THE TOKEN ECONOMY

A Review and Evaluation

ALAN E. KAZDIN

The Pennsylvania State University
University Park, Pennsylvania

PLENUM PRESS · NEW YORK AND LONDON

Library of Congress Cataloging in Publication Data

Kazdin, Alan E
  The token economy.

  (The Plenum behavior therapy series)
  Bibliography: p.
  Includes indexes.
  1. Operant conditioning. 2. Reinforcement (Psychology) I. Title. [DNLM: 1.
Behavior therapy. 2. Reward. WM420 K235t]
BF319.5.06K39   153.8'5   76-44285
ISBN 0-306-30962-9

© 1977 Plenum Press, New York
A Division of Plenum Publishing Corporation
227 West 17th Street, New York, N.Y. 10011

Printed in the United States of America

To my sister, Fran

PREFACE

Applications of operant techniques in treatment and education have proliferated in recent years. Among the various techniques, the token economy has been particularly popular. The token economy has been extended to many populations included in psychiatry, clinical psychology, education, and the mental health fields in general. Of course, merely because a technique is applied widely does not necessarily argue for its efficacy. Yet, the token economy has been extensively researched. The main purpose of this book is to review, elaborate, and evaluate critically research bearing on the token economy. The book examines several features of the token economy including the variables that contribute to its efficacy, the accomplishments, limitations, and potential weaknesses, and recent advances. Because the token economy literature is vast, the book encompasses programs in diverse treatment, rehabilitation, and educational settings across a wide range of populations and behaviors.

Within the last few years, a small number of books on token economies have appeared. Each of these books describes a particular token economy in one treatment setting, details practical problems encountered, and provides suggestions for administering the program. This focus is important but neglects the extensive scholarly research on token economies. The present book reviews research across diverse settings and clients. Actually, this focus is quite relevant for implementing token economies because the research reveals those aspects and treatment variations that contribute to or enhance client performance. The data base now available extends the treatment recommendations that can be made beyond the necessarily limited experience that stems from only one program with a single-client population.

Aside from the general focus and orientation, other features have been included to distinguish this text from others in the area. First, specific influences on client performance that are often omitted or underplayed in other texts are elaborated. These include the influence of self-administration and peer administration of the contingencies, the relationship of economic variables in the token economy and client behavior change, the role of punishment as an adjunctive procedure, the influence of modeling and vicarious processes, the effects of different types of contingencies, and other influences. Second, token reinforcement programs have been extended beyond the usual treatment settings. Currently, many programs focus on socially and environmentally relevant behaviors such as pollution, energy conservation, job procurement, and others. These applications represent a significant extension of reinforcement techniques and are reviewed in detail. Finally, legal issues have become prominent in behavior modification in the last few years. Many court decisions have direct bearing on implementing token economies, particularly in institutional settings. Salient decisions and their implications for implementing token economies are reviewed.

The book begins with a chapter that details the principles of operant conditioning. This chapter serves only as background material for the remainder of the text and may be disregarded by the reader already quite familiar with operant principles. The main text begins by tracing the development of applied operant research in general and the token economy in particular (Chapter 2). This historical focus includes examples of large scale "token economies" that antedated the development of behavior modification. The advantages of token economies over other reinforcement practices also are detailed.

The token economy is not a unitary technique. Token economies vary widely across a range of dimensions including who administers the tokens, how tokens are administered, whether back-up events are used, and if so, what events, the type of contingencies, and so on. The range of options that can be incorporated into the token economy and the effects of these options on client behavior are reviewed (Chapter 3). The token economy has been used with an amazingly large number of populations across several behaviors. Applications and accomplishments with psychiatric patients, the mentally retarded, individuals in classroom settings, delinquents, adult offenders, drug addicts, alcoholics, children and adults in outpatient treatment and in the "natural environment," and several others are reviewed (Chapters 4 and 5).

Several issues bear directly on the effectiveness of a token economy. Major issues are discussed including training individuals to implement a token economy, ensuring that clients respond to the contingencies, and the manipulation of economic variables (Chapter 6). Certainly, one of the most outstanding issues in the field is maintaining behavior changes effected in token economies and ensuring their transfer to nontreatment settings. The technology of maintaining behavioral gains has made important advances in the last few years. With an empirical base, specific procedures can be advanced to increase the likelihood that behaviors will be maintained and transfer to new settings after the token economy has been terminated (Chapter 7).

Although the early chapters in the book stress the accomplishments of the token economy, the criticisms that might be levied against this approach also are detailed. The token economy is critically evaluated across a number of dimensions including the treatment focus, variables contributing to its efficacy, the possible deleterious effects of extrinsic reinforcement, a comparison of the token economy with other treatment procedures, and whether token economies have achieved clinically important changes in behavior (Chapter 8). A relatively new focus in the field is the extension of reinforcement techniques to socially and environmentally relevant behaviors. Applications of reinforcement, particularly those based upon token reinforcement, are detailed along with broader extensions that have attempted to apply operant principles to community and social living situations

(Chapter 9). Involvement of the courts in treatment and rehabilitation, especially in recent years, has generated legal rulings that bear directly upon the token economy as a treatment technique. Ethical and legal issues, major court decisions, and procedures designed to protect client rights are described (Chapter 10). The final chapter of the text briefly summarizes the major accomplishments and salient issues of the token economy (Chapter 11).

Over the last few years, I have had the opportunity to discuss developments in applied behavior analysis with several individuals whose contributions have served as the foundation of the field. These individuals have influenced much of the content of the present text and no doubt its overall orientation. The list of individuals would be too long to acknowledge here and in any case might be omitted to avoid inadvertently implying their endorsement of the present text. As an exception to this, I wish to make a special note of my gratitude to Nathan H. Azrin whose evaluative comments on the manuscript were not only important in achieving the final version of this book but also in stimulating many of my own thoughts about the field in general.

Finally, I wish to acknowledge the support of the Center for Advanced Study in the Behavior Sciences. Final preparation of the book was completed while I was on leave at the Center.

Alan E. Kazdin

The Pennsylvania State University

CONTENTS

Chapter 5
REVIEW OF TOKEN ECONOMIES II 113

Chapter 6
MAJOR ISSUES AND OBSTACLES IN ENSURING
EFFECTIVE PROGRAMS 141

PRINCIPLES OF
OPERANT CONDITIONING

1

The token economy is a type of behavior modification program which relies heavily upon the principles of operant conditioning. There are relatively few major principles of operant conditioning, although there is extensive research bearing on the effective implementation of the techniques derived from the principles. An understanding of the principles and diverse basic research findings is fundamental to the success of token programs. The present chapter provides an overview of the basic principles of operant conditioning. Knowledge of these principles will be assumed throughout the remaining chapters.[1]

The principles of operant conditioning describe the relationship between behavior and environmental events (antecedents and consequences) that influence behavior. In developing behavioral programs, it is important to understand the types of antecedent events and consequences that influence behavior. In many applications of the principles of operant conditioning, emphasis is placed on the *consequences* which follow behavior. For a consequence to alter behavior it must be dependent or contingent upon the occurrence of the behavior. Behavior change occurs when certain consequences are *contingent* upon performance. A consequence is contingent when it is delivered only after the target behavior is performed and is otherwise not available. When a consequence is not contingent upon behavior, it is delivered independently of what the individual is doing. The noncontingent delivery of consequences usually does not result in systematic changes in a preselected behavior because the consequences do not systematically follow that behavior. In everyday life, many consequences such as wages, grades, and physical health are contingent upon behavior.

A contingency refers to the relationship between a behavior (the response to be changed) and the events which follow the behavior. The notion of a contingency is important because reinforcement techniques such as the token economy alter behavior by altering the contingencies which control (or fail to control) a particular behavior. The principles outlined below refer to different kinds of contingent relationships between behavior and the events which follow behavior.

---

[1] For a detailed discussion of the principles of operant conditioning, the reader may wish to consult introductory texts (e.g., Bijou & Baer, 1961; Ferster, Culbertson, & Boren, 1975; Holland & Skinner, 1961; Keller, 1954; Logan, 1969; Rachlin, 1970; Reynolds, 1968; Skinner, 1953a; Williams, 1973).

REINFORCEMENT

The principle of reinforcement refers to an increase in the frequency of a response when it is immediately followed by certain events. The event which follows behavior must be contingent upon behavior. *A contingent event which increases the frequency of behavior is referred to as a reinforcer.* Positive and negative reinforcers constitute the two kinds of events which increase the frequency of a response. *Positive reinforcers* are events which are *presented* after a response is performed and increase the frequency of the behavior they follow. *Negative reinforcers* (also referred to here as aversive stimuli) are events which are *removed* after a response is performed and increase the behavior that preceded their removal. There are two variations of the principle of reinforcement.

Positive Reinforcement

Positive reinforcement refers to an increase in the frequency of a response which is followed by a favorable event (positive reinforcer). The positive or favorable events in everyday language frequently are referred to as rewards. However, it is desirable to distinguish the term "positive reinforcer" from "reward." A positive reinforcer is defined by its effect on behavior. If an event follows behavior and the frequency of behavior increases, the event is a positive reinforcer. Conversely, any event which does not increase the behavior it follows is not a positive reinforcer. An increase in the frequency or probability of the preceding behavior is the defining characteristic of a positive reinforcer. In contrast, rewards are defined as something given or received in return for service, merit, or achievement. Although rewards are subjectively highly valued, they do not necessarily increase the probability of the behavior they follow.

Many events that are evaluated favorably when a person is queried may be reinforcers. Yet, this cannot be known on the basis of verbal statements alone. Moreover, there may be many reinforcers available for an individual of which he is unaware or does not consider as rewards. For example, in some situations verbal reprimands (e.g., "Stop that!") inadvertently serve as a positive reinforcer because they provide attention for a response (Madsen, Becker, Thomas, Koser, & Plager, 1970; O'Leary, Kaufman, Kass, & Drabman, 1970). Even though reprimands sometimes may serve as reinforcers, it is unlikely that anyone would refer to them as rewards. Hence, a reward is not synonymous with a positive reinforcer. Whether an event is a positive reinforcer has to be determined empirically. Does the frequency of a particular behavior increase when the event immediately follows behavior? Only if behavior increases is the event a positive reinforcer.

There are two types of positive reinforcers, namely, *primary* or unconditioned and *secondary* or conditioned reinforcers. Events which serve as primary reinforcers do not depend upon special training to have acquired their reinforcing value. For example, food to a hungry person and water to a thirsty person serve as primary reinforcers. Primary reinforcers may not be reinforcing all of the time. For example, food will not reinforce someone who has just finished a large meal. However, when food does serve as a reinforcer, its value is automatic (unlearned) and does not depend upon previously being associated with any other reinforcers.

Many of the events which control behavior are not primary reinforcers. Events such as praise, grades, money, and completion of a goal have become reinforcing through learning. Conditioned reinforcers are not automatically reinforcing. Stimuli or events which once were neutral in value may acquire reinforcing properties as a result of being paired with events that are already reinforcing (either primary or other conditioned reinforcers). By repeatedly presenting a neutral stimulus prior to or along with a reinforcing stimulus, the neutral stimulus becomes a reinforcer. For example, praise may not be reinforcing for some individuals (Quay & Hunt, 1965; Wahler, 1968). It is a neutral stimulus rather than a positive reinforcer. To establish praise as a reinforcer, it must be paired with an event that is reinforcing such as food or money. When a behavior is performed, the individual is praised and reinforced with food. After several pairings of the food with praise, praise alone serves as a reinforcer and can be used to increase the frequency of other responses. When praise is a neutral stimulus, it can be developed as a conditioned reinforcer by pairing it with another event which is a reinforcer (Locke, 1969; Miller & Drennan, 1970).

Some conditioned reinforcers are paired with more than one other primary or conditioned reinforcer. When a conditioned reinforcer is paired with *many* other reinforcers, it is referred to as a *generalized conditioned reinforcer*. Generalized conditioned reinforcers are extremely effective in altering behaviors because they have been paired with a variety of events (cf. Kelleher & Gollub, 1962). Money and trading stamps are good examples of generalized conditioned reinforcers. They are *conditioned* reinforcers because their reinforcing value is acquired through learning. They are *generalized* reinforcers because a variety of reinforcing events contribute to their value. Additional examples of generalized conditioned reinforcers include attention, approval, and affection from others (Skinner, 1953a). These are generalized reinforcers because their occurrence is often associated with a variety of other events which are themselves reinforcing. For example, attention from someone may be followed by physical contact, praise, smiles, affection, or delivery of tangible rewards such as food and other events.

Tangible generalized conditioned reinforcers in the form of tokens are used frequently in behavior modification programs. The tokens may consist of poker chips, coins, or points and serve as generalized reinforcers because they can be exchanged for a variety of other events which are reinforcing. For example, in a psychiatric hospital tokens may be delivered to patients for attending group activities, grooming and bathing, and other behaviors. The tokens may be exchanged for snacks, cigarettes, and privileges such as watching television and attending social events. The potency of tokens derives from the reinforcers which "back-up" their value. The events which tokens can purchase are referred to as *back-up reinforcers*. Generalized conditioned reinforcers such as money or tokens are more powerful than any single reinforcer because they can purchase a variety of back-up reinforcers.

Two general considerations are important when identifying positive reinforcers. First, an event (praise, candy, pat on the back) may be a positive reinforcer for one person but not for another. Although some events have wide generality in serving as reinforcers (e.g., food, money), others may not (e.g., sour candy). Second, an event may be a reinforcer for one person under some circumstances or at some time but not under other circumstances or at other times. These considerations require carefully evaluating what is reinforcing for a given individual. Because of cultural norms and common experiences of many people, suggestions may be given as to events which probably serve as reinforcers. However, at any given time there is no guarantee in advance that a particular event will be reinforcing.

Reinforcing events referred to above include *stimuli* or specific environmental events such as praise, smiles, food, money which are presented to an individual after a response. However, reinforcers are not limited to stimuli. Allowing an individual to engage in certain *responses* can be used as a reinforcer. Certain responses can be used to reinforce other responses. Premack (1959, 1965) demonstrated that behaviors an individual performs with a relatively high frequency can reinforce behaviors performed with a relatively low frequency. If the opportunity to perform a more probable response is made contingent upon performing a less probable response, the frequency of the latter response will increase. Hence, behaviors of a relatively higher probability in an individual's repertory of behaviors are reinforcers of lower probability behaviors.

Premack (1962) experimentally demonstrated this relationship by altering the probability that rats drank water and engaged in activity (running) by depriving them of either access to water or to an activity wheel. When rats were deprived of water (thus making drinking a high probability behavior), drinking would reinforce running (a low probability response). When the animals were deprived of activity (making running a high probability behavior), running would reinforce drinking. In each case, a lower probability behavior was increased (reinforced) by following it with a high probability behavior. At different times, eating was the higher probability response and running was the lower probability response and vice versa.

On the basis of the laboratory work described above, the *Premack principle* has been formulated as follows: *Of any pair of responses or activities in which an individual engages, the more frequent one will reinforce the less frequent one.* Stated another way, *a higher probability behavior can be used to reinforce a lower probability behavior.* To determine what behaviors are high or low frequency requires observing the behaviors an individual engages in when he is allowed to select activities. Behaviors observed to occur more frequently can be used to follow and reinforce responses observed to occur less frequently. For example, for many children, playing with friends is performed at a higher frequency than is completing homework. If the higher frequency behavior (playing with friends) is made contingent upon the lower frequency behavior (completing homework), the lower probability behavior will increase (cf. Danaher, 1974). Numerous behaviors that an individual performs such as engaging in certain activities, hobbies, or privileges, going on trips, being with friends, and other relatively frequent responses can serve as reinforcers for other behaviors. The requirement of employing the Premack

Principle is that the target response to be altered is of a lower probability than the behavior which will reinforce that response.

Negative Reinforcement

Negative reinforcement refers to an increase in the probability of a response by removing an aversive event immediately after the response is performed. Removal of an aversive event or negative reinforcer is contingent upon a response. An event is a *negative reinforcer* only if its removal after a response increases performance of that response. Events which appear to be annoying, undesirable, or unpleasant are not necessarily negative reinforcers. A negative reinforcer is defined solely by the effect it has on behavior. The qualifications made in the discussion of positive reinforcers apply to negative reinforcers as well. Thus, an event may serve as a negative reinforcer for one individual but not for another. Also, an event may be a negative reinforcer for an individual at one time but not at another time.

It is important to note that reinforcement (positive or negative) always refers to an *increase* in behavior. Negative reinforcement requires an ongoing aversive event which can be removed or terminated after a specific response is performed. Examples of negative reinforcement are evident in everyday experience such as putting on a coat while standing outside on a cold day. Putting on a coat (the behavior) usually removes an aversive state, namely being cold (negative reinforcer). The probability of wearing a coat in cold weather is increased. Similarly, taking medicine to relieve a headache may be negatively reinforced by the termination of pain.

Negative reinforcement requires some aversive event that is presented to the individual before he responds such as shock, noise, isolation, and other events which can be removed or reduced immediately after a response. As with positive reinforcers, there are two types of negative reinforcers, primary and secondary. Intense stimuli which impinge on sensory receptors of an organism such as shock or loud noise serve as primary negative reinforcers. Their aversive properties are unlearned. However, secondary or *conditioned aversive events* have become aversive by being paired with events that are already aversive. For example, disapproving facial expressions or saying the word "no" can serve as an aversive event after being paired with events which are already aversive (Lovaas, Schaeffer, & Simmons, 1965).

Negative reinforcement occurs whenever an individual *escapes* from an aversive event. Escape from aversive events is negatively reinforcing. However, *avoidance* of aversive events is negatively reinforcing, too. For example, one avoids eating rancid food or walking through an intersection with oncoming cars. Avoidance occurs before the aversive event takes place (e.g., becoming sick from rancid food or being injured by a car). Avoidance learning is an area where classical and operant conditioning are operative. Avoidance behavior is sometimes learned by pairing a neutral stimulus (conditioned stimulus) with an unconditioned aversive event (unconditioned stimulus). For example, a frown (conditioned stimulus) from a parent may precede corporal punishment (unconditioned stimulus) of the child which results in crying and escape from the situation (unconditioned response). The child learns to escape from the situation when the adult frowns and thereby avoids corporal punishment. The sight of the frowning parent elicits crying and escape. Avoidance of unconditioned aversive events actually is escape from the conditioned aversive event. Thus, classical conditioning may initiate avoidance behavior. Operant conditioning is also involved in avoidance behavior. Behaviors which reduce or terminate an aversive event, conditioned or unconditioned, are negatively reinforced. The escape from the conditioned aversive event (e.g., frown) is negatively reinforced since it terminates the event. To reiterate, the conditioned aversive event elicits an escape response (classical conditioning) which is negatively reinforced (operant conditioning).

Operant conditioning is involved in yet another way in avoidance learning. The conditioned aversive event serves as a cue signaling that certain consequences will follow. The presence of the conditioned aversive stimulus signals that a certain response (escape) will be reinforced (Reynolds, 1968). A variety of cues control avoidance behavior in everyday life. Indeed, many avoidance behaviors appear to be learned from verbal warnings by others rather than from direct experience with unconditioned aversive stimuli. Warnings merely act as a cue that consequences of a particular sort are likely to follow behavior. An individual does not have to experience physical harm to learn to avoid particular situations.

PUNISHMENT

Punishment is the presentation of an aversive event or the removal of a positive event following a behavior which decreases the probability of that behavior. This definition is somewhat different from the everyday use of the term. Punishment, as ordinarily defined, refers to a penalty imposed for performing a particular act. The technical definition includes an additional requirement, namely, that the frequency of the response is decreased (Azrin & Holz, 1966). Given the technical definition, it should be obvious that punishment does not necessarily entail pain or physical coercion or serve as a means of retribution or payment for misbehaving. Punishment in the technical sense describes an empirical relationship between an event and behavior. Only if the frequency of a response is reduced after a contingent consequence is delivered is punishment operative. Similarly, a punishing consequence is defined by its suppressive effect on the behavior it follows.

There are two different types of punishment. In the first kind of punishment, an aversive event is presented after a response. Numerous examples of this occur in everyday life such as being reprimanded or slapped after engaging in some behavior. In the second kind of punishment, a positive event is withdrawn after a response. Examples include losing privileges after staying out late, losing money for misbehaving, and being isolated from others. In this form of punishment, some positive event is taken away after a response is performed.

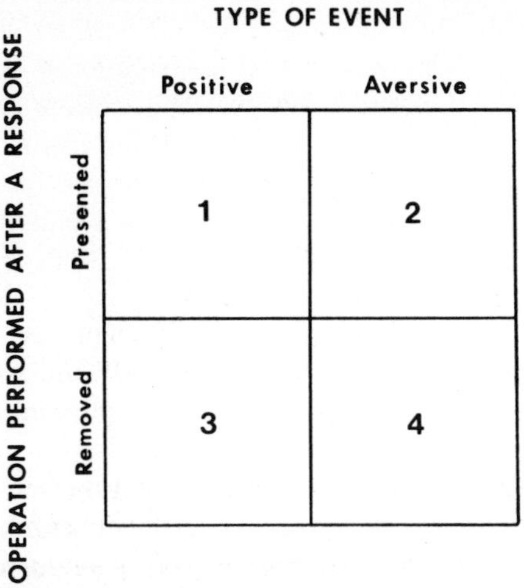

TYPE OF EVENT

Fig. 1. Illustration of the principles of operant conditioning based upon whether positive or aversive events are presented or removed after a response (from Kazdin, 1975; Dorsey Press).

It is important to clearly distinguish punishment from negative reinforcement with which it is often confused. *Reinforcement,* of course, refers to procedures which *increase* a response whereas *punishment* refers to procedures which *decrease* a response. In negative reinforcement, an aversive event is *removed* after a response is performed. In punishment, an aversive event *follows* after a response is performed. The diagram in Fig. 1 clarifies the operations involved in different principles discussed to this point.[2] The diagram depicts the two possibilities which can happen after a response is performed. Something can be *presented* or *removed* after a response is performed (left side of the diagram). The two types of events which

---

[2] This figure or some variation of it is commonly used to describe the basic principles mentioned to this point in the text. While the figure is useful for introductory purposes, technically it is misleading. The figure implies that the presentation or removal of a given event contingent upon behavior has opposite effects. That is, contingently presenting an aversive event decreases behavior whereas contingently withdrawing the same event increases behavior. While this relationship might describe a particular event in relation to some behaviors, this is not necessarily the case. Events which are shown to negatively reinforce behavior do not necessarily suppress behavior, and events shown to suppress behavior do not necessarily negatively reinforce behavior (Azrin & Holz, 1966; Church, 1963). The same relationship holds for the presentation and withdrawal of positive events. Actually, the notion that there are particular *events* that are positive or negative reinforcers ignores much of the experimental literature showing the complex dynamic characteristics of behavior and contingent consequences. Particular events are not invariably positive or negative reinforcers and under different circumstances may serve in both capacities. The effect of presenting a particular consequence after behavior depends upon a number of parameters usually not discussed in applied research (cf. Kish, 1966; Morse, 1966; Morse & Kelleher, 1977).

can be presented or removed are positive and aversive events (top of the diagram). Each cell in the diagram depicts a procedure discussed to this point. Positive reinforcement occupies cell 1; negative reinforcement occupies cell 4; punishment occupies both cells 2 and 3.

EXTINCTION

There is an important principle of operant conditioning that is not represented in Fig. 1. This principle does not involve presenting or withdrawing events in the usual sense. Rather, the principle refers to no longer following behavior with an event that was previously delivered.

Behaviors that are reinforced increase in frequency. However, a behavior which is no longer reinforced decreases in frequency or extinguishes. *Extinction refers to no longer reinforcing a response that has been previously reinforced.* Nonreinforcement of a response results in a gradual reduction or elimination of the behavior. It is important to keep this procedure distinct from punishment. In extinction, no consequence follows the response, i.e., an event is not taken away nor is one presented. In punishment, some aversive event follows a response or some positive event is taken away. In everyday life, the usual use of extinction is in the form of ignoring a behavior that may have been reinforced previously with attention. For example, a parent may ignore whining of a child so that the reinforcer (attention) usually available for the response is no longer presented.

MAXIMIZING THE EFFECT OF POSITIVE REINFORCEMENT

The effectiveness of reinforcement depends upon several factors. These include the delay between performance of a response and the delivery of reinforcement, the magnitude and quality of the reinforcer, and the schedule of reinforcement.

Delay of Reinforcement

Responses which occur in close proximity of reinforcement are more well learned than responses remote from reinforcement (Kimble, 1961; Skinner, 1953a). Thus, a reinforcer should be delivered immediately after the target response to maximize the effect of reinforcement. If reinforcement does not follow the response immediately, another response different from the target response may be performed in the intervening delay period. The intervening response will be immediately reinforced whereas the target response will be reinforced after a delay period. The target response is less likely to change. For example, children are often praised (or punished) for a behavior long after the behavior is performed. If a child straightens his room, a parent would do well to immediately provide praise. If praise is postponed until the end of the day, a variety of intervening responses may occur (including, perhaps, messing up the room).

Immediate reinforcement is important in the early stages of a reinforcement program when the target response is developing. After a response is performed consistently, the amount of time between the response and reinforcement can be increased without a decrement in performance (e.g., Cotler, Applegate, King, & Kristal, 1972). Yet, if a program begins with delayed reinforcement, behavior might not change at all or not as rapidly as when reinforcement is immediate. It is desirable to change from immediate to delayed reinforcement after a behavior is well developed so that behavior is not dependent upon immediate consequences.

Magnitude or Amount of Reinforcement

The amount of reinforcement delivered for a response also determines the extent to which a response will be performed. The greater the amount of a reinforcer delivered for a response, the more frequent the response (Kimble, 1961). The amount can usually be specified in terms such as the quantity of food given, the number of points, or amount of money. Although the magnitude of reinforcement is directly related to performance, there are limits to this relationship. An unlimited amount of reinforcement does not necessarily maintain a high rate of performance. A reinforcer loses its effect when given in excessive amounts. This is referred to as *satiation*. Hence, the effect of magnitude of reinforcement is limited by the point at which the individual becomes satiated. Satiation is especially evident with primary reinforcers such as food, water, and sex. In a short time, each of these reinforcers in excess loses its reinforcing properties and may even become aversive. Of course, satiation of primary reinforcers is temporary because the events regain reinforcing value as deprivation increases. Secondary or conditioned reinforcers such as priase, attention, and tokens are also subject to satiation (e.g., Gewirtz & Baer, 1958; Winkler, 1971b). However, they are less susceptible to satiation than are primary reinforcers. Generalized conditioned reinforcers in particular such as money are virtually insatiable because they have been associated with a variety of other reinforcers. Satiation of generalized reinforcers is not likely to occur until the individual satiates to the other reinforcers with which they have been associated. The more reinforcers for which the generalized conditioned reinforcer such as money can be exchanged, the less likely that satiation will occur.

Quality or Type of the Reinforcer

The quality of a reinforcer is not usually specifiable in physical terms as is the amount of the reinforcer (Kimble, 1961). Quality of a reinforcer is determined by the preference of the client. Reinforcers that are highly preferred lead to greater performance. Preference can be altered by taking a reinforcer such as food and changing its taste. For example, animals show greater performance when a food is sweet than when it is sour or neutral in taste (Hutt, 1954).

For a given client, usually it is not difficult to specify activities which are highly preferred. Behaviors engaged in frequently provide a helpful indication of highly preferred reinforcers. However, preference for a particular reinforcer depends upon satiation. At one point in time a reinforcer may be more effective in changing behavior than another because the client is satiated with one and deprived of another (Premack, 1965). However, as will be discussed later, certain reinforcers tend to result in higher performance than others. Hence, the type of reinforcer alone determines the extent of behavior change.

Schedules of Reinforcement

Reinforcement schedules are the rules describing the manner in which consequences follow behavior. Reinforcers are always administered according to some schedule. In the simplest schedule, the response is reinforced each time it occurs. When each instance of a response is reinforced the schedule is called *continuous reinforcement.* If reinforcement is provided only after some instances of a response, the schedule is called *intermittent reinforcement.*

There are substantial differences between continuous and intermittent reinforcement. Behaviors developed with continuous reinforcement are performed at a higher rate while the behavior is being developed (i.e., while reinforcement is in effect) than behaviors developed with intermittent reinforcement. Hence, while a behavior is being developed it is advisable to administer reinforcement generously, i.e., for every response. The advantage of continuous reinforcement while reinforcement is in effect is offset after reinforcement ceases. In extinction, behaviors previously reinforced on a continuous schedule decrease more rapidly than behaviors previously reinforced on an intermittent schedule. Thus, in terms of maintenance of a behavior, intermittent reinforcement is superior to continuous reinforcement.

The differences in the effects of continuous and intermittent reinforcement are well illustrated by looking at responses made in the presence of vending and slot machines. Vending machines require a response (putting coins into a slot and pulling a lever) which is reinforced (with the desired event such as cigarettes or candy) virtually every time, with exceptions attributed to malfunctioning equipment. Thus, the schedule is designed to be continuous reinforcement. When a reinforcer is no longer provided for putting in coins and pulling a lever, extinction follows the characteristic pattern associated with continuous reinforcement. As soon as the reinforcer (the product) is no longer delivered, extinction is immediate. Few individuals repeatedly place more coins into the machine. Behavior associated with slot machines reveals a different pattern. The response (putting money into a machine and pulling a lever) is reinforced with money only once in a while. If money were no longer delivered, the response would not extinguish immediately. On an intermittent schedule, especially one where the reinforcer is delivered very infrequently, it is difficult to discriminate when extinction begins. The extent to which a response is resistant to extinction is a function of how intermittent the schedule is. Generally, the fewer responses that are reinforced during acquisition, the greater the resistance to extinction.

Intermittent reinforcement can be scheduled in a variety of ways. First, reinforcers can be delivered after the emission of a certain number of responses. This type of delivery is referred to as a *ratio schedule* because the schedule specifies the number of responses required for each reinforcement. Second, reinforcers can be delivered on the basis of the time interval that separates available reinforcers. This is referred to as an *interval schedule,* meaning that the response will be reinforced after a specified time interval.

The delivery of reinforcers on either ratio or interval schedules can be *fixed* (unvarying) or *variable* (constantly changing). Thus, four simple schedules of reinforcement can be distinguished, namely, fixed ratio, variable ratio, fixed interval, and variable interval. Each of these schedules leads to a different pattern of performance.

A *fixed ratio* schedule requires that a certain number of responses occur prior to delivery of the reinforcer. For example, a fixed ratio 2 schedule (FR:2) denotes that every second response is reinforced. A *variable ratio* schedule also requires that a certain number of responses occur prior to delivery of the reinforcer. However, the number varies around an average from one reinforcement to the next. On the average, a specific number of responses are required for reinforcement. For example, a variable ratio 5 schedule (VR:5) means that on the average every fifth response is reinforced. On any given trial, however, the organism does not know how many responses are required.

Performance differs under fixed and variable ratio reinforcement schedules. On fixed schedules, typically there is a temporary pause in responding after reinforcement occurs and then a rapid rise in response rate until the number of responses required for reinforcement is reached. The length of the pause after reinforcement is a function of the ratio specified by the schedule. Larger ratios produce longer pauses. In contrast, variable ratio schedules lead to fairly consistent performance and relatively high response rates.

Interval schedules can be either fixed or variable. A *fixed interval* schedule requires that an unvarying time interval elapse before the reinforcer is available. The first response to occur after this interval is reinforced. For example, in a fixed interval 1 schedule (FI:1), where the number refers to minutes, reinforcement is produced by the first response occurring after 1 minute has elapsed. An interval schedule requires that only one response occur after the interval elapses. Any response occurring before the interval elapses is not reinforced. With a *variable interval* schedule, the length of the interval varies around an average. For example, in a variable interval 4 schedule (VI:4), the reinforcer becomes available after 4 minutes on the average. The first response after the interval elapses is reinforced although the duration of the interval continually changes.

Performance differs under fixed and variable interval schedules. Fixed interval schedules tend to lead to marked pauses after the reinforcer is delivered. Unlike the pattern in fixed ratio schedules, responding after the pause is only gradually resumed. The organism often learns to wait until the interval is almost over. After extensive training, fixed ratio and fixed interval performances may appear very similar. With variable interval schedules, pauses are usually absent and performance is more consistent. In general, the rate of response is higher for variable than for fixed interval schedules.

Across the four simple schedules discussed, it is worth noting that higher rates of response usually are achieved with ratio rather than interval schedules. This is understandable because high response rates do not necessarily speed up delivery of the reinforcer in an interval schedule as they do with ratio schedules. Moreover, variable schedules tend to produce more consistent response patterns, that is, they are not marked by the all-or-none pauses and response bursts of fixed schedules.

The various intermittent schedules are important for developing resistance to extinction. Variable schedules are particularly effective in prolonging extinction. As performance is achieved under a particular schedule, the schedule can be made "leaner" by gradually requiring more responses (ratio) or longer periods of time (interval) prior to reinforcement. With very lean schedules, few reinforcers need to be delivered to maintain a high level of performance. The shift from a "dense" or more generous schedule to a leaner one, however, must be made gradually to avoid extinction. [A more detailed discussion of reinforcement schedules can be obtained from several sources such as Ferster & Skinner (1957), Morse (1966), and Williams (1973).]

SHAPING

Frequently, the development of new behavior cannot be achieved by rein-forcing the response when it occurs because the response may never occur. The desired behavior may be so complex that the elements which make up the response are not in the repertoire of the individual. For example, developing the use of words requires, among other things, the use of sounds, syllables, and their combina-tion. *In shaping, the terminal behavior is achieved by reinforcing small steps or approximations toward the final response rather than reinforcing the final response itself.* Responses are reinforced which either resemble the final response or which include components of that response. By reinforcing *successive approximations* of the terminal response, the final response is achieved gradually. Responses which are increasingly similar to the final goal are reinforced and they increase, while those responses dissimilar to the final goal are not reinforced and they extinguish. Shaping, along with other procedures, is used to develop talking in children. Responses which approach the final goal (e.g., sounds, syllables) are reinforced. Responses which are emitted which do not approach the goal (e.g., screaming, whining) are extinguished along the way toward the final goal.

An obvious example of shaping is training animals to perform various "tricks." If the animal trainer waited until the tricks were performed (e.g., jumping through a burning hoop) to administer a reinforcer, it is unlikely that reinforcement would ever occur. However, by shaping the response, the trainer can readily achieve the terminal goal. Initially, food (positive reinforcer) might be delivered for running toward the trainer. As that response becomes stable, the reinforcer may be delivered for running up to the trainer when he is holding the hoop. Other steps closer to the final goal would be reinforced in sequence, including: walking through the hoop on the ground, jumping through the hoop when it is slightly off the ground, when it is high off the ground, and when it is partially (and eventually completely) on fire. Eventually, the terminal response will be performed with a high frequency, whereas the responses or steps developed along the way are extinguished.

Shaping requires reinforcing behaviors already in the repertoire of the individual which resemble the terminal response or approximate the goal. As the initial approximation is performed consistently, the criterion for reinforcement is altered slightly so that the new response resembles the final goal more closely than the previous response. This procedure is continued until the terminal response is developed.

CHAINING

Behaviors can be divided into a sequence of responses referred to as a *chain*. The components of a chain usually represent individual responses already in the repertoire of the individual. Yet, the chain represents a combination of the individual responses ordered in a particular sequence. For example, one behavioral chain which illustrates the ordering of component responses is going to eat at a restaurant (Reynolds, 1968). Going out to eat may be initiated by a phone call from someone, hunger, or some other event. Once the behavior is initiated, several behaviors follow in a sequence including leaving the house, entering the car, traveling to the restaurant, parking the car, entering the restaurant, being seated, looking at a menu, ordering a meal, and eating. The response sequence unfolds in a relatively fixed order until the chain is completed and the last response is reinforced (e.g., eating). Interestingly, each response in the chain does not appear to be reinforced. Rather, only the last response (the response immediately preceding eating) is followed by the reinforcer (food). Because a reinforcer alters or maintains only the behavior that immediately precedes it, it is not obvious what maintains the entire chain of behaviors leading to the final response.

A concept fundamental to understanding chains is that an event or stimulus which immediately precedes reinforcement becomes a cue or signal for reinforcing consequences. An event which signals reinforcement is referred to as a *discriminative stimulus* (S^D). An S^D sets the *occasion* for behavior, i.e., increases the probability that a previously reinforced behavior will occur. However, an S^D also serves another function. An S^D not only signals reinforcement but eventually becomes a reinforcer itself. The frequent pairing of an S^D with the reinforcer gives the S^D reinforcing properties of its own. This procedure was mentioned earlier in the discussion of conditioned reinforcement. The discriminative stimulus properties of events which precede reinforcement and the reinforcing properties of these events when they are frequently paired with reinforcers are important in explaining how chains of responses are maintained.

Consider the chain of responses involved in going out to eat, described above. (The chain could be divided into several smaller components than those listed above.) A phone call may initiate the response to go to a restaurant to eat. Other behaviors in the chain are performed ending in positive reinforcement (eating). The final response in the chain before reinforcement is ordering a meal. This response is directly reinforced with food. In this chain of responses, the last response performed (ordering a meal) becomes an S^D for reinforcement, since the response signals that reinforcement will follow. The constant pairing of an S^D with the reinforcer (food) eventually results in the S^D becoming a reinforcer as well as a discriminative stimulus. Hence, the response that preceded direct reinforcement becomes an S^D for subsequent reinforcement and a reinforcer in its own right. The response serves as a reinforcer for the previous link in the chain of responses. The response (ordering food) becomes a reinforcer for the previous behavior (looking at a menu). Because looking at a menu now precedes reinforcement, it becomes an S^D. As with other responses, the pairing of the S^D with reinforcement results in the S^D becoming a reinforcer. The process continues in a *backward* direction so that each response in the chain becomes an S^D for the next response in the chain and serves as a reinforcer for the previous response. (The very first response becomes an S^D but does not reinforce a prior response.)

Although the sequence of responses appears to be maintained by a single reinforcer at the end of the chain (food in the above example), the links in the chain are assumed to take on conditioned reinforcement value. To accomplish this requires training from the last response in the sequence which precedes direct reinforcement back to the first response. Since the last response in the sequence is paired immediately and directly with the reinforcer, it is most easily established as a conditioned reinforcer which can maintain other responses. Also, the shorter the delay between a response and reinforcement, the greater the effect of reinforcement. The last response in the chain is immediately reinforced and is more likely to be performed frequently.

PROMPTING AND FADING

Developing behavior is facilitated by using cues, instructions, gestures, directions, examples, and models to initiate a response. *Events which help initiate a response are prompts.* Prompts precede a response. When the prompt results in the response, the response can be followed by reinforcement. When a prompt initiates behaviors that are reinforced, the prompt becomes an S^D for reinforcement. For example, if a parent tells a child to return from school early and returning early is reinforced when it occurs, the instructions (prompt) becomes an S^D. Instructions signal that reinforcement is likely when certain behaviors are performed. Eventually, instructions alone are likely to be followed by the behavior. As a general rule, when a prompt consistently precedes reinforcement of a response, the prompt becomes an S^D and can effectively control behavior.

Developing behavior can be facilitated in different ways such as *guiding* the behavior physically (such as holding a child's arm to assist him in placing a spoon in his mouth); *instructing* the child to do something; *pointing* to the child which directs his behavior (such as making a gesture which signals the child to come inside the house); and *observing* another person (a model) perform a behavior (such as watching someone else play a game). Prompts play a major role in shaping and chaining. Developing a terminal response using reinforcement alone may be tedious and time consuming. By assisting the person in beginning the response, more rapid approximations of the final response can be made.

The use of prompts increases the likelihood of response performance. While a response is being shaped, prompts may be used frequently to facilitate performing the terminal goal. As soon as a prompted response is performed, it can be reinforced. Further, the more frequently the response is reinforced, the more rapidly it will be learned. A final goal usually is to obtain the terminal response in the absence of prompts.

Although prompts may be required early in training, they can be withdrawn gradually or faded as training progresses. *Fading* refers to the gradual removal of a prompt. If a prompt is removed abruptly early in training, the response may no longer be performed. But if the response is performed consistently with a prompt, the prompt can be progressively reduced and finally omitted, thus faded. To achieve behavior without prompts requires fading and reinforcing the responses in the absence of cues or signals. It is not always necessary to remove all prompts or cues. For example, it is important to train individuals to respond in the presence of certain prompts such as instructions which exert control over a variety of behaviors in everyday life.

DISCRIMINATION AND STIMULUS CONTROL

Operant behavior is influenced by the consequences which follow behavior. However, antecedent events also control behavior. Prompts, discussed above, represent a group of controlling events (e.g., instructions, physical guidance, models, cues) which precede and facilitate response performance. Yet, other antecedent stimuli come to exert control over behavior. In some situations (or in the presence of certain stimuli), a response may be reinforced, while in other situations (in the presence of other stimuli) the same response is not reinforced. *Differential reinforcement* refers to reinforcing a response in the presence of one stimulus and not reinforcing the same response in the presence of another stimulus. When a response is consistently reinforced in the presence of a particular stimulus and consistently not reinforced in the presence of another stimulus, each stimulus signals the consequences which are likely to follow. The stimulus present when the response is reinforced signals that performance is likely to be reinforced. Conversely, the stimulus present during nonreinforcement signals that the response is not likely to be reinforced. As mentioned earlier, a stimulus whose presence has been associated with reinforcement is referred to as an S^D. A stimulus whose presence has been associated with nonreinforcement is referred to as an S^Δ (S delta). The effect of differential reinforcement is that eventually the reinforced response is likely to occur in the presence of the S^D but unlikely to occur in the presence of the S^Δ. The probability of a response can be altered (increased or decreased) by presenting or removing the S^D (Skinner, 1953a). The S^D occasions the previously reinforced response or increases the likelihood that the response is performed. When the individual responds differently in the presence of different stimuli, he has made a *discrimination*. When responses are differentially controlled by antecedent stimuli, behavior is considered to be under *stimulus control*.

Instances of stimulus control pervade everyday life. For example, the sound of a door bell signals that a certain behavior (opening the door) is likely to be reinforced (by seeing someone). The sound of the bell frequently has been associated with the presence of visitors at the door (the reinforcer). The ring of the bell (S^D) increases the likelihood that the door will be opened. In the absence of the bell (S^Δ), the probability of opening the door for a visitor is very low. The ring of a door bell, telephone, alarm, and kitchen timer, all serve as discriminative stimuli which increase the likelihood that certain responses will be performed.

Stimulus control is always operative in behavior modification programs. Programs are conducted in particular settings (e.g., the home) and are administered by particular individuals (e.g., parents). Insofar as certain client behaviors are reinforced or punished in the presence of certain environmental cues or of particular individuals and not in the presence of other stimuli, the behaviors will be under stimulus control. In the presence of those cues associated with the behavior modification program, the client will behave in a particular fashion. In the absence of those cues, behavior is likely to change because the contingencies in the new situations are altered.

GENERALIZATION

The effect of reinforcement on behavior may generalize across either the stimulus conditions beyond which training has taken place or across the responses that were included in the contingency. These two types of generalization are referred to as stimulus generalization and response generalization, respectively.

A response which is repeatedly reinforced in the presence of a particular situation is likely to be repeated in that situation. However, situations and stimuli often share common properties. Control exerted by a given stimulus is shared by other stimuli which are similar or share common properties (Skinner, 1953a). A behavior may be performed in new situations similar to the original situation in which reinforcement occurred. If a response reinforced in one situation or setting also increases in other settings (even though it is not reinforced in these other settings), this is referred to as stimulus generalization. *Stimulus generalization refers to the generalization or transfer of a response to situations beyond those in which training takes place.*

Stimulus generalization is the opposite of discrimination. When an individual discriminates in his performance of a response, this means that the response fails to generalize across situations. Alternatively, when a response generalizes across situations, the individual fails to discriminate in his performance of a response. The degree of stimulus generalization is a function of the similarity of new stimulus (or situation) to the stimulus under which the response was trained (Kimble, 1961). Of course, over a long period of time, a response may not generalize across situations because the individual discriminates that the response is reinforced in one situation but not in others. Stimulus generalization represents an important issue in token economies. Although training takes place in a restricted setting (e.g., institution, special classroom, hospital, day-care center, home), it is desirable that the behaviors developed in these settings generalize or transfer to other settings.

An additional type of generalization involves responses rather than stimulus conditions. *Altering one response can inadvertently influence other responses. This is referred to as response generalization.* For example, if smiling is reinforced, the frequency of laughing and talking might also increase. The reinforcement of a response increases the probability of other responses which are similar (Skinner, 1953a). To the extent that a nonreinforced response is similar to one that is reinforced, the similar response too is increased in probability.

The concepts of stimulus and response generalization, as usually used, are more than descriptive terms. These concepts not only denote that there is a spread of the effect of the contingency across either stimulus or response dimensions but also imply an explanation for these effects. Stating the spread of effect in terms of "generalization" implies that the broad effects are due to similarity of stimulus conditions across which a given response transfers (for stimulus generalization) or due to the similarity of the new responses to the one that has been trained (for response generalization). In fact, there is rarely any demonstration that the spread of treatment effects across stimulus or response dimensions is actually based upon the similarity of the stimulus or response conditions to those of training. In applied research, investigators have rarely shown that the spread of effects follows a continuum of similarity across either stimulus or response dimensions. Rather, investigators report that behavior change in one situation transfers to other situations or that a change in one behavior is associated with changes in other behaviors not focused upon in training. Because the spread of effects either across stimulus or response dimensions might not be related to similarity of training and nontraining conditions, there might be technical reasons for avoiding the notion of generalization. Nevertheless, from the standpoint of reinforcement programs in applied settings, the spread of effects either across stimulus conditions or across behaviors has extremely important treatment implications.

CONCLUSION

This chapter provides an overview of the major principles of operant conditioning. These principles provide the basis for the majority of operant conditioning programs in applied settings whether or not these programs are based upon token reinforcement. Understanding the principles is essential for designing effective token economies for at least two reasons. First, as will be evident throughout subsequent chapters, a token economy is not merely based upon delivering positive reinforcers for particular behaviors. Rather, diverse techniques can be used based upon positive and negative reinforcement, punishment, and extinction as part of the token economy. Second, the efficacy of delivering tokens for behaviors depends upon meeting several contingency requirements such as immediately delivering consequences, following generous schedules of reinforcement (at least in the beginning of the program), ensuring the delivery of a sufficient number of tokens to effect change, and so on. Often the ineffectiveness of a given reinforcement contingency can be ameliorated by altering a relatively minor feature of the program. Thus, the parameters that dictate the efficacy of reinforcement in general must be well understood. The importance of the full range of operant principles may be appreciated throughout subsequent chapters which describe the token economy in its diverse forms. As will be evident, token economies include a variety of options many of which entail combining the delivery of tokens with other contingency manipulations.

INTRODUCTION AND HISTORICAL OVERVIEW

2

Throughout history, diverse cultures have recognized the influence of reinforcing and punishing consequences on behavior. Indeed, the systematic application of consequences for behavior has been institutionalized in various cultural practices such as education, child rearing, military training, government and law enforcement, religion, and other practices common to most cultures. For example, consider the widespread use of incentives to develop specific kinds of military behavior. In ancient Greece and Rome during the first century A.D., gladiators received special prizes ranging from wreaths and crowns to money and property for their victorious performance (Grant, 1967). During the same period, charioteers in Rome received freedom from their enslaved status and financial rewards for repeated victory in their duels (Carcopino, 1940). Similarly, ancient Chinese soldiers received colored peacock feathers that could be worn as an honorary symbol of their bravery during battle (Doolittle, 1865). Aztec soldiers in the fifteenth century received titles of distinction and the honor of having stone statues made of them for their triumphant military performance (Duran, 1964). Finally, plains Indians of America in the nineteenth century received attention and approval from their peers for skillful behavior in battle or during a hunt. Meritorious deeds earned the privilege of relating the experiences in public (Eggan, Gilbert, McAllister, Nash, Opler, Provinse, & Tax, 1937). In most contemporary societies, bravery in military performance continues to be rewarded with medals, badges, promotions, and similar incentives.

In education, rewarding consequences have been used as incentives for learning. In the twelfth century, prizes such as nuts, figs, and honey were used to reward individuals who were learning to recite religious lessons (Birnbaum, 1962). Similarly, in the sixteenth century, fruit and cake was advocated by Erasmus to help teach children Greek and Latin (Skinner, 1966). One use of incentives for educational ends that is particularly interesting pertains to the history of pretzels (Crossman, 1975). Development of the pretzel has been traced to the seventh century A.D. A monk in southern Europe made pretzels from remains of dough after baking bread. He shaped these remains into little biscuits to represent children's arms as they are folded in prayer. These biscuits were used to reward children who learned to recite their prayers. The monk referred to these rewards as *petriola* which is Latin for little reward and from which the word *pretzel* is derived.

The select examples pertaining to military training and education exemplify the explicit use of incentives to influence behavior. In everyday life, consequences

19

influence behavior whether or not they are explicitly programmed. In ordinary social interaction, the responses of others contribute to our behavior. For example, if an individual continually fails to reciprocate a friendly greeting, our greeting response is likely to extinguish. Of course, in social institutions and everyday life, the influence of consequences on behavior should not be oversimplified. Behavior is a function of a complex set of variables and learning experiences which may not be obvious by merely looking at the current consequences which follow a particular response.

INVESTIGATION OF THE INFLUENCE OF RESPONSE CONSEQUENCES

Although the influence of consequent events on behavior had been recognized long before the formal development of psychology, the experimental and theoretical basis for this influence has only been empirically investigated since the late 1890s. The earliest programmatic research on the role of consequences on behavior was completed by Thorndike in the late 1890s. Thorndike began research on animal learning using diverse species. His most famous experiments included research on the escape of cats from puzzle boxes. Essentially, Thorndike demonstrated that a hungry animal escaped from a puzzle box to obtain food with increasingly greater efficiency (i.e., shorter response latencies). The responses that were successful in leading to escape were developed and those that were unsuccessful were eliminated. Thorndike's research led to his formulation of various "laws of learning." His most influential law was the law of effect which specified essentially that responses followed by "satisfying consequences" were more likely to be performed and those followed by "annoying consequences" were likely to be eliminated (Thorndike, 1911).

Thorndike was not the only one who investigated response consequences and their influence on behavior. In Russia, at about the same time Thorndike was involved in his research, Pavlov and Bechterev, who were working independently, demonstrated the influence of consequences on behavior. Although Pavlov and Bechterev are appropriately credited with investigation of learning based upon classical conditioning, in fact, their research also encompassed the paradigm now referred to as operant conditioning.

One of Pavlov's accidental encounters with behavior controlled by its consequences occurred in an experiment designed to evaluate respondent conditioning. In this experiment, Pavlov noted that because of a defective apparatus the animal could receive the food (meat powder) by merely shaking a tube through which the food normally passed. The animal quickly learned to shake the tube repeatedly and this response occurred continuously (cf. Pavlov, 1932). Systematic work on the role of consequences in learning was conducted by Ivanov-Smolensky (1927), one of Pavlov's students. A series of experiments was conducted in which a child could obtain food and other rewards contingent upon a response (pressing a bulb) when certain stimuli appeared. Other investigators researched the role of consequences on behavior in laboratory investigations (cf. Bechterev. 1932; Miller & Konorski, 1928).

In the early research on learning, there was some ambiguity in the distinction between the types of learning focused upon by Thorndike and Pavlov. There were several reasons for this ambiguity. Initially, both paradigms seemed to be accounted for by the association or connection of various stimuli and responses. The paradigms appeared to differ only in how the association was made. At a cursory level of analysis, Thorndike's paradigm consisted of associating new responses with a particular stimulus, whereas Pavlov's paradigm consisted of associating new stimuli with a particular response. Second, the different types of learning are not easily distinguished experimentally. The investigation of one type of learning often entails the other (Hilgard & Marquis, 1940; Kimble, 1961). Third, Thorndike and Pavlov accounted for each paradigm within their own theoretical framework. Thus, any distinctions of heuristic value that might be made were obfuscated.

The distinction between Thorndikian and Pavlovian learning was brought into sharp focus by Skinner (1935, 1938). Skinner distinguished between two types of responses and their respective types of conditioning. The two types of responses were *respondents* and *operants.* Respondents refer to reflex responses that are *elicited* such as salivation in response to the presence of food in one's mouth or pupillary constriction in response to bright light. Operants refer to those responses that are *emitted* or performed "spontaneously" by the organism and are not elicited by antecedent stimuli.[1] The two types of conditioning based upon these responses were Type S and Type R conditioning. Type S conditioning refers to the conditioning of respondent behavior. Respondent conditioning (more commonly referred to as classical conditioning) was referred to as Type S because the stimulus which elicits a response is paired with some other stimulus. Emphasis is on the pairing of stimuli (S). Type R conditioning (more commonly referred to as operant conditioning) refers to conditioning of operant behavior. It is referred to as Type R because of the pairing of a response and a stimulus event. In classical conditioning, the reinforcing (unconditioned) stimulus is correlated with another stimulus event (conditioned or neutral stimulus); in operant conditioning the reinforcing stimulus is contingent upon a response.

Skinner's classification of learning into two major types was not unique and can be found in the writings of other authors (e.g., Konorski & Miller, 1937; Miller & Konorski, 1928; Schlosberg, 1937; Thorndike, 1911). Yet, Skinner (1938) emphasized the distinction and discussed the interrelationships and ambiguities between these learning paradigms. More importantly, he began a systematic series of investigations elaborating Type R or operant conditioning (cf. Skinner, 1938, 1956). He investigated diverse processes and experimental manipulations in animal

---

[1] The distinction between respondent and operant behaviors has become less clear than originally posed. For example, respondent behaviors such as heart rate, blood pressure, intestinal contractions, vasomotor reflexes, and others previously considered to be involuntary and not subject to the control by their consequences have been altered through operant conditioning (cf. Kimmel, 1967, 1974; Miller, 1969).

laboratory research including reinforcement, extinction, spontaneous recovery, schedules of reinforcement, discrimination training, response differentiation, and others (Skinner, 1938).

Aside from establishing a substantive area of research, Skinner advanced a methodology for studying behavior referred to as the "experimental analysis of behavior." Specifically, Skinner argued for careful study of overt behavior without making references to inner states of the organism, the use of response frequency as the experimental datum of operant behavior, scrutiny of behavioral processes of individual organisms over extended periods rather than the study of groups of subjects, and a functional analysis of behavior. Skinner's research spawned a well-investigated area of experimental research which consists primarily of laboratory animal investigations of diverse behavioral processes.[2] While this research has served as the basis for many contemporary applications of operant conditioning, extensions of operant principles to human behavior and to clinical populations in general serves as a useful point of departure for tracing the evolution of the token economy in behavior modification.

EXTENSIONS OF OPERANT METHODS TO HUMAN BEHAVIOR: AN OVERVIEW

Conceptual and Experimental Extensions

In the late 1940s and early 1950s, the principles and methods of operant conditioning were extended to human behavior. Initially, the extensions took the form of discussions about the role of reinforcement contingencies in various aspects of behavior and the utility of operant conditioning in conceptualizing behavior generally. For example, Keller and Schoenfeld (1950) noted that "The principles of operant conditioning may be seen everywhere in the multifarious activities of human beings from birth until death" (p. 64). The role of operant principles in various social institutions including government and law, religion, psychotherapy, economics, and education was discussed in *Science and Human Behavior* (Skinner, 1953a). Perhaps the most ambitious extension of operant principles was in the earlier book *Walden Two* (1948) which portrayed a utopian society based upon operant principles.

Although the extensions of operant conditioning to human behavior were broad conceptualizations, the relevance of a behavioral approach for clinical psychology was made explicit. For example, Skinner (1953a) challenged the traditional

---

[2] For a discussion of research in operant conditioning the reader may consult introductory texts (e.g., Williams, 1973), more advanced compendia (e.g., Catania, 1968; Honig, 1966) or the *Journal of the Experimental Analysis of Behavior* (1958–present) which is devoted to operant research. The methodological approach embraced by the experimental analysis of behavior has been described in articles and books (e.g., Sidman, 1960; Skinner, 1953b, 1966).

intrapsychic position of abnormal behavior and the focus of treatment on mental events. According to Skinner (1953a), traditional interpretations of behavior have

> encouraged the belief that psychotherapy consists of removing certain inner causes of mental illness, as the surgeon removes an inflamed appendix or cancerous growth or as indigestible food is purged from the body. We have seen enough of inner causes to understand why this doctrine has given psychotherapy an impossible assignment. It is not an inner cause of behavior but the behavior itself which—in the medical analogy of catharsis—must be "got out of the system" (p. 373).
>
> We have to ask why the response was emitted in the first place, why it was punished, and what current variables are active. The answers should account for the neurotic behavior. Where, in the Freudian scheme, behavior is merely the symptom of a neurosis, in the present formulation it is the direct object of inquiry (p. 376).

Skinner accounted for the effects of psychotherapy in terms of operant principles. He noted that therapeutic change probably resulted from specific reinforcement contingencies managed by the therapist. The therapist reinforces various behaviors, primarily through approval, to effect change. Also, the therapist may extinguish emotional responses by providing a nonpunishing environment in which these responses are expressed. Finally, the therapist helps the patient restructure his own environment to alter the stimuli that adversely control behavior.

The conceptual extensions of operant principles quickly gave way to empirical demonstrations. Initially, the research began with the expressed goal of using operant methods to investigate human behavior. However, even in the early work, the applied implications were evident. An important extension was the research of Skinner and Lindsley who experimented with psychotic patients at the Metropolitan State Hospital in Waltham, Massachusetts in 1953. The research, directed by Lindsley, was designed to extend the methodology of operant conditioning to study the behavior of psychotics. Patients performed a free-operant task (plunger-pulling) for reinforcing events such as candy, cigarettes, projected pictures, music, tokens, escape from a loud noise or a dark room, the opportunity to give milk to kittens, and others (Lindsley, 1956; Skinner, 1954). Patients were seen individually in daily experimental sessions for extended periods (e.g., up to five years) (Lindsley, 1960, 1963). Generally, the research demonstrated the response patterns of the patients were orderly and in many ways resembled the response characteristics of lower organisms.

Interestingly, there were several findings of direct clinical value in this experimental work. First, many patients evinced long pauses between their responses on the task. During these pauses, psychotic symptoms frequently were performed such as pacing, laughing, swearing, staring, and destroying objects. This suggested that assessing response rates and interresponse times might be useful to indirectly assess psychotic symptoms. Other response characteristics of interest were found including lower and more variable response rates for the psychotic patients relative to normal adults and lower organisms. Second, for some patients, an increase in response rate was associated with a reduction of psychotic behavior (Lindsley, 1960).

Plunger-pulling constituted a "competing response" for the symptomatic behavior. This provided a clear demonstration that, at least for some patients, increasing nonpsychotic responses diminished some psychotic behavior. Indeed, for some patients, the reduction of symptoms during the experimental sessions appeared to transfer outside of the experimental environment (Lindsley, 1963). Overall, Lindsley's work clearly demonstrated that the behavior of psychotic patients could be altered as a function of environmental consequences. This general finding was subsequently exploited to alter a wide range of behaviors of psychotic patients.[3]

At approximately the same time that Skinner and Lindsley began exploring operant methods with psychotic patients, Bijou was extending the methodology to the behavior of children at the University of Washington. The primary purpose of Bijou's research, as that of Lindsley, was to develop a methodology to study human behavior. Bijou developed apparatus which allowed free-operant responding (e.g., lever pressing) and studied basic behavioral processes (e.g., Bijou, 1955, 1957a, 1957b, 1958). Bijou's research demonstrated that child behavior could be studied by looking at stimulus and response functions, an approach that had proven useful in infrahuman laboratory research. Aside from his laboratory research, Bijou argued that the methodology of operant conditioning should be employed in the conceptualization of developmental psychology and mental retardation (Bijou, 1959, 1963, 1966). Several other investigators in the 1950s and 1960s applied operant methods to human behavior (e.g., Azrin & Lindsley, 1956; Baer, 1960; Ferster & DeMyer, 1961, 1962). (For a review of laboratory human operant research, see Bijou and Baer, [1966].)

Another extension of operant methods to human behavior was verbal conditioning research in the early 1950s. Verbal conditioning research consisted of altering the verbal behavior of the speaker by following specific responses with contingent events such as comments by the experimenter. The impetus for studying verbal behavior derived from Skinner who applied operant principles to verbal behavior in the 1940s, although his views were not published in detail until the appearance of *Verbal Behavior* (Skinner, 1957).[4] Initially, research began to determine merely whether verbal behavior could be manipulated as a function of consequences provided by the social environment, i.e., the listener, including such consequences as saying "Mmm'hmm," "Good," and "Fine." Several studies using different methods of verbal conditioning were reported (e.g., Ball, 1953; Greenspoon, 1951; Taffel, 1955; Verplanck, 1955).

---

[3] Historically, it is important to note that Lindsley and Skinner were the first to use the term "behavior therapy" (Lindsley, Skinner, & Solomon, 1953; Skinner, Solomon, Lindsley, & Richards, 1954). The term referred to the direct focus on behavior and the use of operant conditioning methodology. Despite the fact that the use of the term antedated other uses (Eysenck, 1959; Lazarus, 1958), it was restricted to unpublished research reports.

[4] Actually, Skinner's ideas were widely available in summary form prior to the appearance of this book. For example, Keller and Schoenfeld (1950) drew upon Skinner's 1948 William James Lectures at Harvard where Skinner's views were explained. Also, Skinner (1953a) provided a brief conspectus of the application of operant conditioning to explain verbal behavior.

The clinical relevance of verbal conditioning studies was recognized relatively early in this research in part because of similarities between conditioning and psychotherapy situations. For example, Krasner (1955) emphasized that a common factor in all psychotherapy was the presence of one person (the therapist) listening, paying attention, and showing interest in the other person (the client). The therapist's expression of interest served as a conditioned reinforcer which effectively directed the client's behavior. By stressing the responses of the therapist, verbal conditioning studies appeared to provide an analog of clinical treatment. Indeed, Krasner (1962) drew attention to the similarities between verbal conditioning and psychotherapy by referring to the therapist as a "social reinforcement machine." This statement did not refer to the dehumanized and mechanical behavior of the therapist but rather only to the fact that the therapist was programmed through training to control client behavior by delivering generalized conditioned reinforcers.

The notion that psychotherapy could be viewed as a conditioning process was bolstered by diverse research. Initially, some verbal conditioning studies with psychiatric patients showed that merely altering verbal behavior in an interview situation affected symptomatically relevant behaviors. For example, conditioning emotional words with psychotic patients was associated with changes in recognizing threatening material, in anxiety, and in adequacy of interpersonal relationships (cf. Krasner, 1962, 1965). Also, reinforcement of self-reference statements in verbal conditioning situations altered self-descriptions and performance on psychological inventories (cf. Kanfer & Phillips, 1970). This research suggested that language was a medium of altering nonverbal behavior and that the verbal conditioning paradigm resembled the dyadic situation of therapy.

The notion that psychotherapy could be viewed from the standpoint of verbal conditioning was strengthened by a study by Truax (1966). In this study, the role of reinforcement in controlling client behavior was examined by analyzing tape recordings of a case handled by Carl Rogers. The analysis revealed that Rogers provided comments reflecting empathy and warmth contingent upon specific responses of the client. The finding was dramatic, of course, because Rogerian therapy, also referred to as nondirective therapy, in fact was controlling client behavior through contingent reinforcement. This demonstration also suggested that conditioning principles are in operation and affect behavior whether or not they are acknowledged by the practitioner.

The similarities and differences between psychotherapy and verbal conditioning have been debated and there is some agreement that these processes are not identical (Heller & Marlatt, 1969; Kanfer, 1968; Luborsky & Strupp, 1962). Yet, verbal conditioning research increased the prominence of operant principles in the context of clinical application. The verbal conditioning studies did not attempt to achieve therapeutic change as much as demonstrate the utility of the operant method in studying verbal behavior. Yet, by the late 1950s and early 1960s, verbal conditioning was directly applied for therapeutic purposes, for example, to alter irrational verbalizations, to reinstate speech in psychotic patients, and to decrease stuttering (e.g., Ayllon & Michael, 1959; Isaacs, Thomas, & Goldiamond, 1960;

Rickard, Dignam, & Horner, 1960; Rickard & Mundy, 1965; Sherman, 1963, 1965).

Clinical Applications

The early extensions of operant conditioning to human behavior were in the domain of the experimental analysis of behavior. That is, the main purpose was to determine whether operant principles and methodology could be extended to human behavior and to investigate the influence of specific contingency manipulations of which behavior was a function. Thus, the purposes were primarily experimental rather than clinical. In the late 1950s and early 1960s, the transition from experimental to clinical research can be discerned.

One of the most influential applications of operant procedures was made by Ayllon (1963, 1965; Ayllon & Michael, 1959) at the Saskatchewan Hospital, in Canada. Ayllon focused on problem behaviors with several psychiatric patients—frequently entering the nurses' office, engaging in violence, spending excessive amounts of time in the bathroom, hoarding magazines, and others. For example, with one of the cases reported, extinction was used to decrease delusional talk (Ayllon & Michael, 1959). The patient continuously talked about her illegitimate child and the men she claimed were always pursuing her. Her psychotic talk had persisted for at least 3 years. Typically, nurses responded to the delusional talk by listening to understand and to get at the "root" of the problem. When extinction was implemented after a baseline period of observation, nurses were instructed not to attend to psychotic talk and to attend to sensible conversation. This resulted in a decrease in psychotic talk.

The role of social consequences in altering the verbal behavior of a psychotic patient was clearly demonstrated in a later report by Ayllon and Haughton (1964). The patient included in this report had a history of 14 years of psychotic verbal behavior that included bizarre statements which referred to "Queen Elizabeth," "King George," and the "Royal Family" (e.g., "I'm the Queen. Why don't you give things to the Queen? . . . How's King George?). To determine whether consequences provided by the staff could control the patient's verbal behavior, staff temporarily ignored neutral talk that was not psychotic (e.g., "It's nice today") and reinforced psychotic verbalizations with attention. In the final phase, the contingencies were reversed so that psychotic verbalizations were ignored and neutral statements were attended to. The results, shown in Fig. 2, reveal that the consequences by the staff clearly dictated the verbal behavior of the client. Ayllon and his colleagues focused on several other behaviors of psychiatric patients including increasing self-feeding and using eating utensils rather than one's hands at meals and reducing somatic complaints, stealing food, hoarding towels, and wearing excessive amounts of clothes (Ayllon & Azrin, 1964; Ayllon & Haughton, 1962, 1964). (For a review and evaluation of this research, see Davison [1969].)

Other investigators reported operant procedures with individual clinical cases. Isaacs *et al.* (1960) reinstated verbal responses in two psychiatric patients by

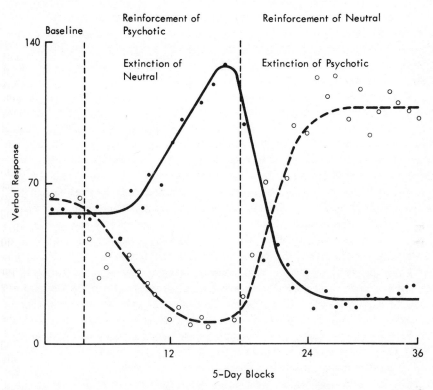

Fig. 2. Verbal responses of a psychotic patient as a function of reinforcement and extinction procedures (from Ayllon & Haughton, 1964; Pergamon Press).

providing gum contingent upon successive approximations toward verbalizations (e.g., a vocalization) and eventually statements. Rickard *et al.* (1960) increased rational talk and decreased psychotic talk by providing attention and ignoring these responses, respectively. Flanagan, Goldiamond, and Azrin (1958) suppressed stuttering by presenting an aversive noise and increased stuttering by removing an aversive noise contingent upon stuttering. Barrett (1962) reduced multiple tics (contracting the neck, shoulder, chest, and abdominal muscles, head nodding, bilateral eyeblinking, and other movements) by either presenting a noise or by terminating music to which the subject listened contingent upon performing a tic. Brady and Lind (1961) partially restored vision of a hysterically blind adult while investigating operant behavior in response to various visual cues. Sulzer (1962) altered excessive alcohol consumption of one client by controlling peer social consequences for drinking and used shaping and social reinforcement to alter studying and social interaction of another client.

In general, early applications of operant conditioning with clinical populations shared several characteristics. First, the applications often were restricted to only one or a few individuals at one time. Second, the procedures frequently were conducted during specific experimental sessions rather than as part of ordinary ward routine. Third, only one or a few reinforcing events were used at one time. Fourth, the focus was restricted to one or a few target behaviors. Fifth, many applications were primarily of methodological interest and were not aimed at effecting therapeutic change *per se*.

There were some obvious exceptions that violated many of the above characteristics. For example, Ayllon and Haughton (1962, Exp. 1) trained self-feeding in 32 psychiatric patients. Normally patients had been fed by the nurses. The nurses reinforced dependent behavior by individually escorting patients to the dining room and feeding them. Thus, patients received attention for not attending the dining room and feeding themselves. To develop self-feeding, patients were told that they had to come to the dining room within a specified time interval (initially 30 minutes and eventually 5 minutes), otherwise they could not eat. Hence, coming to meals would be reinforced with food. Also, staff attention was not provided for refusing to eat as was previously the case. Although some patients missed a few meals, all learned to attend the dining room and feed themselves.

Research with psychiatric patients was not the only impetus for applying operant techniques for clinical purposes. Indeed, perhaps an even greater influence stemmed from research with children. Several clinical applications were conducted by Bijou, Wolf, and their colleagues at the University of Washington, which already had been a center for laboratory operant research with children (cf. Bijou & Baer, 1966). One of the earliest and most influential cases was reported by Wolf, Risley, and Mees (1964) who treated a preschool autistic boy, named Dicky. Dicky refused to wear his eyeglasses which, because of prior eye difficulties and surgery, was likely to result in the complete loss of macular vision. Dicky also engaged in continuous tantrums which included self-destructive behavior. Using the staff and the parents to administer the program, time-out (isolating him in his room or closing the door when he was already in his room) decreased tantrums and problems associated with going to bed. Wearing glasses was shaped gradually by reinforcing contact with empty glass frames (e.g., picking them up, holding them) and eventually wearing the prescription lenses using food and activities as reinforcers. Throwing the glasses, a problem which arose during training, was eliminated by isolating Dicky in his room. Other behaviors were developed such as increasing verbal responses and appropriate eating behaviors. The case of Dicky provided a clear demonstration of the utility of operant techniques with severe clinical problems and constituted an extremely influential report.

Other applications of operant techniques at the University of Washington in the early 1960s included increasing behaviors such as walking, climbing and other play activities, attending and working of children in a classroom setting and decreasing behaviors such as crying in reaction to mildly distressful events and persistent vomiting (Birnbrauer, Bijou, Wolf, & Kidder, 1965; Birnbrauer, Wolf, Kidder, & Tague, 1965; Harris, Johnston, Kelley, & Wolf, 1964; Harris, Wolf, &

Baer, 1964; Hart, Allen, Buell, Harris, & Wolf, 1964; Johnston, Kelley, Harris, & Wolf, 1966; Wolf, Birnbrauer, Williams, & Lawler, 1965).

An example of this series of investigations is provided in a project that increased the social interaction of a 4-year-old girl who isolated herself from other children in a nursery school classroom (Allen, Hart, Buell, Harris, & Wolf, 1964). To develop social interaction, the teacher ignored isolate play and delivered praise when peer play was initiated. During a reversal phase, the child received attention for isolate play and was ignored for peer interaction. As shown in Fig. 3, peer interactions decreased when they were ignored and increased when they were reinforced. After the program was terminated during the last month of school, a high rate of peer interaction was maintained.

Aside from the above demonstrations, several other investigators applied operant techniques to alter child behavior. For example, Williams (1959) demonstrated that tantrums of a 21-month-old child could be eliminated by having the parents ignore tantrums in the home. Zimmerman and Zimmerman (1962) reduced tantrums, baby talk, and not responding appropriately to academic tasks of two emotionally disturbed boys in a classroom setting by providing social reinforcement

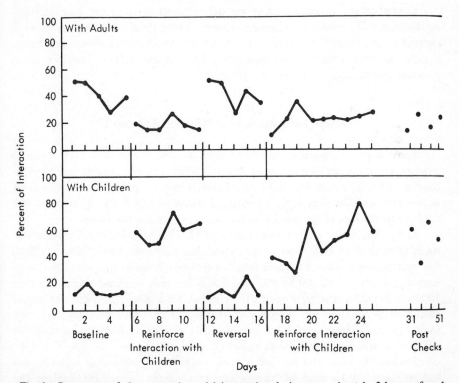

Fig. 3. Percentage of time spent in social interaction during approximately 2 hours of each morning session (from Allen, Hart, Buell, Harris, & Wolf, 1964; Society for Research in Child Development).

for appropriate responses. Rickard and Mundy (1965) reduced stuttering by providing praise, food, and points (exchangeable for toys) for errorless speech across several verbal tasks. Kerr, Meyerson, and Michael (1965) established vocal responses in a previously mute retarded girl. (For a review of early research applying operant techniques with children, see Sherman and Baer [1969].)

The above research provides a brief conspectus of the beginning of the application of operant techniques with clinical populations. The clinical application of operant techniques and the methodology of the experimental analysis of behavior is referred to as *applied behavior analysis*. Essentially, the brief history outlined above represents the development of applied behavior analysis. To develop the basis of the token economy, it is important to discuss more specific influences than clinical applications of operant techniques. Prior to outlining operant research that contributed to the token economy, it is important to review token programs that antedated operant conditioning by several years.

TOKEN ECONOMIES ANTEDATING CONTEMPORARY BEHAVIOR MODIFICATION

Contemporary token programs are derived directly from the principles and methodology of operant conditioning. However, long before these principles were explicitly investigated in psychology, reinforcement systems were used which closely resemble contemporary programs. The present section describes select programs in educational and prison settings.

Joseph Lancaster and the "Monitorial System"

Early token economies may have been relatively common in the school classroom where systems based upon points, merits, demerits, and similar tangible conditioned reinforcers were used to control behavior. An elaborate example of such a program is evident in the educational system of Joseph Lancaster which was implemented across a large number of countries throughout the world.

In England in the early 1800s, Lancaster (1778–1838) developed a system of teaching referred to as the monitorial system (cf. Lancaster, 1805; Salmon, 1904). The system began when Lancaster initiated free education of the poor. This led to large enrollments (e.g., 100 to 300 students in the early years and eventually over 1000) in facilities that were generally crowded. Inadequate funds did not permit having assistants to handle the large number of students who were housed in one room. Thus, Lancaster decided to use students in class as assistants or monitors which permitted teaching the large classes (Kaestle, 1973).

Students in class who excelled in their work were appointed as monitors and had responsibility for most of the functions traditionally reserved for teachers. The monitors were responsible for teaching the students, asking questions, grading answers, noting absences or misbehavior, preparing, distributing, and collecting supplies, and promoting students for improved academic performance. "Teaching monitors," who were responsible for presenting lessons, each had about 10 to 12 students in their group. The groups differed in ability levels and the goal of the system was to promote an individual to higher levels of ability within his own group and eventually to other groups. The students were classified separately in reading, arithmetic, writing, and spelling and promoted to higher groups in these areas when competence was demonstrated. Aside from the teaching monitors, several other types of monitors were used for tasks such as taking attendance, examining students to determine whether they could be promoted, and others. A "monitor general" oversaw activities of the monitors and the distribution of materials to them.

For present purposes, the most significant feature of the Lancasterian system was that it depended heavily upon positive reinforcement, particularly token reinforcement, both for the monitors and the students. The reinforcement system for the students was based upon competition. As students were taught in their groups, while standing in a semicircle around the monitor, each wore a number which indicated his or her relative position in the group in order of merit. As individuals performed in the group, the standing of each was changed as a function of their responses relative to others. For correct responses (e.g., reading) one moved up and for incorrect responses one moved down in the rankings. The top individual in each group earned a rank of 1 and received a leather ticket of merit. Also, he wore a picture pasted on a small board and suspended on his chest to indicate his excellent performance. If his performance was surpassed by someone else in his group, he forfeited the honor of having the ticket and wearing the picture and the highest ranking number. Yet, he received another picture as a reward for having earned first place. Individuals were periodically examined by someone other than their monitor to determine whether they could be promoted to a higher group. When a student knew the lesson and earned a promotion, he received a prize (e.g., toy, picture). His monitor also received a prize of the same value.

Other reward systems were used outside of the group ranking and ticket system. One procedure that was used was to establish an "order of merit" which was a group of individuals who were distinguished by attending to their studies, helping others improve their studies, or endeavoring to check vice. Members of the order were distinguished by a silver medal suspended from their necks with a large plated chain. Thus, these individuals could be readily distinguished at first sight. Another procedure was to reward a few truly exceptional children whose behavior appeared exemplary to the class with engraved silver medals, pens, watches, and books (Lancaster, 1805).

Punishment was relatively infrequent in the system and was reserved for instances of not working or disruptive behavior. For misbehavior the offender would receive a ticket from his monitor. The offense was listed on the card. The card was turned into the head of the school who admonished the offender. Repeated offenses were followed by diverse punishing consequences such as having a wooden log tied to one's neck, thus making turning away from one's lesson difficult, placing one's legs in wooden shackles, having one hand tied behind one's back, or on rare occasions being placed in a sack or basket suspended from the ceiling (cf. Kaestle, 1973).

Lancaster's system spread quickly across Great Britain, several countries in Europe, Africa, Asia, and South America as well as to the United States and Canada. In the United States, the Lancasterian system was implemented shortly after it developed in England. In 1805, New York established a law making free education available for the poor. When the trustees met to enact this law, they realized that a large number of students would have to be educated with extremely limited funds (Salmon, 1904). One trustee had met Lancaster in England and advanced his system as a means to provide education for a large number of students. Aside from New York, the Lancasterian system developed in New Jersey, Pennsylvania, Maryland, and Washington, D.C. Lancaster came to America to further advance his system, established an institute, and published materials about his program.

The New York version of the Lancasterian system is significant because it constituted an early, if not the first, use of token reinforcement in American education and also because it extended Lancaster's methods to more closely resemble contemporary token economies. As in the original method, monitors conducted class and students competed within small groups for tickets exchangeable for prizes or money (Ravitch, 1974). A few distinguishing features of the system are worthy of note. Initially, although the number of tickets provided for performance was relatively fixed, the teacher had discretionary power over 1000 tickets per month to provide special rewards for deserving students. Second,

Table 1. Punishment (Ticket) System Used in New York City Schools in the Early 1800s (Ravitch, 1974)

Offense	Fine in tickets
Talking, playing, inattention, out of seats, etc.	4
Being disobedient or saucy to a monitor	4
Disobedience of inferior to superior monitor	8
Snatching books, slates, etc., from each other	4
Monitors reporting scholars without cause	6
Moving after the bell rings for silence	2
Stopping to play, or making a noise in the street on going home from school	4
Staring at persons who may come into the room	4
Blotting or soiling books	4
Monitors neglecting their duty	8
Having dirty face or hands, to be washed and fine	4
Throwing stones	20
Calling ill names	20
Coming to school late, for every quarter of an hour	8
Playing truant, first time	20
Ditto, second time	40
Fighting	50
Making a noise before school hours	4
Scratching or cutting the desks	20

punishment in the system relied upon fines rather than the usual penalties practiced by Lancaster. Disruptive and inappropriate classroom behaviors were associated with specific costs in terms of tickets or fines (see Table 1).

The ticket system of reward and punishment in the New York City Schools was withdrawn in the 1830s. The trustees believed that cunning behavior rather than meritorious behavior was rewarded.[5] Also, problems appeared to arise including jealousies associated with earning rewards and the stealing of tickets. The system was replaced with a program in which students earned certificates which commended appropriate performance. The certificates, signed by the trustees, were only administered semiannually and were helpful to students in procuring subsequent employment.

An interesting sidelight about the use of reinforcement in the New York system is that the trustees attempted to eliminate corporal punishment which was commonly accepted as a method of discipline. In the mid 1840s, testimonial certificates to principals were delivered if they did not resort to corporal punishment for one or more years. However, few principals qualified (Ravitch, 1974).

The Excelsior School System

The use of token reinforcement for classroom management appears to have become relatively widespread in the United States in the late 1880s somewhat independently of Lancaster's monitorial system of education. One program described in 1885 was referred to as the "Excelsior School System" (Ulman & Klem, 1975). In this system, students earned merits in various denominations for performance of appropriate behavior including being punctual, orderly, and studious. Students could earn a token referred to as an "excellent" for commendable performance or a different token referred to as a "perfect" for perfect performance each day. Excellents and perfects were exchangeable for one or two merits, respectively. When merits were accumulated, they were exchanged for yet other tokens worth varying amounts of merits. If a large number of merits were earned, the student could exchange them for a special certificate by the teacher attesting to his performance.

An interesting feature of the program was that parents were encouraged to ask the students how many merits they earned in school and received feedback from the school. The Excelsior School System was sold commercially (including tokens of different denominations, instructions, certificates, and slips to parents). The program was reported to have been successful for almost 10 years, across thousands of teachers, and across several states. Direct testimonials were provided with a description of the program (Ulman & Klem, 1975).

---

[5] It is interesting to note that contemporary authors have discussed the concern shared by many that reinforcement programs will teach greed and avarice similar to the concern expressed here in the early programs (cf. Kazdin, 1975a; O'Leary, Poulos, & Devine, 1972).

Alexander Maconochie and the "Mark System"

A particularly interesting application of a token system was made by Alexander Maconochie (1787–1860) who established a program for the management and rehabilitation of several hundred prisoners at Norfolk Island (Australia) (Maconochie, 1847, 1848). Maconochie was critical of the brutal and inhumane methods of treating prisoners. He believed that prison should focus primarily on the rehabilitation of prisoners by developing behaviors that would ensure their adjustment to society. He devised a system whereby prisoners could obtain release on the basis of their behavior rather than on the basis of merely serving time.

Maconochie developed a "mark system" as the primary basis of rehabilitation. Sentences were converted into a fixed number of "marks" that individuals had to earn to obtain release. The number of marks needed by a prisoner was commensurate with the severity of the crime committed. Prisoners traversed through a system of different stages. Although some initial restraint and deprivation were included in the program, the main treatment consisted of inmates earning marks for work and for appropriate conduct. The marks could be exchanged for essential items including food, shelter, and clothes. Disciplinary offenses were punished by fines and by withdrawal of privileges rather than by the cruel and unusual punishments that had been commonly practiced (e.g., flogging, being bound in chains).

A given level of accumulated marks determined whether one earned release. The marks toward this final goal had to be accumulated over and above those marks spent on essential items or lost for offenses. After a prisoner had earned a large number of excess marks toward the terminal goal and, thus, had shown appropriate conduct for a protracted period, he joined other prisoners (in groups of five or six) and performed tasks as a member of a group. The group was utilized so that individuals would become responsible for each other's conduct. To this end, offenses performed by one member of a group resulted in a fine for all group members. Eventually, marks were earned on a collective basis for the group.

Maconochie's system was beset with an extremely large number of obstacles (Barry, 1958). For example, the prison in which he implemented his program included particularly intractable inmates who had been sent for their multiple convictions. Extreme administrative and political opposition ensured that major conditions required in the program (e.g., guaranteeing release of prisoners contingent upon earning a predetermined number of marks) could not be enforced. Perhaps the most serious obstacle was that prison life during the time of Maconochie's work was based heavily upon brutal punishment and explicitly coercive treatment. Thus, his program by comparison was often viewed as indulging the prisoners rather than providing the punishment and social revenge usually accorded them.

Despite these difficulties, there was some indication that his method was associated with superior outcome, as suggested by a recidivism rate claimed to be approximately 2% in five years (Maconochie, 1847). Other prisons throughout the British colonies showed a recidivism rate between 33% and 67%. Although the records do not permit an uncritical acceptance of these figures, evaluation of

Maconochie's achievements by colleagues, contemporaries, and by the verdict of history attest to the efficacy of his procedures (Barry, 1958; West, 1852).

Maconochie's achievements from the standpoint of subsequent developments in prison reform and rehabilitation are marked and cannot be conveyed in this brief treatment of his overall views. From the standpoint of behavior modification, and more specifically, token economies, his achievements are clear. His system entailed a token program where an individual's behavior and rate of progress determined the consequences he attained. Prisoners were given clear guidelines of the required behaviors and predetermined consequences for specific offenses (visually displayed to the prisoners) rather than arbitrary consequences imposed by fiat. Treatment focused on developing positive behaviors rather than merely punishing undesirable behaviors. Also, treatment included approximations of social living, through a group contingency, to help develop behaviors that would facilitate community adjustment after release. The major achievements by Maconochie in developing a large-scale treatment system earned him the title of "inventor of the mark system in prison discipline" by several writers (cf. Barry, 1958).[6,7]

CONTEMPORARY FOUNDATIONS OF TOKEN ECONOMIES

The early token economies provide clear examples of relatively complex and sophisticated programs. Although they provide a historical basis for contemporary applications, their role has been minimal in stimulating current practices. Operant conditioning research and its extensions to human behavior in laboratory and clinical settings serve as the direct roots of current token economies.

Laboratory research on token reinforcement began as early as the 1930s. This original research clearly established the capacity of tokens to bridge the delay between performance and receipt of back-up reinforcers. In the infrahuman research on token reinforcement, chimpanzees were trained to work for tokens (e.g., by lever-pressing) that could be exchanged for food (Cowles, 1937; Kelleher, 1957a, 1957b; Wolfe, 1936). Initially, animals could be trained to operate a vending machine with tokens. Insertion of a token was associated with food delivery. After the value of tokens had been established, the tokens could be used to reinforce responding on a laboratory apparatus.

The early research demonstrated several interesting features of token reinforcement. First, token reinforcement was shown to be effective in altering behavior, and in some cases, as effective as food reinforcement not mediated by tokens. Second, animals would work for tokens that could be exchanged either immediately or after a delay period. Third, animals could discriminate different types of tokens,

---

[6] Maconochie's significance in the area of behavior modification in general has been recognized by Eysenck (1972).

[7] It is worth noting that the label "inventor" probably is inaccurate given the large number of precursors to the specific features of system (cf. Barry, 1958; West, 1852). Certainly, Maconochie was the greatest proponent of the mark system of treatment and has to his credit the first well-publicized (and controversial) attempt to implement such a program in prisons.

namely, those that could be exchanged for back-up events and those that could not be exchanged and respond differentially. Fourth, the mere possession of tokens apparently served as a stimulus which controlled behavior. When animals were required to save tokens and could not immediately exchange them for back-up events, they tended to work less to procure additional tokens. That is, there appeared to be some effects of satiation of token earnings.

While the infrahuman research provides a clear laboratory basis for the use of token reinforcement, it did not immediately stimulate the use of tokens for clinical purposes. Eventually, however, continued research revealed that various character-istics of token reinforcement were suited to some of the demands of applied work. In the late 1950s and early 1960s laboratory research with humans increasingly investigated the use of tokens. The use of tokens served to ameliorate a problem of administering more commonly used reinforcers such as food. Primary reinforcers such as food are more subject to satiation than are conditioned reinforcers. Children could be trained to respond for protracted periods for tokens that were eventually exchanged for back-up events such as food without the threat of satiation (cf. Bijou, 1958).

Several investigators used tokens in laboratory studies with children. Ferster and DeMyer (1961, 1962) used tokens to reinforce key-pressing of autistic children. The responses earned coins which could be used to operate a number of devices that delivered back-up events including a vending machine with food, candy, toys, and trinkets, a pigeon in a box that performed for 30 seconds, a pinball machine, a record, milk or juice, and other events. Similarly, Lovaas (1961) allowed children to select toys that were placed in a transparent box. A toy could be earned by opening the box which required a preselected number of tokens. Bijou (Bijou & Baer, 1966) also utilized token reinforcement in experiments with children. Tokens could be exchanged for toys, candies, trinkets, and mechanical and electrical games, and other events. In Bijou's procedure, sometimes different types of tokens were used, viz., one color token which was redeemable for back-up events and another color which was not redeemable for back-up events. The latter token provided feedback to the child that his behavior was appropriate.

The transition from laboratory to applied research utilizing token reinforce-ment is apparent in the work with psychiatric patients and with children. In the case of psychiatric patients, Ayllon and Haughton (1962) conducted three experi-ments evaluating the effect of different contingencies related to self-feeding. In an initial experiment, patients learned to come to meals on their own rather than having to be coaxed or cajoled by the nurses. In a second experiment, Ayllon and Haughton required patients to deposit a penny in a slot of a collection box to gain entrance into the dining room. The coins were delivered noncontingently outside of the dining room by nurses prior to a meal. Verbal prompts were used to aid patients in obtaining and depositing the pennies to gain entrance. All 38 patients in this study learned to deposit coins to obtain food reinforcement.

In the final experiment, Ayllon and Haughton made the delivery of pennies to the patients contingent upon "social responses." Specifically, patients had to engage in a cooperative task which consisted of pressing two separate buttons simultaneously. The two buttons had to be pressed by separate patients; the buttons were separated so that one patient could not press both buttons at the same time. When the buttons were pressed simultaneously, a light and buzzer came on and a nurse delivered a penny to each patient who pressed a button. As in the previous study, the penny could be exchanged immediately for access to the dining room. In general, the experiments by Ayllon and Haughton (1962) demonstrated that patients could learn that tokens were required to utilize a back-up event and could learn a response to earn tokens. Essentially, this report directly preceded large-scale applications where tokens could be earned for several behaviors and could purchase a variety of back-up events.

Additional research utilizing token reinforcement which eventually led to applied work was completed by Staats. Staats began programmatic investigations with children of diverse learning disabilities in 1959 (cf. Staats, 1970). He developed a methodology to study reading behavior as an operant. Children worked individually at a laboratory apparatus in which stimulus presentation, response recording, and consequence delivery were partially automated (cf. Staats, 1968a). The reinforcing system involved a marble dispenser which delivered a marble when activated by a button press. The marble served as a token and could be used in exchange for plastic trinkets, edibles (candy, pretzels), or a penny by placing it in another dispenser. Individuals could also place the marbles in transparent tubes which could accumulate different quantities of marbles. Above the tube was the back-up reinforcer (a toy previously selected by the child) that could be earned when the tube was full.

Interesting findings were derived from Staats' early research. He and his colleagues demonstrated that responses to a complex verbal task produced learning curves analogous to operant conditioning with more simple behaviors of children and with animals. Thus, the viability of the operant methodology with complex behaviors such as reading was shown. More specific to ultimate development of the token economy, the research demonstrated that a token system could sustain child performance for a large number of sessions (Staats, Minke, Finley, Wolf, & Brooks, 1964). The token system proved to be superior to social reinforcement alone. Indeed, preschool children would work for only a relatively short time (e.g., 15 or 20 minutes) when they were praised for correct reading responses. The children quickly became bored and restless. Yet when additional reinforcers were introduced including tokens (exchangeable for toys), edibles, and trinkets, they worked over twice as long without distraction (Staats, Staats, Schultz, & Wolf, 1962).

The initial work completed by Staats and his colleagues was designed primarily to systematically develop a research methodology to investigate reading acquisition and various parameters of reinforcement rather than to construct an effective reading repertoire *per se* (cf. Staats *et al.,* 1964; Staats, Finley, Minke, & Wolf, 1964; Staats *et al.,* 1962). Of course, there was obvious applied significance suggested by the preliminary results. First, diverse children had been studied including normal children as young as 2 years old, educable and trainable retardates, behavior problem, and culturally deprived children. Second, the responses altered were significant because they developed reading, writing, and arithmetic skills over a protracted training period and were sometimes shown to be associated with increases in intelligence test performance (Staats, 1968b).

The applied value of the token system became increasingly apparent. For example, in one study, a 14-year-old boy was exposed to the reading program for a period of over four and one-half months (Staats & Butterfield, 1965). The boy increased his vocabulary and reading achievement scores and was less of a problem in school from the standpoint of deportment and academic performance. The applied value of this work was extended by training nonprofessionals (volunteer students and housewives) to conduct the program with several subjects (Staats, Minke, Goodwin, & Landeen, 1967). Instructor-technicians were trained to alter the behavior of junior high school students with reading deficits, many of whom were emotionally disturbed or mentally retarded. Reading responses including identifying new words, reading the words in paragraphs and stories, and answering questions based upon material read were increased by providing tokens backed-up by pennies or specific items the individual selected as a reward.

In a subsequent investigation, the token reinforcement system administered by nonprofessional technicians was extended to a large number of black, ghetto children enrolled in regular or special education classes (Staats, Minke, & Butts, 1970). Not only were reading responses developed, but subjects trained on the reading task showed greater increases in grade point average and less truancy during the training period than control subjects who did not receive training.

Overall, the work by Staats and his colleagues established the efficacy of token reinforcement in developing complex responses included in reading. The investigations in the early 1960s suggested a technology of token reinforcement that might be extended. While Staats extended the laboratory method to various populations with the use of nonprofessionals, token economies had already developed.

As noted earlier, in the late 1950s and early 1960s, clinical applications of operant techniques generally had been restricted in the number of individuals, the target behaviors, and the reinforcers that were incorporated into the contingencies. Also, many applications in clinical settings had been restricted to experimental sessions and were not conducted in ordinary or routine living conditions. These applications did not immediately suggest a way to apply reinforcement on a large scale. Nevertheless, select findings from early applications provided hints on how to structure the contingencies. For example, the use of token reinforcement in laboratory settings suggested that generalized conditioned reinforcers could sustain performance for protracted periods.

The large-scale extension of reinforcement for clinical purposes was achieved with the development of the token economy. The initial development of the token economy is to be credited to Ayllon and Azrin who designed and implemented the first program in 1961 at Anna State Hospital in Illinois. Their objective was to design a total environment that would motivate psychiatric patients (Ayllon & Azrin, 1968b). They constructed a comprehensive program by rearranging the hospital environment so that a large number of reinforcers in the setting could be brought to bear to alter patient behavior. The programmed environment had to be arranged in such a way that the administration of reinforcement could be carefully controlled and monitored. Because tokens were tangible conditioned reinforcers, their delivery, receipt, and expenditure could be monitored.

Tokens were used as a generalized conditioned reinforcer not only because they provided a convenient way to incorporate other reinforcers but also because they provided a way to structure the relationship of the staff and patients. By using tokens, the behavior of the staff and patients could be engineered in such a way as to create a total contingency-based environment. Importantly, the token economy provided a major advance in extending applications of reinforcement beyond one or a few patients or behaviors.

The program developed by Ayllon and Azrin exerted immediate influence on other treatment, rehabilitation, and educational programs. The influence of their program preceded publication of their initial results (Ayllon & Azrin, 1965). Actually, prior to 1965 several investigators visited their program or communicated with them about it. These communications led directly to several other programs with psychiatric patients, the mentally retarded, delinquents, and other populations (cf. Ayllon & Azrin, 1968b). Independently of the program of Ayllon and Azrin with psychiatric patients, Bijou, Birnbrauer, Wolf, and their colleagues developed a token economy for mentally retarded children (e.g., Birnbrauer et al., 1965a; Birnbrauer et al., 1965b). The program which was initiated in 1962 at Ranier School in Washington was extremely influential because it demonstrated the viability of the token economy in changing academic behavior and classroom deportment. (The token programs at Anna State and at Ranier School will be discussed later.)

REINFORCEMENT PROGRAMS IN APPLIED SETTINGS

Tracing the development of token reinforcement in the laboratory and applied research should not imply that tokens constitute the only type of reinforcer employed in clinical applications. A large number of reinforcers have been used in laboratory and applied research (Bijou & Sturges, 1959; Kazdin, 1975a). However, it is especially important to single out token reinforcement because it has permitted a larger extension of programs than ordinarily is the case with other type of reinforcing events. It is useful to consider briefly the type of reinforcing events that are used in programs in applied settings to appreciate the advance that the token as a reinforcer provides.

Food and Other Consumable Items

Food and consumable items have been used in several studies. Because food is a primary reinforcer, it is extremely powerful. Studies have used such events as meals, cereal, candy, ice cream, drinks, and others. Consumable items such as cigarettes and gum also have been used. These events have less of a universal appeal as reinforcing events than food but they share some properties with food such as a relative dependence upon deprivation and satiation states. Food has been used frequently in applied programs. For example, Risley and Wolf (1968) used food (such as ice cream, candy, and coke) along with praise to develop functional speech in autistic children who were echolalic (i.e., inappropriately mimicked conversation of others and excessively repeated songs and television commercials). Children were trained to answer questions, to initiate requests and comments, to name objects with an increased vocabulary, and to use grammatically correct sentences.

While food and consumables have been effective in a large number of programs, these events have distinct limitations. First, as noted earlier, these events depend heavily upon deprivation and satiation states of the individual. Thus, for the events to be effective, some deprivation usually is required. Even if this can be accomplished (e.g., by using food as a reinforcer immediately prior to a meal or by using highly desirable food such as ice cream and candy), satiation can become a problem. The client may not perform for the reinforcing event for protracted periods because of satiation.

A second limitation of food and consumables is that their effectiveness depends upon the specific event used. Although food generally is a primary reinforcer, specific foods used in a particular program may not be reinforcing. For example, even such foods such as ice cream may depend upon the flavor for a particular individual. Thus, if a single or limited number of foods is used, the possibility exists that the event will not be effective for a number of clients. Moreover, preferences within a given individual change over time so that a single food or consumable item may have short-lived reinforcing properties.

A third potential problem with food and consumables is in their administration. The delivery and consumption of food after a response sometimes interrupts ongoing behavior. For example, if a teacher distributes candy to her students while they are working attentively on an assignment, each individual might be distracted momentarily from the task. Although the purpose of reinforcement is to augment attentiveness to classwork, the consumption of the reinforcer may temporarily distract the students.

A fourth and related problem is that some items may be difficult to dispense immediately because they are cumbersome. Although staff can carry pockets full of candy, other foods such as beverages and ice cream generally are not readily carried. Of course, the setting in which food is used dictates the ease with which a particular type of food can be administered.

A related problem is that food is not easily administered to several individuals in a group immediately after behavior is performed. Since the administration of food to several individuals takes some time (e.g., selecting the quantity of food, putting spoon or fork into individual's mouth, or passing a piece to each individual), it is not particularly well suited to group situations where everyone receives the reinforcer. Hence, many programs using food have been conducted on an individual basis rather than in groups (Ayllon, 1963; Fuller, 1949; Isaacs *et al.*, 1960; Risley & Wolf, 1968; Sherman, 1965; Sloane, Johnston, & Harris, 1968).

Social Reinforcers

Social reinforcers such as verbal praise, attention, physical contact, facial expressions, proximity, and similar events are conditioned reinforcers. Numerous studies have shown that attention from a parent, teacher, or attendant exerts considerable control over behavior (Kazdin, 1975a). For example, Madsen, Becker, and Thomas (1968) used praise to improve attentive behavior in two elementary classrooms. Using a reversal design, different phases were evaluated including classroom rules (telling students rules for behaving in class), ignoring inappropriate behavior (not attending to those disruptive behaviors which interfere with learning), and praising appropriate (e.g., praising children who worked on assignments, answered questions, listened, and raised their hands to speak). Providing rules to the class and ignoring disruptive behavior did not have a noticeable effect on performance. Disruptive behavior decreased only in the last phase when appropriate behavior was praised.

Social consequences have a variety of advantages as reinforcers. First they are easily administered by attendants, parents, and teachers. A verbal statement or smile can be dispensed quickly. The complications of delivering food reinforcement are not present with praise and attention. Obviously, little preparation is involved before delivering praise. Administration takes little time so there is no delay in reinforcing a number of individuals almost immediately. Indeed, praise can be delivered to a group as a whole such as a class.

Second, the delivery of praise need not disrupt the behavior that is reinforced. A person can be praised or receive a pat on the back while engaging in appropriate behavior. Performance of the target behavior can continue. Third, praise is a generalized conditioned reinforcer because it has been paired with many reinforcing events and, thus, is less subject to satiation than are food and consumable items. Fourth, attention and praise are "naturally occurring" reinforcers employed in everyday life. Behaviors developed with social reinforcement in a treatment or training program may be more readily maintained outside of the setting than behaviors developed with other reinforcers. Social reinforcers in everyday life may continue to provide consequences for newly acquired behavior.

Despite the advantages, social reinforcement has one major drawback. Praise, approval, and physical contact are not reinforcing for everyone. Because the reinforcement value of praise and attention has to be learned, one can expect to find individuals who do not respond to events which are normally socially reinforcing. Indeed, for some individuals praise may be aversive (Levin & Simmons, 1962). Because social reinforcers are employed in everyday life, it is important to establish social events as reinforcers by pairing praise, approval, and physical contact with events that are already reinforcing.

Activities, Privileges, and Other High Probability Behaviors

Aside from food and praise or attention, high probability behaviors have been used as reinforcing events. When provided with the opportunity to engage in a variety of behaviors, an individual chooses certain activities with a relatively higher frequency than others. Behaviors of a relatively higher frequency in an individual's repertoire of responses can reinforce behaviors of a lower probability, a relationship referred to as the Premack principle. Engaging in preferred activities or earning various privileges serve as reinforcers because they have a relatively higher probability than other behaviors which serve as target responses. To determine whether a particular activity is high in probability requires observing the frequency that an individual engages in the activity when given the opportunity. Several programs in applied settings have used the Premack principle effectively (Kazdin, 1975a). For example, Lattal (1969) used swimming (high probability behavior) to reinforce tooth brushing of boys in a summer camp. After baseline observations were made of the number of boys who brushed their teeth, only those boys who brushed were permitted to go swimming. Virtually everyone brushed their teeth daily during this phase.

High probability behaviors offer distinct advantages as reinforcers. In most settings, activities and privileges are readily available. For example, in the home, access to television, peers, or the family automobile are likely to be high probability behaviors depending upon the age of the person. At school, access to recess, free time, games and entertaining reading materials may serve a similar function. In hospital and rehabilitation facilities, engaging in recreation, leaving the ward, access to desirable living quarters or personal possessions, and sitting with friends at meals also can be used. In short, activities and privileges which can be made contingent upon performance usually are available in any setting. Hence, "extra" reinforcers (e.g., candy, money) need not be introduced into the setting.

There are potential limitations in using high probability behaviors as reinforcing events. First, access to an activity cannot always immediately follow a low probability behavior. For example, in a classroom setting, activities such as recess or games cannot readily be used to immediately reinforce behavior. Usually activities and privileges have some scheduling limitations. Hence, in some cases there will be a delay of reinforcement. If the reinforcer is not delayed, another problem arises, namely, interruption of continuous performance of the target behavior.

A second consideration is that providing an activity is sometimes an all or none enterprise so that it is either earned or not earned. This can limit the flexibility in administering the reinforcer. For example, in institutions for psychiatric patients or delinquents access to overnight passes and trips to a nearby town are sometimes used as reinforcing activities. These activities cannot easily be parceled out so that "portions" of them are earned. They have to be given in their entirety or not at all. If a client's behavior comes very near the performance criterion for reinforcement but does not quite meet the criterion, a decision has to be made whether to provide the reinforcer. A solution is to shape behavior by setting initially low criteria to earn the activity. Gradually the criteria for earning the reinforcer are increased. Another alternative is to incorporate many privileges and activities into the contingency system (O'Brien, Raynes, & Patch, 1971). Different behaviors or varying degrees of a given behavior can be reinforced with different privileges.

A third consideration in using high probability behaviors as reinforcers is that relying on one or two activities as reinforcers runs the risk that some individuals may not find them reinforcing. Preferences for activities may be idiosyncratic so that different activities need to be available. Providing free time is desirable if individuals can choose from a variety of activities (e.g., Osborne, 1969).

A final consideration in using activities and privileges is that in many institutions, activities must be freely available to the clients. Activities which might be made contingent upon performance are delivered independently of the client's performance. The ideology of presenting activities and other potentially reinforcing events (e.g., meals, sleeping quarters) noncontingently was developed to ensure that individuals would not be deprived of basic human rights. Institutionalized clients usually are deprived of many amenities of living simply by virtue of their institutionalization. Withholding or depriving individuals of the already limited number of available reinforcers is viewed as unethical and illegal in light of recent court decisions. In some settings, activities already provided to the clients cannot be given contingently. (The issues of deprivation of privileges in institutional settings and patient rights will be discussed in Chapter 10.)

Informative Feedback

Providing information about performance can serve as a reinforcer. Feedback is a conditioned reinforcer because it usually is associated with the delivery of other events which are reinforcing. Feedback is implicit in the delivery of any reinforcer because it indicates which responses are appropriate or desirable from the standpoint of those who provide reinforcement. Thus, when reinforcers such as food, praise, activities, or points are provided, a client receives feedback or knowledge of how well he is doing. However, feedback can be employed independently of these other reinforcers. Individuals can be informed of their behavior or of the extent to which their behavior has changed. Feedback refers to *knowledge of results* of one's performance without necessarily including additional events that may be reinforcing in their own right.

Feedback was used with a psychiatric patient who had a severe knife phobia (Leitenberg, Agras, Thompson, & Wright, 1968). The patient had obsessive thoughts about killing others when using a kitchen knife and became unable to look at or come into contact with sharp knives. She was told that practice in looking at the knife would help reduce her fear. The patient was told to look at a sharp knife displayed in a small compartment until she felt any discomfort or distress. Feedback was provided indicating how many seconds the patient kept the compartment open and exposed herself to the knife. Feedback steadily increased time of self-exposure to the knife. Adding praise (e.g., "That was great") to the feedback slightly augmented the effect of feedback alone. By the end of the study, the patient was able to use a knife to slice vegetables for use in the ward.

Feedback can be readily employed in most settings particularly where some performance criterion is explicit such as academic achievement or productivity on the job. In other situations, performance criteria can be set (e.g., number of hallucinatory statements of a psychotic patient, number of cigarettes smoked) and daily feedback can be delivered to provide information comparing performance with the criterion. When feedback is used, a criterion for performance is essential

(Locke, Cartledge, & Koeppel, 1968). By using feedback alone, extrinsic reinforcers which are not delivered as part of the routine need not be introduced.

There is a major potential limitation in applying feedback for therapeutic purposes. The effectiveness of feedback alone has been equivocal. In some studies, providing information about performance has not altered behavior (Kazdin, 1973a; Salzberg, Wheeler, Devar, & Hopkins, 1971; Surratt & Hopkins, 1970). Because feedback is implicit in the delivery of other reinforcers, these other reinforcers should be used along with explicit feedback.

Token Reinforcement

The delivery of tangible conditioned reinforcers (tokens) that are exchangeable for a variety of back-up events has distinct advantages over other reinforcing events (Ayllon & Azrin, 1968b; Kazdin & Bootzin, 1972; Sherman & Baer, 1969). First, tokens are potent reinforcers and can often maintain behavior at a higher level than other conditioned reinforcers such as praise, approval, and feedback (Birnbrauer *et al.*, 1965; Kazdin & Polster, 1973; Locke, 1969; O'Leary, Becker, Evans, & Saudargas, 1969; Staats *et al.*, 1962; Zimmerman, Stuckey, Garlick, & Miller, 1969). Second, tokens bridge the delay between the target response and back-up reinforcement. If a reinforcer (e.g., an activity) cannot immediately follow a target behavior, tokens which can be used to purchase a back-up reinforcer later can be delivered immediately. Third, since tokens are backed up by a variety of reinforcers, they are less subject to satiation than are other reinforcers. If a client is no longer interested in one or two back-up reinforcers, usually there are many other reinforcers which are of value. For example, if an individual is satiated with food, nonfood items can be purchased with tokens. Fourth, tokens can be easily administered without interrupting the target response. Since the reinforcer does not require consumption (such as food) or performance of behaviors which may be incompatible with the target response (such as participating in a special activity), it does not disrupt behavior. Fifth, tokens permit administering a single reinforcer (tokens) to several individuals who ordinarily may have different reinforcer preferences. Individual preferences can be exercised in the exchange of tokens for back-up reinforcers. Hence, there is less concern with the reinforcers having value to only a few individuals in the setting. Sixth, tokens permit parceling out other reinforcers (e.g., activities) which might have to be earned in an all-or-none fashion. The tokens can be earned toward the purchase of the back-up reinforcer. Finally, from a practical standpoint, tokens are usually easier to administer than some other back-up events, easier to carry on one's person for both the staff and clients, and provide an immediate quantitative basis to evaluate how well one is doing in the program.

There are potential disadvantages in employing tokens. First, the tokens and the back-up events for which they can be exchanged usually introduce artificial events into the setting. If reinforcers not usually available are introduced for the program, they must at some point be withdrawn. Tokens constitute a reinforcing event not available in most settings (excluding tokens as money and grades).

Because the delivery of tokens is clearly associated with reinforcement of desirable behavior, they may exert stimulus control over that behavior. Clients learn that the presence of tokens signals that desirable behavior is reinforced and the absence of tokens signals that desirable behavior is not likely to be reinforced. Once tokens are withdrawn, desirable behavior may decline. Specific procedures need to be implemented to withdraw the token program without a loss of behavior gains. In some settings, conditioned reinforcers normally available such as grades, money, and praise can be substituted for tokens. In any case, a potential problem is removing the token system after behavior gains have been made and transferring control of behavior to naturally occurring events such as privileges, activities, and social reinforcers. (Chapter 7 discusses techniques to maintain changes after a token reinforcement program is withdrawn.)

Other disadvantages of the use of tokens pertain to handling tokens and monitoring their use. One problem is that individuals may obtain tokens in unauthorized ways. For example, clients may steal tokens from each other or perform behaviors (e.g., sexual favors) to earn tokens from other clients rather than perform target behaviors as part of the program (Liberman, 1968). If tokens can be obtained without performing the target responses, their effect on behavior will decrease. Stealing has been combatted by individually coding tokens for each client in a program. Another problem is the loss of tokens on the part of clients. If the client has responsibility for handling the tokens, the possibility exists that the tokens will be lost. Finally, the procedures needed to monitor the administration of tokens are much more complex than those needed to monitor other reinforcers because of the complexity of the interactions involving tokens (earnings, expenditures, and fines). Thus, token systems often provide more work for staff who manage the system.

All reinforcers have limitations with respect to specific purposes, settings, clients, and constraints of the situation. However, it is important to keep in mind two major points which recommend token reinforcement. First, tokens tend to be extremely effective as a reinforcer and over the long run tend to maintain behaviors better than other events. Superiority of tokens over other events derives from the capacity of tokens to incorporate all other reinforcers that might be used in their own right such as food, activities, and attention from others (e.g., spending time with someone), all of which can be purchased with tokens. Second, the use of token reinforcement permits the extension of reinforcement techniques on a larger scale than is ordinarily available with other reinforcers. A large number of target behaviors, clients, and back-up events can be incorporated into a single system. The system allows great diversity in designing programs and in delivering or withdrawing consequences for behavior, as will be evident in the next chapter.

CONCLUSION

Reinforcement techniques have been in use in some form throughout history. Of course, the terminology and technical and methodological sophistication that characterize contemporary applications are quite recent. The development of operant conditioning initially in laboratory settings and eventually in clinical and educational settings underlies contemporary applied behavior analysis. Investigators gradually extended operant conditioning to human behavior initially with psychiatric patients, children, and the mentally retarded. Laboratory extensions with clinical populations led to a focus on increasingly clinically relevant behavior. Responding to laboratory tasks was shown in select cases to alter behaviors that were clinically significant. Also, specific extensions focused on target behaviors of direct clinical import. Although there were select applications with increasingly clinical relevance, the programmatic applications with psychiatric patients by Ayllon and his colleagues and with children by Wolf and his colleagues are the major landmarks in the history of applied behavior analysis.

The development of token reinforcement has a related history gradually evolving from animal laboratory research, through extensions to human populations, and finally in the form of token economies. Interestingly, rather complete token economies bearing resemblance to contemporary programs were in wide use in several countries over 150 years ago. While these programs are of historical interest, they did not serve as the basis of current programs. The token economy, as it is now known, was formally developed by Ayllon and Azrin at Anna State Hospital. The program began in the early 1960s and exerted immediate influence on other treatment, rehabilitation, and educaitonal programs even before Ayllon and Azrin published their first report. Many contemporary applications have been modeled after this first program. However, over the years the token economy has developed considerably to include a wide range of variations and techniques. The next chapter describes the basic requirements of a token economy, provides examples of contemporary programs in different settings, and elaborates the options and variations available.

THE TOKEN ECONOMY: PROGRAM OPTIONS AND VARIATIONS

3

The present chapter describes the essential features of the token economy and details examples of specific programs in different settings. There are relatively few essential features of a token economy. Yet, several program variations, options, and ancillary procedures are used so that token economies vary considerably in their characteristics. Program variations and options frequently used in token economies and evidence pertaining to their effects on client behavior are also reviewed.

THE BASIC TOKEN PROGRAM AND ITS REQUIREMENTS

The token economy, as any other behavior modification program, has several basic requirements. The target behaviors need to be identified and assessed, experimental procedures used to evaluate the contingencies must be selected, and so on. Aside from these requirements, there are various features peculiar to a token system including selection of the token or medium of exchange, selection of back-up reinforcers, and specification of the contingencies so that performance of the target behaviors is translated into points and points can be translated into back-up events.

The Token

The medium of exchange that serves as the token must be selected. A wide range of tokens has been used including poker chips, tickets, stars, points on a tally sheet or a counter, checkmarks, punched holes on cards, currency or coins individually devised for the program, foreign coins, plastic credit cards, computer cards, and others. The items used for tokens may have to meet specific requirements depending upon the setting, population, and in some cases special problems that arise.

One requirement is that the token is a unique event that cannot be easily duplicated or counterfeited by the clients. For example, simple checkmarks on a card might not be appropriate for many populations because the marks can be duplicated and clients can introduce tokens into the system. Another requirement of tokens is that they are easily administered. In some programs, tokens are carried by the staff so that they can administer tokens immediately as they see appropriate behavior. Thus, tokens that can be carried easily sometimes are selected.

Tokens sometimes are individually marked so that they are not transferable between clients. The need to individually mark tokens has developed in part because of reports that some clients steal tokens from other clients or from a token bank (source of supply) (Gates, 1972). If the tokens are individually marked, the incentive for theft is reduced. (Interestingly, clients have still been reported to steal individualized tokens from others even though they cannot be spent [Lachenmeyer, 1969].) Another reason to individualize tokens is that token earnings and expenditures can be more easily monitored for a given client if all the tokens used by a given person are marked.

Currency devised specifically for the program often is used in token economies. Paper money usually specifies the number of tokens earned and allows a place for the names of the client who earned the token and the staff member who administered the token, the date of administration and the behavior which was reinforced (cf. Logan, 1970). Using currency that has to be completed by a staff member has the advantage of being the same for all clients until it is marked by the staff member which individualizes it for a given client. Another advantage of currency is that it can be stamped in various denominations. Also, in many programs, there are different types of tokens which vary according to the specific behavior that is reinforced. For example, one type of behavior may be reinforced with one type of token or a token of one color, and another behavior may be reinforced with another type of token. This is done to trace the amount of tokens earned for specific behaviors. If currency is used, it is easy to monitor the specific behaviors that are reinforced and to use the same token for all behaviors. If necessary, the behavior that was reinforced can be marked or checked on the token itself. A sample of paper money developed in a token economy is provided in Fig. 4.

Another type of token that is useful is punched holes in specially coded cards (Aitchison, 1972; Datel & Legters, 1970; Tanner, Parrino, & Daniels, 1975). Clients

Fig. 4. Example of the type of "paper money" used in a token-economy system (from Logan, 1970; Society for the Experimental Analysis of Behavior).

carry cards (or sometimes booklets) which are sectioned. A punched hole in a given section of the card or over a select number indicates the behaviors performed and/or the number of tokens earned. In some cases, the punches are special shapes (e.g., diamond shape) so that they are not easily reproduced by the clients. Client earnings are indicated by punched holes on a card. Expenditures are indicated either by a larger punched hole which eliminates the smaller hole that indicated earnings (Aitchison, 1972) or by punching a different section of the card (Foreyt, 1976). An advantage of the system of punching cards or booklets is that the tokens are easily administered. Also, in many systems where the punches are not altered when they are redeemed (i.e., are marked or circled rather than repunched), the card or booklet serves as a permanent record of an individual's earnings and can be used for data analyses over time.

In some token programs, a token, in the usual sense, is not used. For example, in a few programs a credit card system is used to handle all transactions related to token earnings (cf. Ayllon, Roberts, & Milan, 1977; Lehrer, Schiff, & Kris, 1970). Clients carry a credit card which identifies them by name and other data. When tokens are earned, the earnings are recorded on a slip of paper which is imprinted with the client's information in the fashion that credit card transactions normally occur at department stores and gas stations. The credit card system records the type of transaction made (e.g., points earned, lost, or spent for particular behaviors or back-up items). The client then receives a weekly statement indicating daily earnings and charges and the code to specify what each transaction reflected (Tanner *et al.,* 1975). As with credit cards in national economies, some individuals can overspend from their earnings and must be monitored to avoid illegal spending (Gates, 1972; Lehrer *et al.,* 1970).

Back-up Reinforcers

The events that back up the tokens have been diverse across token programs depending in part upon the setting in which the program is conducted and the population. In any given program, usually a variety of back-up events are used to enhance the value of the tokens. However, as discussed later, there are several exceptions to this. The back-up events often include consumables (e.g., food, cigarettes), high probability behaviors (e.g., leaving the ward, visiting with friends, engaging in leisure or free time, watching television), money, and select items such as clothes, cosmetics, small appliances, and luxuries as part of client living quarters which are provided on a purchase or rental basis. In institutional settings for diverse populations, access to visitors, private living quarters, recreational activities, extra time with a physician, psychologist, or social worker, passes off the grounds, and similar events are used as back-up events. In classroom settings, extra recess, free time, games, snacks, small toys and prizes, and school supplies often back up the value of tokens. In most settings, there is a "store" or canteen devised so that clients can exchange tokens for a large variety of back-up events. In institutions, a canteen or commissary serves in this capacity. The large variety of events that usually back up the tokens may be listed in the form of a menu which allows patients to clearly see those items available.

Aside from the usual back-up events listed above, many programs have introduced innovative and relatively complex events that can be purchased with tokens. For example, in some programs clients can select items from department or discount store catalogs and earn points toward these items (Cohen & Filipczak, 1971). In a program for prisoners, an interesting variation of this procedure was used. Inmates earned trading stamps which could be spent on items selected from a catalog by their relatives on the outside (Ayllon *et al.*, 1977).

An extremely interesting back-up event was used for young retarded adults in a day-care program designed to develop vocational skills (Welch & Gist, 1974). Among the many back-up events available was a "token economy insurance policy" which was designed to help the individual in case an extended illness prevented attendance at work. The client could purchase an insurance policy on a monthly basis which would allow him some token earnings even if he had an extended illness (verified by a physician).

In most token programs prices for back-up events are set so that an individual can spend points earned for minimal performance. While minimal performance does not lead to wealth, it should allow exposure to some sources of reinforcement. Back-up events ordinarily assume a wide range of values so that there is an incentive for earning higher amounts of tokens for the more highly prized events.

In some token economies, clients can only earn enough tokens to purchase essential back-up reinforcers on a short-term basis and an excess of tokens cannot be accumulated. More commonly, clients can earn more tokens than necessary for expenditures and thus accumulate a surplus of tokens. (As discussed later, the surplus of tokens has important implications for the therapeutic effects of token economies.) In most programs, individuals are permitted to save tokens, perhaps toward some expensive back-up event, or for time off from the token economy and the contingencies which control their behavior. A record is kept of excess earnings and clients may make deposits in much the same way as one would in a bank. When savings are accumulated, as has been the case in programs in hospitals, institutions, prisons, and classrooms, the surplus tokens an individual earns are merely recorded. In some programs, tokens can be exchanged for money and individuals are encouraged to save tokens so that they may have money upon their release (e.g., Karacki & Levinson, 1970).

In a few programs, banking arrangements for token earnings parallel banking services ordinarily provided in national economies involving both savings and checking services. Individuals may be able to write personalized checks for back-up events, a service which costs tokens (Welch & Gist, 1974). Even more elaborate banking services were provided in a token economy with delinquent youths (Cohen & Filipczak, 1971). Tokens earned primarily for academic performance could be exchanged directly for money (1 token = 1 penny). For a fee, an individual could open a savings account for their earnings. The system required a minimum deposit (of 1000 points equivalent to $10.00) and provided a bank book. Interest was provided for money in the savings account and early withdrawals (before a period of three months) cost a small "service fee." Loans were provided for various problems that arose in the token economy. For example, incoming clients could take out an "orientation loan" to purchase essential back-up events until pay day. This loan had a 1% interest charge. Other loans could be taken for educational fees (tuition for classes), emergency situations, or another unanticipated expense. Although complex banking arrangements can be found in several token programs, these programs are exceptions. In most programs, savings merely constitute a record of earnings which exceed expenditures and are not associated with interest or checking and loan privileges.

Specification of the Contingencies

The introduction of the program includes precise specification of the responses that are to be reinforced and punished and the specific token values associated with each response. In addition, the specific back-up events, the prices in terms of tokens, and the periods in which back-up events are made available are specified. To introduce the program, client manuals are sometimes written to detail the contingencies. The contingencies may be described orally or displayed in a conspicuous place. The manner of introducing the program, verbal, written, or perhaps no formal introduction at all, in part depends upon the level of the clients.

It is important to provide details of the program for several reasons. First, the contingent delivery of events in many treatment and rehabilitation settings differs from usual practices and thus diverges from the experiences of the clients. Second, an attempt is made to show that the contingencies are clearly specified in advance to the clients and that token earnings will not be delivered by fiat or whim on the part of the staff. Third, by establishing clear rules, performance may be facilitated. If clients are aware of the responses and the consequences associated with them, performance sometimes is enhanced to a greater extent than merely providing consequences without a clear specification of the behaviors and consequences (Ayllon & Azrin, 1964). Token reinforcement has effectively altered client behavior even when the contingencies are not made explicit (Kazdin, 1973g). In some programs, the contingencies are relatively complex because there are several behaviors that can earn or lose tokens and back-up events with a wide range of prices. The contingencies might not be easily memorized so a manual or posted set of guidelines provides a constant source of reference to both the clients and staff.

Aside from describing the contingencies, introducing the program requires establishing the value of the tokens. For some populations, it is sufficient to explain that tokens are earned for behavior and, once earned, can be exchanged for various events. After the explanation, tokens take on immediate value which is maintained by their actual exchange for other reinforcers. For individuals whose behavior is not controlled by instructions about the contingencies and the stated value of tokens, tokens may be administered noncontingently. Immediately after they are delivered, they can be exchanged for a back-up reinforcer (e.g., Ayllon & Haughton, 1962, Exp. 2; Ayllon, Layman, & Burke, 1972; Keilitz, Tucker, & Horner, 1973). For example, a psychiatric patient may be given a few tokens immediately prior to entering a dining room. After having the tokens for only a few moments, an attendant at the door may take them as the patient is allowed access to food. Informing the patients that they need tokens for access to reinforcing events and providing them noncontingently usually is sufficient to establish the value of the tokens.

EXAMPLES OF TOKEN ECONOMIES IN THREE SETTINGS

The basic requirements for token economies are illustrated in three examples which include a program in a psychiatric hospital (Ayllon & Azrin, 1965), a home-style cottage facility for pre-delinquents (Phillips, 1968), and an elementary school classroom (McLaughlin & Malaby, 1972a). The basic features of the programs will be described in detail. The specific experiments conducted in these and other programs will be reviewed in the next two chapters which detail the evidence pertaining to token economies across diverse settings and treatment populations.

Psychiatric Ward

A very influential token program was implemented by Ayllon and Azrin (1965) in a ward of over 40 psychiatric patients. Although the majority of patients were diagnosed as schizophrenic, some other diagnoses were included (mentally retarded and chronic brain syndrome). Patients earned tokens for performing various jobs on and off the ward. Job performance was selected as the target response because it was considered useful to the patient and because it could easily be observed. Completion of a job could be easily detected because of the change made in the physical environment (e.g., clean dishes or floor) and thus could be readily followed by specific consequences. The on-ward jobs included tasks related to the dietary, secretarial duties, housekeeping, laundry, grooming other patients, serving as a waitress, and performing special services. Table 2 provides a sample of only some of the jobs available. Included in the table is a list of self-care jobs that were reinforced with tokens. While patients did not all receive tokens for the same ward job, all could earn tokens for self-care behaviors. In addition, off-ward jobs were reinforced including assisting with dietary, clerical, laboratory, and laundry tasks.

The tokens could be exchanged for a large number of back-up reinforcers including selection of living quarters on the ward (among rooms differing in the amount of privacy), activities on and off the ward, and items at the commissary. The reinforcers available appear in Table 3. From the table, it is obvious that the program encompassed a variety of tangible items that could be purchased or high probability events that could be rented. Some of the reinforcers constituted increased access to events that were already available in small quantities on a noncontingent basis such as audience with a chaplain, physician, or psychologist.

Table 2.   Types and Number of On-Ward Jobs

Types of jobs	Number of jobs	Duration	Tokens paid
DIETARY ASSISTANT			
1. Kitchen Chores	3	10 min	1
Patient assembles necessary supplies on table. Puts one (1) pat of butter between two (2) slices of bread for all patients. Squeezes juice from fruit left over from meals. Puts supplies away. Cleans table used.			
2. Pots and Pans	3	10 min	6
Patient runs water into sink, adds soap, washes and rinses all pans used for each meal. Stacks pans and leaves them to be put through automatic dishwasher.			
WAITRESS			
1. Meals	6	10 min	2
Empties trays left on tables and washes tables between each of four (4) meal groups.			
2. Commissary	3	10 min	5
Cleans tables, washes cups and glasses used at commissary. Places cups and glasses in rack ready for automatic dishwasher.			
SALES CLERK ASSISTANT			
1. Commissary	3	30 min	3
Assembles commissary items. Displays candy, cigarettes, tobacco, cosmetics, dresses and other variety store items so that they can be seen by all. Prepares ice, glasses and cups for hot and cold beverages. Asks patient what she wishes to buy. Collects the tokens from patient and tells the secretary the name of the patient and the amount spent. Puts commissary supplies away.			
SECRETARIAL ASSISTANT			
1. Tooth Brushing	1	30 min	3
Assists with oral hygiene. Writes names of patients brushing teeth.			
2. Commissary			
Assists sales clerk assistant. Writes names of patients at commissary, records number of tokens patient spent. Totals all tokens spent.			
WARD CLEANING ASSISTANT			
1. Halls and Rooms	24	30 min	3
Sweep and mop floors, dust furniture and walls in seven rooms and hall.			
2. Special	1	30 min	4
Cleans after incontinent patients.			

continued

Table 2. (*continued*)

Types of jobs	Number of jobs	Duration	Tokens paid
ASSISTANT JANITOR			
1. Supplies	1	10 min	1
Places ward supplies in supply cabinets and drawers.			
2. Trash	3	5 min	2
Carries empty soft drink bottles to storage area, empties waste paper baskets throughout the ward and carries paper to container adjacent to building. Carries mop used during the day outside to dry.			
LAUNDRY ASSISTANT			
1. Hose	1	15 min	1
Match and fold clean anklets and stockings.			
2. Pick Up Service	1	60 min	8
Sorts dirty clothing and linens and puts items into bags marked for each item.			
GROOMING ASSISTANT			
1. Clothing Care	1	15 min	1
Patient sets up ironing board and iron. Irons clothing that belongs to patients other than self. Folds clothing neatly. Returns ironed clothing, iron and ironing board to nurses station.			
2. Bath	2	45 min	4
Patient assists with baths, washing, shampooing and drying. Cleans tub after each bath.			
RECREATIONAL ASSISTANT			
1. Walks	1	20 min	3
Assists ward staff when taking group of patients on walks. Walks in front of group.			
2. Exercise	1	20 min	3
Operates record player and leads patients in exercises.			
SPECIAL SERVICES			
1. Errands	1	20 min	6
Leaves the ward on official errands throughout the hospital grounds, delivering messages and picking up supplies and records pertaining to the ward.			
2. Tour Guide	1	15 min	10
Gives visitors a 15-min tour of the ward explaining about the activities and token system. Answers visitors questions about the ward.			
SELF-CARE ACTIVITIES			
1. Grooming			1
Combs hair, wears: dress, slip, panties, bra, stockings and shoes (three times daily).			

Table 2. (*continued*)

Types of jobs	Number of jobs	Duration	Tokens paid
2. Bathing			1
Takes a bath at time designated for bath (once weekly).			
3. Tooth Brushing			1
Brushes teeth or gargles at the time designated for tooth brushing (once daily).			
4. Exercises			1
Participates in exercises conducted by the exercise assistant (twice daily).			
5. Bed Making			1
Makes own bed and cleans area around and under bed.			

The tokens could be exchanged daily either at the commissary or, in the case of items that had to be rented, at the nurse's station. The metal disks that served as tokens could be exchanged without staff intervention since they operated turnstiles that regulated entrance to specified areas and operated a TV set.

Ayllon and Azrin (1965, 1968a, 1968b) completed several investigations evaluating the effect of token reinforcement in determining selection and performance of particular jobs plus a variety of specific features of the program. (These experiments will be treated separately in the next chapter.)

Home-Style Facility for Pre-Delinquents

A token program for pre-delinquent youths has been implemented at a home-style community-based facility in Kansas. The facility, referred to as Achievement Place, serves youths under 16 years of age who have committed various offenses such as theft, fighting, school truancy and have been processed through Juvenile Court. The boys are referred to as pre-delinquent by the court because they were considered to be likely to advance to more serious crimes if steps were not taken to alter their behavior. Achievement Place houses a small number of boys and a married couple, referred to as teaching parents. The teaching parents are responsible for administering the program designed to develop a wide range of behaviors which ultimately result in return of the individual to the community and natural living environment. As will be reviewed later, the program at Achievement Place has been evaluated in a large number of experiments and the program has evolved as a function of that research.

The basic program, as initially reported, included a token economy that focused upon social, self-care, and academic behaviors (Phillips, 1968). The behaviors that earned and lost points are summarized in Table 4. Information in this table was displayed on a bulletin board so that the residents were familiar with the contingencies. The table reveals that most of the behaviors could be reinforced either once or many times per day. The tokens (points accumulated on 3 by 5

Table 3.   List of Reinforcers Available for Tokens (Ayllon & Azrin, 1965)

	Number of tokens daily		Tokens
I. Privacy		**IV. Devotional opportunities**	
Selection of room 1	0	Extra religious services on ward	1
Selection of room 2	4	Extra religious services off ward	10
Selection of room 3	8	**V. Recreation opportunities**	
Selection of room 4	15	Movie on ward	1
Selection of room 5	30	Opportunity to listen to a live band	1
Personal chair	1	Exclusive use of a radio	1
Choice of eating group	1	Television (choice of program)	3.
Screen (room divider)	1	**VI. Commissary items**	
Choice of bedspreads	1	Consumable items such as candy, milk, cigarettes, coffee, and sandwich	1–5
Coat rack	1		
Personal cabinet	2		
Placebo	1–2		
II. Leave from the ward		Toilet articles such as Kleenex, toothpaste, comb, lipstick, and talcum powder	1–10
20-min walk on hospital grounds (with escort)	2		
30-min grounds pass (3 tokens for each additional 30 min)	10	Clothing and accessories such as gloves, headscarf, house slippers, handbag, skirt	12–400
Trip to town (with escort)	100	Reading and writing materials such as stationery, pen, greeting card, newspaper, and magazine	2–5
III. Social interaction with staff			
Private audience with chaplain, nurse	5 min free	Miscellaneous items such as ashtray, throw rug, potted plant, picture holder, and stuffed animal	1–50
Private audience with ward staff, ward physician (for additional time—1 token per min)	5 min free		
Private audience with ward psychologist	20		
Private audience with social worker	100		

Table 4. Privileges That Could Be Earned Each Week with Points (Phillips, 1968)

Privileges for the week	Price in points
Allowance	1000
Bicycle	1000
TV	1000
Games	500
Tools	500
Snacks	1000
Permission to go downtown	1000
Permission to stay up past bedtime	1000
Permission to come home late after school	1000

Table 5   Behaviors and the Number of Points That They Earn or Lose (Phillips, 1968)

Behaviors that earned points	Points
1. Watching news on TV or reading the newspaper	300 per day
2. Cleaning and maintaining neatness in one's room	500 per day
3. Keeping one's person neat and clean	500 per day
4. Reading books	5 to 10 per page
5. Aiding house-parents in various household tasks	20 to 1000 per task
6. Doing dishes	500 to 1000 per meal
7. Being well dressed for an evening meal	100 to 500 per meal
8. Performing homework	500 per day
9. Obtaining desirable grades on school report cards	500 to 1000 per grade
10. Turning out lights when not in use	25 per light

Behaviors that lost points	Points
1. Failing grades on the report card	500 to 1000 per grade
2. Speaking aggressively	20 to 50 per response
3. Forgetting to wash hands before meals	100 to 300 per meal
4. Arguing	300 per response
5. Disobeying	100 to 1000 per response
6. Being late	10 per min.
7. Displaying poor manners	50 to 100 per response
8. Engaging in poor posture	50 to 100 per response
9. Using poor grammar	20 to 50 per response
10. Stealing	10,000 per response

index cards that the boys carried) could be exchanged for various privileges. A list of some of the basic reinforcers is provided in Table 5. There were other reinforcers that were not always available or were unavailable to all individuals and are not listed in the table. For example, choosing where one would sit when riding in the car or serving as a manager to supervise the work of other boys were auctioned to the highest bidder of tokens. An interesting feature of the program is that the reinforcers were bought on a weekly basis. At the end of each week, the boys could trade their points for specific privileges during the next week.

As the program evolved, changes were made in the way in which tokens were exchanged for back-up reinforcers (Phillips, Phillips, Fixsen, & Wolf, 1971). Initially, back-up reinforcers could be bought on a daily or weekly basis. Individuals who came to the facility were first placed on a daily system whereby tokens earned on a given day could be exchanged for privileges on the next day. This minimized the delay between earning and exchanging tokens. Second, individuals who purchased back-up events on a weekly basis were required to earn a minimum level of points on a given day to be able to use the privilege (e.g., watching TV) on the next day. This ensured maintenance of high-level performance even though the back-up reinforcer had been purchased. Finally, individuals could buy basic privileges of the house as a package. These privileges included the use of tools, telephone, radio, recreation room, and the privilege of going outdoors. These privileges had to be purchased before any other privilege could be bought because their utilization was difficult to monitor on the part of the teaching parents.

A large number of behaviors have been altered as part of the program well beyond those listed in the table of behaviors that earned tokens. In addition, the token program at Achievement Place has developed innovative variations of the basic contingencies such as peer delivery of token reinforcement, self-government, and others, which will be detailed later.

Elementary School Classroom

Token economies are frequently established in classroom settings. In the usual program, attention to the lesson or studying as well as academic performance are reinforced with tokens. For example, in the program by McLaughlin and Malaby (1972a) diverse behaviors were reinforced with points in a classroom of fifth and sixth grade students. The behaviors that earned or lost points are listed in Table 6. The behaviors encompassed general deportment, studying, and completing math, spelling, language, and handwriting assignments. The back-up reinforcers

Table 6. Behaviors and the Number of Points That They Earned or Lost (McLaughlin & Malaby, 1972a)

Behaviors that earned points	Points
1. Items correct	6 to 12
2. Study behavior 8:50–9:15	5 per day
3. Bring food for animals	1 to 10
4. Bring sawdust for animals	1 to 10
5. Art	1 to 4
6. Listening points	1 to 2 per lesson
7. Extra credit	Assigned value
8. Neatness	1 to 2
9. Taking home assignments	5
10. Taking notes	1 to 3
11. Quiet in lunch line	2
12. Quiet in cafeteria	2
13. Appropriate noon hour behavior	3

Behaviors that lost points	Points
1. Assignments incomplete	Amount squared
2. Gum and candy	100
3. Inappropriate verbal behavior	15
4. Inappropriate motor behavior	15
5. Fighting	100
6. Cheating	100

Table 7. Weekly Privileges (McLaughlin & Malaby, 1972a)

Privilege	Price in points	
	Sixth	Fifth[a]
1. Sharpening pencils	20	13
2. Seeing animals	30	25
3. Taking out balls	5	3
4. Sports	60	40
5. Special writing on board	20	16
6. Being on a committee	30	25
7. Special jobs	25	15
8. Playing games	5	3
9. Listening to records	5	2
10. Coming in early	10	6
11. Seeing the gradebook	5	2
12. Special projects	25	20

[a]Privilege costs were determined separately for the two grades. Also, fifth-grade children were in the room less time per day than were the sixth-grade children.

consisted of privileges normally available in the classroom. They were selected by having the children enumerate and rank in order of preference those activities in the classroom that they considered to be privileges. The privileges were priced so that the more preferred privileges cost more tokens than the less preferred privileges. The privileges and their prices are listed in Table 7. Daily assignments were corrected by the students (in some medium other than that used to complete the assignment such as ink). Students were responsible for keeping records of their own transactions (earnings and losses) which were verified by a separate record of the teacher. One student served as a banker who subtracted points from a student's point total when a privilege was purchased and initialed the transaction.

PROGRAM VARIATIONS, OPTIONS, AND THEIR EFFECTS

In the previous discussion, the basic requirements of a token economy were described and exemplified. Aside from the essential ingredients which define a token program, there are several optional features and variations that are used. The present discussion reviews major optional features that either have been shown to contribute to the efficacy of the program or represent widely utilized variations of the basic economy. The features include variations in the types of contingencies, the methods of administering the consequences, the range and extent of back-up reinforcers, the use of token withdrawal or response cost, the use of leveled programs, and peer- or self-administration of the token reinforcement contingencies.

TYPES OF CONTINGENCIES

The type of contingency refers to the response criterion used as a basis for delivering reinforcing consequences. In token economies, one of three types of contingencies is usually used, namely, individualized, standardized, and group contingencies.

Individualized contingencies refer to those contingencies in which the criterion for reinforcement or the specific consequences selected are designed for a particular client. If several clients are in the program, the contingencies might be individualized for each of them. Thus, different behaviors (e.g., talking versus remaining quiet), different criteria for the same behaviors (e.g., number of correct responses on an academic assignment) or different back-up events (e.g., food or special privileges) might be used across individuals in the program.

The advantage of an individualized contingency may be obvious. Because it is designed for a particular individual, the performance characteristics of the individual client are taken into account. The initial level of behavior already apparent in the client's repertoire, the unique reinforcers for which he or she will respond, and the rate of altering the criteria for reinforcement as the client approaches a terminal goal is oriented to the client's progress. Individualization of the contingencies maximizes the likelihood that the parameters of reinforcement are administered for a given person in an optimal fashion. The constraints of utilizing individualized contingencies are equally obvious. In settings where there are low staff-client ratios, administration of individualized programs is prohibitive. Staff simply cannot monitor different behaviors or a single behavior with different performance criteria across individuals.

As will be evident in the review of token programs, there are situations well suited to individualized contingencies. For example, programs conducted with a child or spouse in the home or in outpatient behavior therapy have utilized individualized token reinforcement contingencies. In these situations, someone can closely monitor and administer the program and respond to changes in client performance. In institutional settings, individualized programs are used occasionally. Clients may be removed from the ward for special treatment sessions in which select behaviors are focused upon. In general, there are few programs where several individuals are simultaneously placed on individualized contingencies.

Standardized contingencies refer to contingencies that apply widely across several individuals in a given program. The criterion for reinforcement is the same for each of the clients. Thus, the standardized contingency is not designed for a particular individual's behavior. Rather, it is used in group situations where performance of a particular response on the part of all clients is desired. The contingency specifies that a certain number of tokens are provided for a given behavior and each individual who meets the standard criterion receives the tokens. For example, attending activities and completing jobs on the ward in a psychiatric hospital or completing homework and complying with instructions in a classroom usually are encompassed in standardized contingencies. Most token programs rely heavily upon standardized contingencies. Indeed, the three programs detailed at the beginning of this chapter relied heavily upon standardized contingencies.

Standardized contingencies are relatively easily administered in settings where there are low staff-client ratios. Although consequences are administered on an individual basis, the criterion for reinforcement does not vary across clients. Thus, staff need only remember that when a given behavior is performed (e.g., making one's bed), a certain unvarying number of tokens is delivered. Ideally, standardized contingencies are used in situations where everyone in the setting is expected to perform a particular response (e.g., routine behaviors) and the response is already in the repertoire of many clients. If the response is not in a given client's repertoire, it may have to be shaped gradually. An individualized rather than standardized contingency may be more well suited to this objective.

Group contingencies refer to contingencies in which the criterion for reinforcement is based upon performance of the group as a whole. Whereas the standardized contingency applies the same criterion to each client individually, the group contingency considers the behavior of all individuals as a unit, i.e., a single organism. There are several variations of the group contingency, some of which have been particularly well investigated in the classroom (cf. Greenwood, Hops, Delquadri, & Guild, 1974; Litow & Pumroy, 1975; McLaughlin, 1974).

In the most commonly used variation of the group contingency, the performance of the group as a whole determines the consequences that are administered. Essentially, performance of each individual determines the consequences received by the group and the collective performance of the group determines the consequences received by the individual. In one variation of this, the average performance of the group may serve as the basis for reinforcement. For example, the performance of all group members for a target behavior (e.g., arithmetic problems of students) is averaged and the consequences are delivered if the average performance meets a predetermined criterion (cf. Hamblin, Hathaway, & Wodarski, 1974).

An excellent illustration of a group contingency was provided by Schmidt and Ulrich (1969) who controlled excessive noise in a fourth grade classroom. The students were told that when a timer sounded at the end of a 10-minute period, they would receive two extra minutes of gym plus a two-minute break to talk if the class had not been noisy. The noise level registered automatically by a sound-level meter which was monitored by an observer. If the noise level surpassed the criterion level, the timer was reset. Noise decreased when the group contingency was in effect. The program clearly illustrates a contingency where the performance of an individual influenced the group and performance of the group influenced the consequences delivered to the individual.

In another variation of the group contingency, the group is divided into subgroups each of which functions on a separate group contingency. As in the previous variation, an individual can still earn or lose for the group and the collective behavior of the group determines what consequences the individual receives. However, the subgroups compete against each other. The consequences are delivered to the subgroup or team with the better performance.

For example, Maloney and Hopkins (1973) evaluated the effect of a team contingency in developing writing skills for elementary school students attending a remedial summer school session. Students were matched according to their baseline writing skills and randomly assigned to one of two teams. When the contingency was in effect, students received points for writing behaviors such as the use of different adjectives, action verbs, and novel sentence beginnings within a given composition. The team that earned the higher number of points of those possible from an accumulation of all individual members was allowed to go to recess five minutes early and receive a small piece of candy. To ensure that excellent performance was reinforced, both teams could win on a given day if performance met a prespecified high criterion. The team contingency increased the use of specific writing responses.

The efficacy of the team contingency has been demonstrated in several studies (e.g., Barrish, Saunders, & Wolf, 1969; Medland & Stachnik, 1972). Evidence suggests that dividing a group into teams accelerates performance over and above use of the group as a single unit (Harris & Sherman, 1973). Thus, when possible it might be useful to rely upon teams to accelerate performance.

In other variations of the group contingency, performance of select individuals (or a subgroup of individuals) serves as the criterion for reinforcement. For example, the consequences delivered to the group might be determined by the individual with the best performance on a given day or by the average performance of a select number of individuals with the best performance. In a classroom situation, the consequences provided might be determined by the individual who performs least disruptively or achieves the highest academic performance on a given day (cf. Drabman, Spitalnik, & Spitalnik, 1974; Hamblin *et al.*, 1974). The individual whose performance is the best on a given day might change on a daily basis. Similarly, the consequences for the group can be provided on the basis of the worst performance by a given individual. Another variation that has been used is to randomly select an individual from the group each day and to use his or her performance as the criterion to administer the reinforcer to everyone else in the group (Drabman *et al.*, 1974).

Frequently, group contingencies are combined with other types of contingencies. For example, in a token program in a psychiatric hospital, patients were placed into small decision-making groups (Greenberg, Scott, Pisa, & Friesen, 1975). The task set for the group was to develop their own treatment programs for their members and to make recommendations to the staff that would ensure a group member's progress toward eventual discharge from the hospital. The recommendations were made in the form of proposals that were evaluated by the staff. The proposals were differentially graded so that proposals designed to develop behaviors that were likely to result in discharge were reinforced with tokens. A group contingency was used so that each patient received tokens as a function of the overall behavior of the group, i.e., developing treatment proposals. A patient determined the points earned for the group by contributing to the proposals at group meetings. Aside from the group contingency, standardized contingencies were also used so that an individual could earn points for himself for such behaviors as working on and off the ward, grooming, and engaging in social behavior.

The relative effectiveness of different types of contingencies has been evaluated in relatively few studies. The comparison most commonly reported is between group contingencies and standardized or individualized contingencies (McLaughlin, 1974). The results have been equivocal. While some programs have shown that the target behavior changes to a greater extent under a group contingency than under a standardized contingency based upon individual performance (e.g., Brown, Reschly, & Sabers, 1974; Hamblin et al., 1974; Long & Williams, 1973), others have not (e.g., Axelrod, 1973; Drabman et al., 1974; Grandy, Madsen, & De Mersseman, 1974; Herman & Tramontana, 1971). In general, unambiguous comparisons of individualized and group contingencies have been difficult to achieve in some studies because the amount and schedule of providing reinforcing consequences or the sequence with which conditions are presented are confounded with the type of contingencies (e.g., Long & Williams, 1973).

There is an interesting difference between individualized or standardized contingencies and group contingencies. Programs have reported that group contingencies sometimes generate interactions between clients that are not present with other types of contingencies. The interactions sometimes include activities to exert pressure on select individuals to behave in a way to maximize the group's earnings. For example, investigators have reported that peers make threatening verbalizations and gestures or have chastised target subjects for not earning the reinforcers (Axelrod, 1973; Harris & Sherman, 1973; Packard, 1970; Schmidt & Ulrich, 1969). Not all of the interactions generated by a group contingency are aversive. For example, in classroom situations, peers have been found to help each other in a variety of ways such as helping "slower" students with academic assignments when a group contingency is implemented (Hamblin et al., 1974; Wodarski, Hamblin, Buckholdt, & Ferritor, 1973).

As with many other options available in token programs, the type of contingency used is a function of the specific goals and practical constraints of the program. Each of the different types of contingencies might be used in a given program. Perhaps, more commonly, individualized contingencies are combined with either standardized or group contingencies. For example, in a psychiatric hospital, standardized contingencies are commonly used to increase the performance of routine ward behaviors of all of the patients. Individualized contingencies might also be included to alter the specific bizarre or idiosyncratic behaviors of individual patients. While the standardized contingencies might constitute the majority of the contingencies in this situation, the individualization ensures a specific treatment focus.

Group contingencies provide an extremely convenient way to implement a program (McLaughlin, 1974). For example, when students receive consequences on a group basis, tokens, checkmarks, and points can be placed or removed from the board rather than administered separately to each individual (Axelrod, 1973; Kazdin & Forsberg, 1974). The ease of administration of group contingencies may explain why teachers sometimes prefer them over individualized contingencies (Drabman *et al.,* 1974; Rosenbaum, O'Leary, & Jacob, 1975).

An advantage of group contingencies may be the peer monitoring of behavior that sometimes results. If peers monitor behavior and enforce the contingencies, behavior is less likely to come under the narrow stimulus control of the staff members who administer the program (Kazdin, 1971b). In many programs, staff and peers seem to be reinforcing incompatible or opposite behaviors (e.g., nondisruptive versus disruptive behavior in a classroom). For example, with delinquents, reports have suggested that staff reinforce conformity with institutional rules whereas peers reinforce rule violations (Buehler, Patterson, & Furniss, 1966; Lachenmeyer, 1969). Similarly, with psychiatric patients, staff may reinforce behaviors reflecting independence whereas patients reinforce behaviors among their peers reflecting dependence (Wooley & Blackwell, 1975). Group contingencies might align staff and peer consequences toward the same objectives. If the group is involved in the contingency, socially reinforcing peer consequences for undesirable behavior are unlikely (Sulzbacher & Houser, 1968).

ADMINISTRATION OF CONSEQUENCES

The type of contingencies refers primarily to the different criteria that are used for delivering consequences. The *criteria* that are used for providing reinforcing consequences can be distinguished from the *manner* by which the consequences are administered or dispensed. Two ways of administering consequences can be distinguished.

Consequences can be *individually administered* which denotes that each individual in the program receives tokens. The administration of the consequences is independent of the type of contingency that is used. For example, if the client's behavior is governed by an individualized, standardized, or group contingency, this leaves unspecified how the consequences are finally administered. In fact, in most programs an individual receives consequences no matter what the criteria are for reinforcement. For example, in a program with a group contingency, the group members usually receive consequences earned by the group on an individual basis. As an example, Axelrod (1973) utilized individualized and group contingencies across different phases in two special education classes. Under each condition, the tokens were administered on an individual basis. Thus, in the group contingency, the behavior of the class as a whole served as the criterion for receiving tokens. However, when the tokens were earned, each individual received the select amount of tokens and could then exchange tokens for individually selected back-up events such as candy and toys.

Consequences can also be *group administered*. Group administration entails providing a reinforcing event to the clients as a single unit rather than on an individual basis. For example, Kazdin and Forsberg (1974) provided tokens to the class as a whole based upon the group's performance. Rather than administering tokens to each individual for the performance of the group, the tokens earned by the class as a whole were placed on the board. When a certain number of tokens accumulated, the class earned a special back-up event or activity that was administered to the group (e.g., recess).

The type of contingency (individualized, standardized, and group) can be combined with the various methods of administering consequences (individual or group). Figure 5 illustrates the combinations. Cell 1 depicts programs where the criterion for reinforcement and manner of administering tokens are individualized. Cell 2 shows the programs where the criterion for reinforcement is standardized across all clients and they receive their tokens individually on the basis of meeting the standard. The majority of programs fall within cell 2 where a common standard for several behaviors is reinforced and one's earnings are determined by the number of performances which meet the standard. Cell 3 constitutes the majority of group contingency programs where the criterion for reinforcement is determined by the performance of the group as a unit but the consequences are delivered on an individual basis. Cells 4 and 5 represent infrequently used program options. In programs based upon these cells, consequences are provided to the group as a unit if each individual has met his or her individually set performance criterion (Cell 4) or if each individual has met the standard criterion (Cell 5). These variations constitute complex types of group contingencies that are not usually used as the main program. The final type of program, as specified in Cell 6, involves the delivery of group consequences on the basis of group performance and has been referred to in the initial discussion of group-administered consequences. Programs using group contingencies and group-administered consequences are relatively easy to administer although they have not been used very frequently.

There is a major program variation involving primarily the method of administering consequences that is not represented in Fig. 5. This variation involves designing a contingency for a single individual (or select subgroup of individuals) and having the individual share the earned consequences with others. The contingency is individualized because the response requirements for reinforcement are

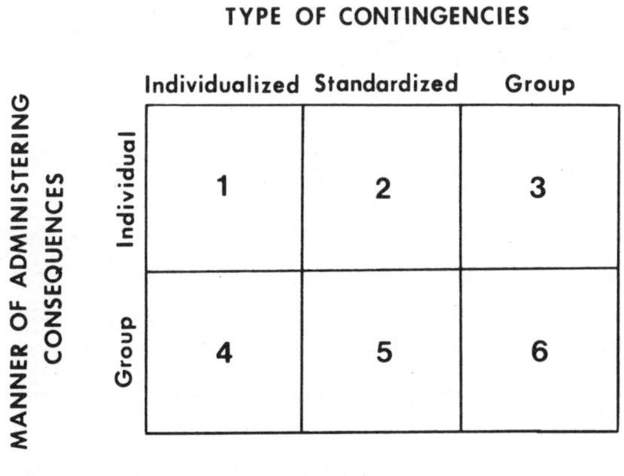

TYPE OF CONTINGENCIES

Fig. 5.   Program variations based upon the type of contingencies and the manner of administering consequences (original figure; no copyright).

suited to the particular client's behavior. Yet, the consequences are administered to the group. This program does not fit into Cells 1 and 4, the most closely related variations, because the group members who share the consequences earned by the target subject do not necessarily have specific contingencies that govern their own behavior. This type of contingency might be viewed as *consequence sharing* because the group's formal involvement is entailed only by sharing the reinforcing consequences earned by one individual.

Consequence sharing is used in many situations where a single client is focused upon in a group and it is desirable to incorporate peer group influences into the contingency. For example, an individualized contingency with consequence sharing was used in a classroom program designed to develop sitting and remaining quiet in an extremely disruptive first grade boy (Kubany, Weiss, & Sloggett, 1971). The boy was monitored by a timer that ran only when he was performing appropriately. When a prespecified amount of time had accumulated, he earned treats including trinkets and candies. These were placed in a "sharing jar" where they remained until the end of the day. At the end of the day, he took one of the treats for himself and distributed the remaining ones to the rest of the class. Thus, the class earned treats based solely on the contingency designed for the target subject. In addition to earning treats for others, the boy also earned tokens (gummed stars) for himself. These were placed on a sheet to plot his progress in earning stars but were not exchangeable for other events. As a result of the group contingency, combined with the individual token program, disruptive behavior decreased substantially.

A number of token programs with psychiatric patients, the mentally retarded, and "emotionally disturbed" and "normal" children have demonstrated the efficacy of reinforcer sharing to alter the behavior of select clients (Feingold & Migler, 1972; Jones & Kazdin, 1975; Morgan, 1975; Patterson, 1965; Rosenbaum *et al.,* 1975; Wolf, Hanley, King, Lachowicz, & Giles, 1970). An interesting feature of consequence sharing is that the peers often directly support the behavior of the target subject. For example, in a program for a psychiatric patient, one peer for whom tokens were earned prompted the patient, helped her perform self-care behaviors, and assisted with ward jobs to increase the likelihood of the patient's completion of the target responses (Feingold & Migler, 1972).

RANGE AND EXTENT OF BACK-UP REINFORCERS

With few exceptions (mentioned below), the token economy depends heavily upon the extent and reinforcing capacities of the back-up reinforcers. Thus, an important area across which programs may vary is the number and kinds of events for which tokens may be exchanged. Despite the obvious import of back-up reinforcers, relatively few studies have systematically evaluated the relationship between the events which back up the tokens and behavior. Some basic questions have been addressed. For example, the contingent delivery typically results in greater behavior change than the noncontingent delivery of the back-up reinforcers (Arann & Horner, 1972; Bushell, Wrobel, & Michaelis, 1968). However, a number of variations available in selecting back-up events have not been explored. The variations include the number and kinds of back-up reinforcers made available and, indeed, whether back-up reinforcers are even provided at all.

In many programs, tokens have been used successfully *without* specific back-up reinforcers (Ayers, Potter, & McDearmon, 1975; Deitz & Repp, 1974; Haring & Hauck, 1969; Jens & Shores, 1969; Repp, Klett, Sosebee, & Speir, 1975; Smith, Brethower, & Cabot, 1969; Sulzer, Hunt, Ashby, Koniarski, & Krams, 1971). Token reinforcement contingencies without back-up events usually are less effective than contingencies with back-up events (Drabman, 1973; Hall, Axelrod, Tyler, Grief, Jones, & Robertson, 1972, Exp. 2; Zimmerman *et al.,* 1969), although there are exceptions (Ribes-Inesta, Duran, Evans, Felix, Rivera, & Sanchez, 1973). However, it is important to note in passing some effective programs where tokens are not backed by specific reinforcers and their effects on behavior.

For example, a token program conducted in the home was designed to develop performance of household chores of a 10-year-old girl (Hall *et al.,* 1972, Exp. 2). The girl did not perform routine tasks such as cleaning her room, hanging up her clothes, making her bed, doing odd jobs, and similar tasks. After baseline observations, the girl was told that she would receive a specified number of points for each chore completed. The points earned were recorded each evening and graphically displayed. Although the points could not be exchanged for back-up reinforcers, the girl increased her performance of chores markedly over baseline performance. In subsequent phases, points could be exchanged for pennies, later for the uniform of an organization to which she belonged, and finally toward the purchase of Christmas gifts. When tokens were exchangeable for back-up reinforcers, the completion of household chores increased above the level achieved when no back-up reinforcers were provided.

Similarly, in a sheltered workshop setting, multiply handicapped clients were told about a token economy prior to its implementation (Zimmerman *et al.,* 1969). Prior to the program, clients were told how many points they would receive for the work they completed had the program been in effect. No tokens were actually given nor were the back-up reinforcers available. Yet, client work performance increased relative to baseline. In a subsequent phase, points finally were delivered and exchangeable for back-up reinforcers which resulted in further increases in performance.

The studies in which tokens or points are shown to be effective without back-up reinforcers might be viewed as demonstrating the effect of feedback or knowledge of results on performance (Kazdin, 1975a). Several studies in applied settings have shown the effect of merely informing someone about their performance on behavior (e.g., Drabman & Lahey, 1974; Jens & Shores, 1969; Leitenberg *et al.,* 1968). In the majority of programs, tokens without back-up reinforcers have little or no effect on behavior (e.g., Long & Williams, 1973). Thus, the question might arise as to the value of excluding back-up reinforcers. In many programs, it may be desirable to attempt to alter behavior with less potent reinforcers such as tokens without back-up events or feedback alone. If back-up reinforcers are not used for the tokens, the problems associated with withdrawing the tokens and maintenance of the target behavior appear to be less (cf. Drabman, 1973; Jones & Kazdin, 1975).

In the usual program, of course, back-up reinforcers are provided for the tokens. Programs differ greatly on the variety of back-up events provided, depending in part upon the events that can be brought to bear as contingent consequences in a given environment. Programs usually have several events to back up the tokens. However, this is not always essential to an effective economy. For example, in one token program, psychiatric patients received a ticket for completing various jobs such as washing, cleaning, and mopping, as well as self-care behaviors such as grooming and bedmaking (Arann & Horner, 1972; Hollander & Horner, 1975). The ticket was backed by a single reinforcing event, viz., access to lunch. Only those patients who completed the target responses received the ticket. The number of patients completing jobs and self-care behaviors as a result of this contingency increased markedly.

A comment needs to be made about the contingent delivery of food. The use of meals as a reinforcing event is relatively uncommon. Increased sensitivity to the rights of patients has called into question the ethics and legality of using food deprivation as an incentive (see Chapter 10). The reason lunch was selected as a back-up reinforcer in this particular program was that the patients were in an open psychiatric ward. They could come and go as they wished. Thus, events usually available in a institutional setting such as access to television or passes off the grounds had to be made available to these patients as a matter of course i.e., noncontingently. If events were not presented to these patients in their ward, they had the option of visiting another ward where they were available. In any case, the example illustrates the use of tokens where only one back-up reinforcer was provided.

The advantage in using only one back-up reinforcer is that it obviates the need to establish prices for a large number of reinforcers and to keep track of expenditures. Of course, the potential disadvantage is that individuals will not consistently perform for the single back-up reinforcer due to satiation or simply to the lack of reinforcing value of that event for some clients. There is a way to combine the practical advantages of administering only one back-up reinforcer and the advantage of diversity of reinforcers. This can be accomplished by having as a back-up reinforcer for the tokens a specified amount of free time. During free time, clients can select events that are individually reinforcing. Because the events that an individual selects to occupy his free time can vary from moment to moment or daily, the likelihood of satiation to free time is small.

A token economy which relied upon free time as a back-up reinforcer was conducted in an inner-city junior high school class (Long & Williams, 1973). To control disruptive behavior, classroom rules were invoked such as remaining in one's seat, bringing materials to class, and working quietly. The rules themselves did not control behavior and a token program was implemented. The teacher displayed a rotary-type file with 18 cards (numbered from 18 to 1). For any disruption or rule violation during the day, the teacher turned the card back to the next smaller number. At the end of the day, the number displayed on the card reflected the number of minutes of free time that the children earned. Free time included several activities as selected by the individual student such as talking with friends, playing games, working on other assignments than the lesson, reading magazines and comics, playing records, and others. Thus, the program actually included several back-up events although these were obtained by only earning one consequence, free time. Earning free time in this fashion led to a dramatic increase in task-relevant behavior in the classroom, a finding replicated in other classroom studies (Glynn, Thomas, & Shee, 1973).

In most programs, a large number of events are used to back up the tokens. It would be difficult to catalog all of the events used. Because events will be mentioned in the next chapters as part of the review of token programs, only general comments are provided here. Initially, many programs rely on consumable rewards such as food, cigarettes, toiletries, privileges such as renting a special place to sleep and dine in the institution, special events such as attending social activities, going home for a weekend, having visitors or talking with friends, and the privilege of delivering reinforcing and punishing consequences (e.g., delivering or withdrawing tokens) to peers.

There has been relatively little research on the selection of back-up reinforcers by the clients who participate in the program (Ruskin & Maley, 1972) and the kinds of events that should be incorporated to maximize patient behavior change (Winkler, 1971b). As a general rule, it would seem advisable to include in the economy diverse back-up events to maintain the high value of tokens and to avoid satiation to any particular event. At present, there are few guidelines for selecting reinforcers based upon research developed from token economies. Recent evidence evaluating the role of economic variables in token economies suggests that certain events are particularly likely to be useful to sustain high levels of client earnings. (This evidence will be detailed later in Chapter 6 which discusses the role of economic variables.)

TOKEN WITHDRAWAL OR RESPONSE COST

Typically, token economies are based primarily upon positive reinforcement, i.e., the delivery of tokens for specified target behaviors. A major program option is whether punishment should be incorporated into the system and, if so, in what form. Although token programs often incorporate some form of punishment into the system, including time out from reinforcement, reprimands, overcorrection, and others, response cost is very commonly used to suppress behavior. Response cost refers to a procedure in which a positive reinforcer is lost or some penalty is invoked. Unlike time-out, there is no specified time limit to the withdrawal of the reinforcer. The most commonly used form of response cost is the withdrawal of tokens or fines (Kazdin, 1972b).

Despite the diverse forms of punishment that are used in applied settings, it is important to single out response cost as a procedure for at least three reasons. First, response cost programs have been used in their own right as a special variation of token economies. That is, some programs have been designed solely on the basis of withdrawing tokens for inappropriate behaviors rather than delivering tokens for appropriate behaviors. Second, token reinforcement and response cost are somewhat unique in that reinforcement and punishment can be administered along the same reinforcer dimension (i.e., tokens). Incorporating punishment in an ongoing token economy requires relatively few additions to the program other than specifying the behaviors that cost tokens. Thus, from the standpoint of convenience, response cost has special utility in a token economy. Finally, although somewhat less pertinent to the present discussion, response cost has been used relatively frequently as a therapy technique independently of the token economy as, for example, in outpatient treatment (Kazdin, 1972b).

Response Cost Programs

Relatively few token programs have been based entirely upon response cost. However, response cost programs have been reported as very effective. For example, in a classroom setting, response cost was used to reduce whining, crying, and complaining that frequently occurred when an "emotionally disturbed" boy was given academic assignments (Hall, Axelrod, Foundopoulos, Shellman, Campbell, & Cranston, 1972). The interesting feature of this program is the specific response cost procedure used. Slips of paper bearing the boy's name were given to him at the beginning of the lesson. Unlike most token reinforcement or response cost programs, the slips had no specific back-up value. However, removal of the slips contingent upon whining, crying, or complaining led to a dramatic suppression of these behaviors. The effect of the program, shown in Fig. 6, was demonstrated in a combined reversal and multiple-baseline design across academic situations (reading and math periods).

More commonly, when response cost is used, a specific back-up reinforcer augments the value of the tokens. For example, Sulzbacher and Houser (1968) employed response cost to suppress use of the "naughty finger" or references to it in a classroom program. When points (backed by minutes of recess) were lost for these responses, they were dramatically reduced.

In general, response cost has been used in its own right to alter a variety of behaviors such as cigarette smoking (Elliott & Tighe, 1968), overeating (Harmatz & Lapuc, 1968), alcohol consumption (Miller, 1973), speech disfluencies (Kazdin, 1973a; Siegel, Lenske, & Broen, 1969), repetitive verbal behavior (Doleys & Slapion, 1975), and others.

Several investigators have compared the efficacy of response cost (token withdrawal) and token reinforcement programs. Iwata and Bailey (1974) compared token reinforcement with response cost in a special education class. Within the class, some students earned points for obeying classroom rules such as remaining seated during the lesson, raising hands to signal for the teacher's help, not talking to others, and in general, paying attention; other students lost tokens given to them at the beginning of the lesson for violating classroom rules. At predetermined intervals, the teacher approached each child and provided a token in his or her cup on

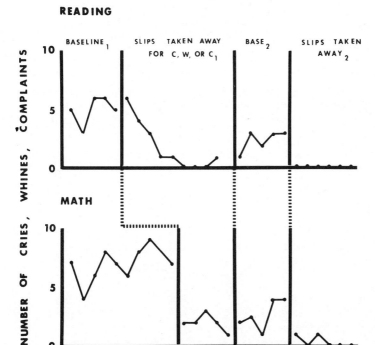

Fig. 6. The frequency of Billy's complaints during the 30-minute reading and arithmetic periods. Baseline$_1$—Prior to experimental conditions. Slips Taken Away for Crying, Whines, or Complaints$_1$—Teacher removes slips bearing child's name following complaining behaviors. Baseline$_2$—Return to baseline conditions. Slips Taken Away$_2$—Same as original Slips Taken Away phase. (From Hall, Axelrod, Foundopoulos, Shellman, Campbell, & Cranston, 1972; Pergamon Press.)

the desk or removed a token previously placed in the cup noncontingently at the beginning of the lesson. Tokens could be exchanged daily for a snack and periodic surprises (toys and candy). After a reversal phase, the conditions were crossed so that students previously on the token reinforcement contingency were switched to response cost and vice versa. The results indicated that paying attention, adhering to classroom rules, and performing arithmetic lessons improved equally across both response cost and reinforcement conditions.

Two interesting features of the study are worth noting. At the end of the study, when children were given their choice to be under the cost or reinforcement condition, preferences for the two systems across the majority of children did not differ. Although the two systems did not differentially affect the children, an interesting finding emerged in relation to teacher behavior. Administration of the token reinforcement condition was associated with an increase in the teacher's rate of approval over baseline levels; this was not found for the response cost condition.

Thus, the two systems for administering tokens may differentially affect aspects of the teacher's behavior and perhaps have long-term implications for the efficacy of a given program. At present, the differential effects of cost and reinforcement programs on staff behavior has not been sufficiently well investigated to make clear statements.

Other research has also shown that response cost and token reinforcement systems are equally effective in altering client behavior (Bucher & Hawkins, 1973; Hundert, 1976; Kaufman & O'Leary, 1972; Panek, 1970), although there are exceptions. For example, in one study in a classroom setting, withdrawing tokens for inappropriate verbalizations was not as effective as delivering tokens for the absence of such verbalizations (McLaughlin & Malaby, 1972b). In another study with pre-delinquent boys, providing a given number of tokens for correctly answering quiz questions based upon a daily television news program was not as effective as withdrawing the same number of tokens for incorrect answers (Phillips *et al.*, 1971, Exp. 4).

Token Economies Including Response Cost

The use of response cost as the sole means of administering the program is the exception rather than the rule. More commonly, token reinforcement contingencies are merely supplemented with fines. Fines serve to complement the main contingencies that rely on positive reinforcement. Thus, the main question is not the effect of response cost versus token reinforcement but token reinforcement with and without the addition of response cost.

One of the most thorough evaluations of the role of response cost in a token program was provided in a study of pre-delinquents who lived at a home-style facility (Phillips *et al.*, 1971). In one experiment alluded to above, an attempt was made to increase the boys' knowledge of current national and international events by having them watch television news. Specific contingencies were designed to evaluate the information gained by providing consequences for performance on daily news quizzes that were administered after the news. For present purposes, the interesting comparisons were based on the different effects of reinforcement alone, response cost alone, and reinforcement combined with response cost. Across several phases (each of the above conditions was presented in at least two different phases), the combination of reinforcement and response cost was more effective than either procedure used alone, both in the accuracy of performance on the quizzes and the number of boys who attended the news sessions. This study suggests rather clearly that response cost can add directly to the token reinforcement contingencies.

The beneficial effects of response cost in a token program were also illustrated in a study in a sheltered workshop setting (Weisberg, Lieberman, & Winter, 1970). A 20-year-old retarded male received tokens (backed by an opportunity to play the piano, a high probability behavior) for completing work, complying with special requests, and performing other work-related behaviors. To eliminate a bizarre stereotypic hand-to-face gesture, a response cost contingency was invoked. One token was lost for each gesture. In an ABAB design, the response cost contingency was shown to dramatically reduce gestures. Interestingly, the initial token program had reduced a number of inappropriate behaviors such as tantrums, wandering, and others. However, response cost added to the general effect by specifically suppressing a remaining bizarre behavior. This result suggests that response cost can be a useful addition to a token reinforcement system. Whether the behaviors that are fined could be altered by reinforcing appropriate rather than fining inappropriate behavior remains to be determined.

A number of other token programs have shown that response cost contingencies provide an effective adjunct to eliminate such diverse behaviors as aggressive statements (Phillips, 1968), antisocial behaviors (Burchard & Barrera, 1972), episodes of noise and violence (Winkler, 1970), out-of-seat behavior (Wolf *et al.,* 1970), bizarre verbalizations (Kazdin, 1971a), self-injurious behaviors (Myers, 1975), violation of ward rules (Liebson, Cohen, & Faillace, 1972), and others.

Use of Response Cost

Given the existing evidence, response cost appears to be a useful technique to eliminate select behaviors in a token economy. The desirability of using a response cost system as the sole basis of the program cannot be unequivocably evaluated at this time. Generally, the desirability of using a punishment system rather than a reinforcement system can be questioned. Thus, it might be useful to consider response cost as an adjunct to an overall reinforcement system. The reason for this recommendation is that punishment sometimes is associated with undesirable side effects in applied and laboratory settings (Kazdin, 1975a). Indeed, in some applications, when cost is added to token reinforcement contingencies, individuals react adversely to the system by avoiding the situation or with adverse verbal reactions (Boren & Colman, 1970; Meichenbaum, Bowers, & Ross, 1968), although these instances appear to be exceptions (Kazdin, 1972b). The possibility that response cost, when relied upon exclusively, has less beneficial effects on staff behavior than token reinforcement is another consideration (Iwata & Bailey, 1974).

Perhaps, the main reason for recommending response cost as an adjunct rather than an end in its own right is that response cost is a punishment procedure, and as such, emphasizes what the client is not supposed to do. Punishment alone is not likely to result in the development of prosocial behaviors. On the other hand, reinforcement for specific target behaviors is likely to develop prosocial responses as alternatives for the behaviors to be suppressed.

LEVELED PROGRAMS

Many token economies have contingencies that do not change for a given client. These economies might be referred to as maintenance programs insofar as they attempt to maintain a given level of performance. The program is a quasi-permanent part of the system to which the clients are exposed. Of course, a potential criticism of such programs is that there is no means of ensuring that the clients will be taken off the system and that some permanent gains will be achieved. In other programs, the contingencies are repeatedly changed as a function of client behavior to ensure progress toward some terminal level of performance. However, programs which focus on an individual's behavior and shape responses according to the individual's rate of progress often are too cumbersome to administer for very many individuals or across very many behaviors of a few individuals in a given setting.

There is a way to structure the token program so that the contingencies are adjusted to client performance and develop behavior toward some goal but are not too cumbersome to implement widely across a number of clients. The program is structured to achieve different levels through which clients must pass. A goal of the leveled system is to have clients pass through different levels until they reach the highest level at which point behaviors are performed without token reinforcement (Atthowe, 1973; Kazdin & Bootzin, 1972).

The levels of a token program refer to different stages. Clients may begin the program at an initial level or stage and, depending upon their improvement and sustained performance of the target behaviors, progress to higher levels. Initial levels require the performance of few behaviors and also offer relatively few reinforcers. However, after the client meets the requirements of the level for a certain period of time, he is able to progress to the next level that requires more complex behaviors and provides added privileges.

In a sense, each level is a separate subprogram with new contingencies. The rationale for use of subprograms is that as the client progresses, each one is more appropriate to his current performance level than the previous program or level. From the client's standpoint, the incentive for progressing from one level to the next is that increasingly desirable reinforcers are made available at higher levels. Finally, in some cases, at the highest level the client's behavior is no longer subjected to token reinforcement contingencies. The client has access to all of the available reinforcers and must maintain appropriate behavior without specific reinforcment contingencies.

The notion of levels is not new. Education might be regarded as a common leveled system where one progresses to higher levels (grade levels) depending upon behavior in the previous level. Also, completion of different levels (e.g., high school versus college or college versus graduate school) is associated with different consequences (occupational level and salaries).

Levels have been used frequently in token programs in psychiatric hospitals where the goal of treatment is to discharge the patient. Early contingencies may focus on basic self-care skills and participation in various activities. Contingencies associated with higher levels may focus on looking for a job in the community, visiting relatives, staying at home rather than the hospital, and other behaviors more likely to be associated with successful discharge.

One of the first leveled token economies was described by Atthowe and Krasner (1968) in their program for psychiatric patients. Diverse patient behaviors were reinforced including self-care, attending activities, and interacting with others.

Tokens were exchangeable for cigarettes, money, privileges such as watching television and passes off the ward, as well as special events idiosyncratic to the tastes of individual patients. Essentially, there were two levels of the system. Most individuals participated in the general program or lower level. However, an elite group functioned somewhat independently of the token system at a higher level. These patients carried a "carte blanche" which entitled them to all privileges within the general system plus added privileges. To obtain this higher level of the program, patients had to work for long periods on job assignments, consistently perform appropriate behaviors for protracted periods, and had to accumulate a large number of tokens. The incentive for progressing to this level was the ability to function without tokens, added privileges, and possibly increased status on the ward.

A similar type of leveled system has been used with pre-delinquent youths participating in a token economy at Achievement Place (Phillips *et al.,* 1971). In this system, the youths earned tokens for a large number of behaviors including maintaining their rooms, attending activities promptly, and completing academic assignments. Tokens could be exchanged for a variety of privileges such as receiving an allowance, watching television, eating snacks, going on special trips, and others. Each individual who entered the system was placed in the token economy. However, the treatment program was leveled to facilitate the return of the youth to his home and community life. If the individual functioned successfully in the token system, as defined by earning high token levels for several weeks, he was advanced to the *merit system*. On the merit system, all privileges were free and only social consequences were supplied for behavior. If appropriate behavior continued, the individual became a candidate for the next level or *homeward-bound system*. His family was prepared for his return. As in most of the leveled systems, inappropriate behavior could result in a return to a previous level.

In a program for hospitalized drug addicts, a leveled system was devised to facilitate the patients' return to society (Melin & Gotestam, 1973). The three levels of the program included: (1) detoxification, (2) treatment, and (3) rehabilitation. In the detoxification phase, patients lived in relatively sterile living quarters and could not participate in any ward activities. However, clients were required to dress themselves and make their beds. After traces of drugs were no longer found in the urine, individuals were admitted to the treatment phase. In this phase, individuals were required to arise early, dress, make their beds, attend a morning conference, engage in routine ward behaviors, and attend physical therapy. In return for the points earned for these behaviors, individuals could live in semiprivate rooms and go for walks off the ward. After progress in this phase, as defined by judgment of the staff and token earnings of the patient, the patient could progress to the rehabilitation phase. In this phase, the patient could earn points for off-ward activities such as procuring a job in the hospital or in town. Also, new back-up reinforcers were available including living in a private room, leaving the ward unaccompanied, leaving for longer periods, and receiving visitors in one's room. To progress from one level to another, specific criteria were set such as earning a specific total number of tokens and maintaining a minimum weekly level of earnings. Failure to meet the criteria for a given level led to a return of the patient to the previous level.

There are several features of the leveled token program that appear to be advantageous. First, by leveling the program, client progress is programmed into the system. Clients are provided with an incentive to progress through the treatment program. This is important because, as discussed later, there is a tendency for clients to reach a plateau in their performance and token earnings. Second, the leveled system provides a means to shape increasingly complex behaviors across a large number of clients without having highly individualized contingencies. In most rehabilitation settings, where staff-client ratios are extremely low, individualized contingencies and close monitoring of behaviors that are being shaped are not feasible. The leveled system provides, as it were, a group shaping program and is useful when a group (ward, classroom) is exposed to the program. Third, the leveled system can be structured so that tokens are withdrawn completely while the clients are in the setting. The final level can include the gradual elimination of the large number of contingencies that otherwise govern client behavior. If the individual can function free or relatively free of the contingencies, he can be released from the facility. In contrast, if performance is not maintained, the individual can remain in the facility with further attempts to develop independence from the specific contingencies.

Despite the desirable features of a leveled token system, there is no clear evidence evaluating its specific therapeutic benefits. For example, whether having many or only a few levels or using a leveled system at all produces short- or long-range gains in client behavior vis-à-vis a nonleveled system remains to be studied. The use of leveled systems can be justified on the basis of shaping, i.e., making increasingly greater performance demands and reinforcing successive approximations of the final response, and the practical appeal of having a few levels for all clients rather than individualized contingencies which could only be implemented for a few.

PEER ADMINISTRATION OF THE CONTINGENCIES

An interesting program option pertains to who administers the reinforcement contingencies. Administration of the contingencies can refer both to who selects the criteria for reinforcement and who actually delivers the tokens. In the majority of programs, staff including aides, teachers, and parents both design the contingencies and deliver tokens. However, a number of programs have shown that peers and the client himself can administer tokens and sustain performance.

The use of peers as reinforcing agents has been investigated extensively in the program for pre-delinquents at Achievement Place. In one report, boys at the facility served as therapists for their peers who made articulation errors (Bailey, Timbers, Phillips, & Wolf, 1971). Two boys who had pronunciation errors such as substituting inappropriate sounds in some words or omitting various sounds from others (e.g., saying *motho* for *mother* and *probem* for *problem*) attended special therapy sessions. Their peers conducted the sessions. One peer function was to judge whether a word was correctly pronounced. At various points in the program, the client earned points for correct pronunciation or lost points for errors depending upon the judgment of the peers. Peers themselves received points depending upon their identification of errors and correct speech and for the correct pronunciations made by the client. Peers were effective in increasing correct pronunciation in the clients.

Another use of peers in the administration of the contingencies at the same facility was a managerial system (Phillips, 1968; Phillips, Phillips, Wolf, & Fixsen, 1973). This system consisted of one of the clients serving as a reinforcing agent by delivering or withdrawing tokens to develop behavior in his peers. The individual who served as a manager either purchased this privilege with tokens or was elected by his peers. Once the manager was decided, he was responsible for ensuring that the target behaviors were performed. Several studies have evaluated the effect of peer managers in maintaining clean bathrooms in the facility. The manager selected the peers to do the cleaning and rewarded or fined them based upon his judgment of the job. When the bathroom was checked by the teaching parents, the manager earned or lost points based upon how well the bathrooms had been cleaned. Thus, although a peer administered the contingencies that governed other residents, the teaching parents maintained control over the contingencies of the peer-manager. The peer-manager system of administering contingencies appears to be quite effective in controlling behavior. Indeed, the managerial administration of reinforcement contingencies led to higher levels of performance than group contingencies administered by the teaching parents (Phillips *et al.*, 1973; Exp. 1).

In yet another use of peers, a semi-self-government was devised at Achievement Place in which individuals could determine whether one of their peers was guilty of violating a rule or misbehaving in general (e.g., borrowing clothes or personal belongings from others without their permission, excessive teasing or bickering) and what consequences should be associated with the behavior (Fixsen, Phillips, & Wolf, 1973). The youths who resided at the facility met to discuss rule violations with the teaching parents each evening at a "family conference." At the meeting, the violation would be discussed and one's peers would decide whether the alleged rule violator was guilty. If guilty, the peers would also vote on an appropriate penalty for the violation. The teaching parents could intervene primarily through discussion in cases where penalties appeared either much too lenient or severe for the violation. However, the primary means of deciding the penalty was through votes of the clients themselves. To ensure that rule violations were called to the attention of the teaching parents, at one point in the program, points were administered for reporting misbehavior and for assigning the consequences associated with the violation.

Peer-administered contingencies also have been used effectively in classroom situations. Solomon and Wahler (1973) had peers reinforce appropriate behavior (working on assignments, appropriate talking, and remaining at their desks). After baseline observations of disruptive behavior, students were asked if they wished to help their disruptive peers. These volunteer students were trained by discussing basic operant principles and by practicing recognition of appropriate and inappropriate behavior on videotapes of baseline performance. During the program, these

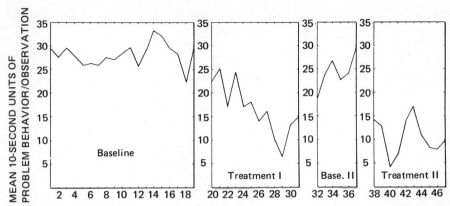

Fig. 7. Mean number of problem behavior units produced by target subjects overall baseline and treatment periods (from Solomon & Wahler, 1973; Society for the Experimental Analysis of Behavior).

students were instructed to attend to appropriate behavior in the target subjects. As shown in Fig. 7, the average rate of problem behavior for ten target subjects decreased during the phase in which their peers attended to appropriate behavior. During a reversal phase when the peers were told to resume their baseline inter-actions (i.e., not attend to appropriate behavior), problem behavior of the target subjects increased. In the final phase, treatment was reinstated and problem behavior again decreased.

In an interesting use of peers in a second grade classroom, the child who administered tokens varied daily (Winett, Richards, & Krasner, 1971). The child selected on a given day circulated among the students and delivered tokens (poker chips backed by trinkets, school supplies, and tickets to movies). The child himself received the maximum tokens on the day he was selected as a monitor. Several other studies have shown that peers can effectively administer reinforcement contingencies in the classroom (Axelrod, Hall, & Maxwell, 1972; Drabman, 1973; Patterson, Shaw, & Ebner, 1969).

Peers have been used relatively infrequently as reinforcing agents in token economies. Thus, the relative efficacy of peer versus staff-administered contin-gencies has not been well studied. Drabman (1973) compared the effects of teacher- and peer-administered contingencies in token programs implemented in four small classrooms (6 children each). The students, most of whom were diagnosed as schizophrenic or unsocialized aggressive children, were residents in a children's psychiatric hospital. In the teacher-administered program, the teacher delivered tokens for adhering to classroom rules pertaining to deportment (e.g., sitting in one's seat, not distracting others, coming to class on time). In the peer-administered program, the students elected a peer who administered points to the students. The peer administered consequences to each student by placing poker chips in each child's canister on the teacher's desk and individually informing the students how many tokens they earned for appropriate behavior. The peer who served as a reinforcing agent received tokens from the teacher on the basis of his performance.

The results indicated that teacher- and peer-administered systems were equally effective in reducing disruptive child behavior.

If peer managers are to be used, they must be adequately trained to administer the contingencies. Without adequate training and incentives for administering the contingencies correctly, managers may be substantially less effective than staff. For example, Greenwood, Sloane, and Baskin (1974) instructed select peers in a behavioral problem classroom how to deliver points and social approval to the students. Instructions generally were not effective in developing the contingent delivery of consequences on the part of the managers. However, the delivery of points to the managers for their appropriate administration of the contingencies markedly improved their performance. In general, peers may be useful as a supplement to a token reinforcement program provided that their behaviors can be monitored.

There are potential advantages that recommend using peer-administered contingencies. First, as noted earlier in the discussion of group contingencies, stimulus control exerted by the staff sometimes is a problem in token programs (Kazdin, 1973e). If peers administer some of the contingencies, somewhat broader stimulus control is likely to result (Johnston & Johnston, 1972). Second, peers provide reinforcing and punishing events (e.g., praise and reprimands) which might differ in their effects from similar events administered by staff. It may well be that consequences delivered by a staff member are somewhat less effective than consequences delivered by a peer, at least for some populations. If behaviors are likely to occur almost exclusively in the presence of peers (e.g., aggressive acts), peers might be the only ones who can feasibly implement the program. Third, peers often wish to function in the capacity of a staff member and indeed will even purchase with tokens the right to administer consequences to their peers (Phillips, 1968; Phillips et al., 1973). Finally, if peers administer tokens to each other, social interaction might be increased which may be a therapeutic goal in its own right (Abrams, Hines, Pollack, Ross, Stubbs, & Polyot, 1974).

SELF-ADMINISTRATION OF CONSEQUENCES

The self-administration of reinforcing or aversive consequences has received increased attention in behavior modification (Goldfried & Merbaum, 1973; Thoresen & Mahoney, 1974). Clients are trained to self-reinforce or self-punish themselves in outpatient treatment for a variety of behaviors.[1] In many token

[1] The literature commonly uses the terms self-reinforcement and self-punishment. Because of the accepted use of these terms in general, and the specific use of these terms by the authors whose work is reviewed in this section, they will be retained. However, recent papers have carefully analyzed both the terminological and conceptual problems in referring to the self-administration of consequences as self-reinforcement (Brigham, 1977; Catania, 1975). Few studies referred to as "self-reinforcement" demonstrations have met the criteria associated with reinforcement and have permitted the individual to freely determine the self-administered consequences without constraints (Kazdin, 1975a).

economies, clients have administered tokens to themselves in place of staff- or peer-administered contingencies.

Applications of self-reinforcement can include different procedures. First, the client can determine the response requirements for delivering the reinforcer and how much of the reinforcer (i.e., tokens) should be delivered. When the individual determines the criteria for reinforcement, this is referred to as *self-determined reinforcement* (Glynn, 1970). Second, the client can administer the reinforcers to himself for achieving a particular response criterion. The criterion may not necessarily be self-determined. When the individual administers the reinforcers to himself, this is referred to as *self-administered reinforcement.* Whether an individual administers reinforcers to himself or whether someone else does is not the crucial element in defining self-reinforcement. The singularly important component is whether the individual can determine when to deliver the reinforcers and for what behaviors. Self-reinforcement probably is best achieved when the individual can self-determine and self-administer the consequences which minimizes the external constraints on the contingency.

Several studies conducted primarily in the classroom have shown that individuals can reward themselves with tokens and that their behavior usually improves as a result (e.g., Bolstad & Johnson, 1972; Felixbrod & O'Leary, 1973, 1974; Glynn, 1970; Lovitt & Curtiss, 1969; McLaughlin & Malaby, 1974a). As an example, Glynn and Thomas (1974) permitted elementary school students to self-determine and self-administer checkmarks (points) to themselves for on-task behavior. Intermittent tape-recorded signals ("beeps") were played aloud at various randomly ordered intervals. Each child placed a check on his card if he considered himself to be attending to the task at the moment the signal sounded. At the end of the lesson, individuals received free time according to how many checks they had. To help the children discriminate on-task performance at any given point in time, charts at the front of the room were displayed by the teacher noting whether the students should be working on their in-seat assignment or paying attention to the teacher. The self-administration of points increased on-task behavior over baseline levels.

Although several studies have shown that children in classroom settings can provide consequences for their behavior, a major question is whether self-administered and externally-administered consequences are differentially effective. Studies have provided direct comparisons of self- and other-administered consequences. Felixbrod and O'Leary (1973) compared experimenter versus self-determined criteria for reinforcement with elementary school students. Students attended individual sessions where they completed arithmetic problems. Self-determined reinforcement subjects were told they could determine how many correct problems were required for a given amount of tokens; experimenter-determined reinforcement subjects (yoked to the self-determined subjects) were unable to determine the ratio of correct problems to token earnings. The results indicated that these conditions were equal in their effects of performance accuracy and time spent on the task. Both conditions were superior to a nonreinforcement control group.

Similarly, Frederiksen and Frederiksen (1975) evaluated teacher- and self-determined tokens in a class of junior high school students. During the self-determined phase, the teacher approached each student at the end of a half hour and asked whether they had earned a token (i.e., behaved appropriately). If they replied affirmatively, the teacher administered a token. In an ABAB design, teacher- and self-determined tokens were equally effective in changing on-task and disruptive behavior. Other studies have shown that experimenter- or teacher-determined and self-determined token reinforcement are equally effective (Felixbrod & O'Leary, 1974; Glynn, 1970; Glynn et al., 1973).

In some of the studies, conclusions about the relative efficacy of self- and externally determined consequences cannot be made without ambiguity. When comparisons of self- and externally determined consequences are made in ABAB designs, conditions are frequently confounded with sequence effects (e.g., Frederiksen & Frederiksen, 1975; Glynn et al., 1973; Kaufman & O'Leary, 1972). Also, other factors that can markedly affect behavior such as the amount of reinforcement, the delay between behavior and receipt of tokens, and uncontrolled and unmonitored teacher behavior are confounded with teacher- versus self-administered conditions and obfuscate any pattern or lack of pattern in the data (e.g., Frederiksen & Frederiksen, 1975).

In some comparisons, experimenter- or teacher-determined reinforcement has been superior to self-determined reinforcement because when students are allowed to reward themselves, they become increasingly lenient over time. For example, in one token program, "emotionally disturbed" children in a psychiatric hospital were permitted to self-determine their point earnings for appropriate classroom behavior (Santogrossi, O'Leary, Romanczyk, & Kaufman, 1973). Although disruptive behavior remained low for a few days after the consequences were changed from teacher to self-determined reinforcement, the students began to administer tokens to themselves noncontingently and disruptive behavior increased markedly. Other studies have found increased leniency when individuals self-determine their consequences relative to experimenter- or teacher-determined consequences (e.g., Felixbrod & O'Leary, 1973, 1974; Frederiksen & Frederiksen, 1975). As an exception, Glynn et al. (1973) found that the majority of children provided themselves with too few tokens for their performance during self-reinforcement phases.

Increased leniency or noncontingent delivery of reinforcers is not a necessary consequence of self-reinforcement contingencies. Drabman, Spitalnik, and O'Leary (1973) developed accurate self-evaluation and self-reward in a token economy to control disruptive behavior in a special adjustment class of 9- to 10-year-old boys. Students were trained to evaluate and rate their own behavior accurately. The teacher randomly selected some children, checked their evaluations, and reinforced accuracy of self-ratings (i.e., agreement with the ratings she had made). The number of children checked was gradually decreased over time so that eventually the children were evaluating themselves without checks. The children did not become increasingly lenient in the period (12 days) that they had complete control over self-reward free of the possibility of being checked or monitored by the teacher. Thus, for at least short periods, self-reinforcement is possible without increased leniency. Other research has replicated the utility of reinforcing accuracy of self-evaluation as a procedure to ensure maintenance of appropriate behaviors when clients self-reinforce (Turkewitz, O'Leary, & Ironsmith, 1975).

Aside from self determining and self-administering tokens, clients have been involved in their own programs in other ways. Sometimes clients observe their own behavior even though they do not control the actual delivery of tokens. Knapczyk and Livingston (1973) had junior high school students in a special education class record their own correct responses on daily reading assignments and record their accumulated token (money) earnings. The teacher was responsible for determining the criteria for token reinforcement and for administering the tokens. The token program markedly increased reading accuracy. There was no difference in the effects of token reinforcement when the students or the teacher recorded the behavior.

Self-observation has sometimes been used as a behavior change technique in its own right in applied settings (Kazdin, 1974b). For example, when children observe and record their own disruptive or inattentive behavior in the classroom, these behaviors sometimes improve even though extrinsic consequences are not provided (Broden, Hall, & Mitts, 1971). Thus, self-observation in token programs may warrant additional research.

Overall, self-administration of consequences appears to be a viable option in a token program. To use this effectively, research suggests that some means are required to ensure that individuals can easily evaluate whether the appropriate response has occurred and that the consequences are delivered contingently. It is unclear whether individuals can administer consequences to control their own behavior without explicitly programmed consequences for adherence to or violation of the contingent relationship between behavior and the consequences beyond relatively brief periods (Anderson & Alpert, 1974; Drabman et al., 1973).

The self-delivery of tokens may be a useful asset to programs otherwise administered by staff alone. As with other options discussed, an advantage of allowing self-reward as an adjunct to the program is that stimulus control over the target behavior may be broadened beyond a few staff-monitored situations. Also, as will be discussed in a subsequent chapter, self-administration of the contingencies may play a role in maintaining behavior after the token reinforcement contingencies are withdrawn.

CONCLUSION

The token economy has several basic requirements such as selecting the target responses, selecting the medium of exchange to serve as a token, enumerating the back-up events, and describing the contingencies so that performance of the target responses is translated into token values and so that token values can be translated into back-up events. The basic requirements provide tremendous room for variation. For example, programs vary widely on the tokens and back-up events used. Indeed, in some token economies "tokens" as such are not used. Rather, credit in a "banking" account serves as the conditioned reinforcer. Most token programs, of course, provide a conditioned tangible event for the client to retain and use for purchase of back-up events. Occasionally, programs do not utilize back-up events but merely provide tokens that cannot be exchanged for other reinforcers. This, too, is an exception because research indicates that tokens with back-up value lead to greater behavior change than those with no back-up value. In many cases, however, it may be unnecessary to use an elaborate token system involving a wide range of back-up events.

The selection of tokens and back-up events is dictated by a number of considerations. The medium that serves as the token may be determined by the requirements for monitoring client earnings and expenditures, the ease of delivery, the likelihood of theft of tokens or use of counterfeit tokens, and other exigencies of the setting or population. The back-up events are determined by the preferences of the population, items, privileges, and activities normally available in the setting, and the range of events that may be delivered on a contingent basis. Variations of the basic ingredients of a token economy and their effects on behavior have not been carefully studied. The medium of exchange and specific back-up reinforcers usually are not topics for empirical research in their own right.

Aside from the basic ingredients of a token economy, a wide range of options and variations can be incorporated into the program. These variations include the type of contingencies that are used (individualized, standardized, and group), the methods of administering consequences (individual or group administration), the range of back-up events, the use of response cost as an alternative or adjunct to token reinforcement contingencies, leveled programs where clients traverse through stages leading to superior levels of performance, and peer- and self-administration of the contingencies. The range of alternatives available contributes to the diversity of token programs.

Research on the effect of specific program variations on behavior has been sparse. Thus, the particular option selected for the program often is based upon convenience rather than empirical evidence. When evaluated, many of the variations frequently are shown not to differentially influence the efficacy of the program. For example, programs based upon the presentation or withdrawal of tokens, on group or standardized and individualized contingencies, or on self- or staff-administered contingencies, and similar options usually have been shown to be equally effective. Occasionally, one of the program options is particularly effective in altering a problem that arises in the program such as the failure of a client to respond to a given contingency. The use of program options to resolve particular problems will be discussed in a later chapter. The next two chapters review the token economy research across diverse treatment populations and target behaviors.

REVIEW OF
TOKEN ECONOMIES I

4

The token economy has been applied across a wide range of populations and treatment, rehabilitation, and educational settings. This chapter reviews token programs with psychiatric patients, the mentally retarded, and individuals in classroom settings. Token economies with these populations constitute the vast majority of programs. Because of the extensive literature in each of these areas, the review will only highlight exemplary programs of historical and contemporary interest and illustrate recent advances. The next chapter reviews token programs with delinquents, adult offenders, drug addicts and problem drinkers, outpatients, and other populations.

PSYCHIATRIC PATIENTS

Use of the token economy with psychiatric patients, in part, stimulated the widespread application of token programs. As noted earlier, the token economy of Ayllon and Azrin (1965) attracted widespread attention and the techniques seen in that program were applied broadly elsewhere. Currently, token economies are commonly employed in psychiatric facilities in open and closed wards and across diverse patient groups, although, of course, chronic schizophrenics have received the greatest attention (see Carlson, Hersen, & Eisler, 1972; Gripp & Magaro, 1974; Kazdin, 1974c, 1975d; Kazdin & Bootzin, 1972; Milby, 1972, for reviews of programs with psychiatric patients).

For purposes of review, it is useful to consider programs under three categories, namely, general ward-wide programs that focus on adjustment to hospital living, programs that focus on bizarre or "symptomatic" behaviors, and programs that attempt to develop social behaviors. Obviously, there is overlap among these categories because most programs in the hospital attempt to develop behaviors in these and other categories. However, by reviewing programs according to the major emphases of treatment, the diversity of token economies and their outcomes can be more fully described.

General Ward-Wide Programs

The majority of token economies in psychiatric hospitals have focused on a wide range of behaviors considered adaptive within the hospital including self-care, performance of jobs on and off the ward, attendance to and participation in activities, and general activity on the ward. Certainly, the series of studies by Ayllon and Azrin (1965, 1968b) stand out as significant from a historical standpoint. In several studies, Ayllon and Azrin (1965) focused upon the job preferences of female chronic schizophrenic patients at Anna State Hospital in Illinois. Patients could perform any of a large variety of jobs such as serving as a sales clerk assistant at the commissary, or as an assistant for any of several dietary, secretarial, janitorial, nursing, recreational, and grooming activities. In addition, patients received tokens for grooming, bathing, tooth brushing, exercising, and making their own beds. Among token economies, this one provided one of the most extensive lists of back-up reinforcers which included renting living quarters with bedspreads, choice of eating group, leaving the ward, social interaction, recreational and devotional opportunities, and a large number of commissary items, as noted in the previous chapter.

Ayllon and Azrin conducted several studies that showed the importance of token reinforcement in maintaining job performance on and off the ward. The reinforcement contingencies were effective in altering voluntary job preferences by manipulating the number of tokens that could be earned. Additionally, the noncontingent delivery of tokens led to a decrease in job performance, thus demonstrating the importance of the response-reinforcement relationship. The dramatic effect of tokens in this early token economy is illustrated in Fig. 8. which shows the effect of token reinforcement contingent upon job performance in an ABA design. The results show the effect of eliminating tokens altogether on job performance of 44 patients who participated in this experiment. Clearly, the presentation and withdrawal of token reinforcement controlled patient performance.

After the work of Ayllon and Azrin, several other token programs were reported. Schaefer and Martin (1966) altered the "apathy" of psychiatric patients, as defined by engaging in only one behavior (e.g., standing) without the simultaneous performance of a concomitant behavior (e.g., talking). Half of the patients on the ward received tokens contingent upon various responses related to personal hygiene, social interaction, and work performance while the remaining patients received tokens noncontingently. Over a three-month period, patients receiving tokens contingently showed a significantly greater reduction in apathy, as rated on behavioral checklist data, than the control patients. This study is significant for focusing upon overall activity and for evaluating the program with patients randomly assigned to groups.

Another early program with psychiatric patients was reported by Atthowe and Krasner (1968) on a closed ward of a Veterans Administration hospital. This program was designed to encompass a wide variety of behaviors of hospital living including self-care, attending activities, and social interaction. The tokens were exchangeable for privileges and items at a canteen. Several interesting aspects of this program included the ability of patients to earn their way off the system by performing appropriately for protracted periods and accumulating a large number of tokens, increased discharge and lower readmission rates relative to preprogram rates, and an overall increase in activity on the ward. Although the program was evaluated in an AB design and causal statements could not be made about the effect

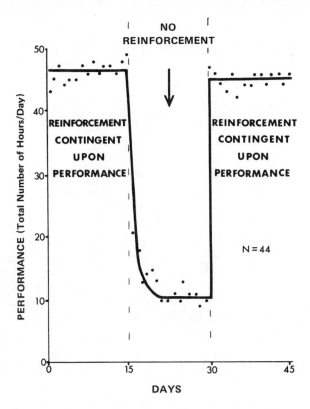

Fig. 8. The total number of hours of on-ward job performance (from Ayllon & Azrin, 1965; Society for the Experimental Analysis of Behavior).

of the contingencies *per se,* the broad effects on behavior suggested by the data stimulated other programs to encompass a wide range of behaviors.

An example of such a program was provided by Winkler (1970) whose program for chronic psychiatric patients encompassed a large number of behaviors including getting up, making one's bed, dressing, attending and completing morning exercises, and completing work. Other behaviors were associated with fines including unnecessary episodes of noise and violence (e.g., tantrums, screaming, banging) as well as idiosyncratic behaviors of individual patients. The program was shown to be responsible for behavior change in several experiments.

Not all programs for psychiatric patients have been for adult chronic schizophrenic patients. For example, Aitchison and Green (1974) reported a program with male and female adolescents admitted to a psychiatric hospital. The primary reason for admission was a lack of parental control over behavior. Tokens were provided for attendance at school and therapy, ward maintenance, promptness in taking medication, and others. The effect of tokens was evaluated in room-cleaning behaviors such as making one's bed, sweeping the floor, hanging up one's clothes, and emptying the wastebasket. In an ABAB design, the effect of tokens on room cleaning was evaluated with a ward of patients (51–59 patients over the course of the experiment). After initial baseline and a brief period in which the patients were instructed to clean their rooms, the point system was implemented. As shown in Fig. 9, the program markedly increased room-cleaning behaviors. Withdrawing and reinstating the contingencies showed the effect of the program. In the last phase depicted in Fig. 9, points were still delivered and the high rate of cleaning was sustained.

A plethora of programs have demonstrated the efficacy of token reinforcement contingencies on behaviors related to adaptive and independent functioning in the hospital including self-care, attending and participating in diverse activities designed to be therapeutic, taking medication, and working on jobs (Allen & Magaro, 1971; Arann & Horner, 1972; Chase, 1970; Cohen, Florin, Grusche, Meyer-Osterkamp, & Sell, 1972; Ellsworth, 1969; Gericke, 1965; Glickman, Plutchik, & Landau, 1973; Greenberg *et al.,* 1975; Grzesiak & Locke, 1975; Heap, Boblitt, Moore, & Hord, 1970; Hersen, Eisler, Smith, & Agras, 1972; Lloyd & Garlington, 1968; Maley, Feldman, & Ruskin, 1973; McReynolds & Coleman, 1972; Steffy, Hart, Craw, Torney, & Marlett, 1969; Suchotliff, Greaves, Stecker, & Berke, 1970). While the majority of programs for psychiatric patients have been conducted in traditionally composed institutional settings, similar programs have been used in community-based and day-treatment facilities as alternatives to hospitalization (cf. Henderson, 1969; Henderson & Scoles, 1970; Liberman, 1973).

Complex and Therapeutically Relevant Behaviors and "Symptoms"

In the early token programs for psychiatric patients, attention focused primarily on routine ward behaviors. There was some justification for this focus.

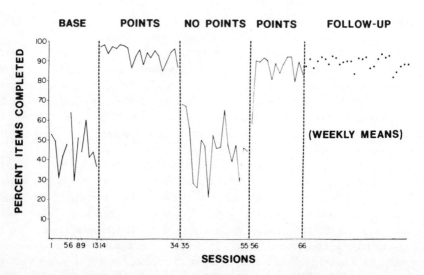

Fig. 9. Percent of completed room-maintenance items for all subjects under each treatment condition (from Aitchison & Green, 1974; Pergamon Press).

Initially, the purpose of many programs was of a demonstrational nature. Behaviors selected such as job completion, bed making, and attendance to activities were easily observed and could readily reflect the effect of the contingencies. Second, institutionalization often fosters maladaptive and dependent behaviors (Paul, 1969). Token reinforcement programs could affect ward functioning and general activity and seemingly overcome deleterious effects of institutionalization. Third, development of routine ward behaviors was shown to be related to the reduction of bizarre and symptomatic behaviors of therapeutic significance.

Several studies demonstrated the role of a general program in altering bizarre behaviors. O'Brien and Azrin (1972) showed that a high rate of screaming of a female schizophrenic patient decreased by providing tokens for adaptive behaviors such as social skills, housekeeping, and grooming. These behaviors were not incompatible with screaming. However, the authors suggested that positive behaviors on the ward "functionally displaced" screaming in the patient's response repertoire.

Hersen, Eisler, Alford, and Agras (1973) provided tokens to three hospitalized neurotically depressed males. Tokens were delivered for work, occupational therapy, and personal hygiene. An increase in token-earning behaviors was associated with a reduction in depression as measured by behavioral ratings. In an ABA design, depression was shown to systematically vary inversely with performance of routine ward behaviors.

Similarly, Anderson and Alpert (1974) provided tokens to a schizophrenic patient who had a high rate of hallucinations (defined by overt ritualistic and compulsive mannerisms that accompanied the reported hallucinations). The hallucinations occurred between 10 and 55 seconds of each wakeful minute and interfered with the patient's completion of routine activities. Tokens were provided for decreases in the latency of performing routine behaviors including eating meals, dressing, and completing bathroom activities. Marked improvements in routine behaviors were associated with reductions in hallucinations, as demonstrated in an ABAB design.

Increasingly, programs focusing on ward-wide behaviors have assessed behaviors of symptomatic significance. These studies also have shown that reinforcing routine behaviors are associated with changes in bizarre behaviors and measures of symptoms on psychological inventories (DiScipio & Trudeau, 1972; Gripp & Magaro, 1971; Grzesiak & Locke, 1975; Schwartz & Bellack, 1975). Indeed, it is reasonably clear that the effects of token reinforcement extend well beyond changes in routine behaviors that serve as target responses. For example, token economies focusing on routine behaviors show patient improvements in cooperativeness on the ward, communication skills, social interaction, and mood states as well as a reduction in the use of medication (Maley et al., 1973; Shean & Zeidberg, 1971).

Many token economies that focus on routine behaviors have reported increased discharge and/or lower readmission rates than custodial treatment (Atthowe & Krasner, 1968; Gorham, Green, Caldwell, & Bartlett, 1970; Heap et al., 1970; Hollingsworth & Foreyt, 1975; Lloyd & Abel, 1970; McReynolds & Coleman, 1972; Rybolt, 1975; Shean & Zeidberg, 1971; Steffy et al., 1969). To many investigators, these data argue for the importance of general ward-wide contingencies and their effects on symptomatically relevant behaviors. Of course, the importance of discharge as a measure of program success cannot be overstressed. However, the measure is sensitive to a variety of demands including changes in hospital philosophy, "push" efforts by the staff, and so on, independently of changes in the psychological status of the patient (cf. Gripp & Magaro, 1974; Kazdin & Bootzin, 1972).

Several programs have focused upon target "symptoms" directly rather than reflecting change as an inadvertent effect of the program. Wincze, Leitenberg, and Agras (1972) altered delusional talk of ten paranoid schizophrenics. Feedback for delusional talk (i.e., "that was incorrect") produced transitory effects whereas token reinforcement for nondelusional talk produced clear changes. Token reinforcement was more effective in therapy sessions where nondelusional talk was more systematically evoked and followed with consequences than on the ward.

In a similar program, Patterson and Teigen (1973) increased rational talk of a patient who failed to answer questions correctly about her personal history. The patient had previously participated in a program in which special privileges and praise were contingent upon rational as opposed to paranoid and delusional talk in individual interviews (Liberman, Teigen, Patterson, & Baker, 1973). Because irrational talk remained, tokens (exchangeable for a variety of privileges on the ward) were provided for correct answers to questions in a multiple-baseline design. Verbalizations improved and were partially maintained several weeks after discharge. Other investigations have shown that irrational verbalizations can be suppressed either by reinforcing rational talk with tokens or by withdrawing tokens for bizarre verbalizations (Kazdin, 1971a; Meichenbaum, 1969).

Other symptoms have been focused upon aside from verbalizations. Reisinger (1972) developed a token program for an "anxiety-depressive" institutionalized patient who had been hospitalized for 6 years. The patient frequently cried without apparent provocation, and appeared withdrawn, fearful, and generally depressed. Episodes of crying and smiling were observed. After baseline rates were obtained, tokens were delivered for smiling and were withdrawn for crying. The effect of these procedures can be seen in Fig. 10 (second phase). After smiling improved and crying decreased, the contingencies were withdrawn completely and a reversal phase was invoked in which crying was *reinforced* and smiling was *fined*. The original token reinforcement and cost procedures were reinstated and praise for smiling was added. In the final weeks of the program, tokens were eliminated and crying was ignored while smiling was praised which effectively maintained performance.

Social Behaviors

A significant response class for psychiatric patients is social behavior. Of course, a precondition for social interaction for many individuals may entail elimination of specific bizarre behaviors. However, the elimination of deviant behaviors alone does not ensure the appearance of appropriate social behaviors.

Social behavior of psychiatric patients has been successfully altered using other reinforcers than tokens. For example, Kale, Kaye, Whelan, and Hopkins (1968)

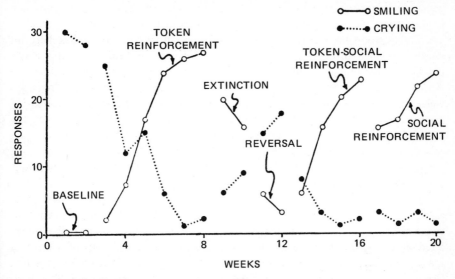

Fig. 10. Smiling responses and crying episodes under each treatment condition (from Reisinger, 1972; Society for the Experimental Analysis of Behavior).

developed greeting responses (e.g., "Hi" or "Hello, Mr. [*staff member's name*] ") in three chronic schizophrenic patients who rarely interacted with staff or peers. Using cigarettes as a reinforcer, staff reinforced greeting responses. To maintain behavior, cigarettes were delivered on an increasingly intermittent schedule. Also, Milby (1970) demonstrated that contingent staff attention (going in close proximity to, looking at, nodding, and talking approvingly) increased social behavior (talking to, working or playing with another individual) of three undifferentiated schizophrenic patients.

Token reinforcement has been used to alter conversation. Leitenberg, Wincze, Butz, Callahan, and Agras (1970) treated a socially withdrawn patient by providing contingent token reinforcement for conversing with staff. Neither instructions nor noncontingent reinforcement had increased the time devoted to conversation. Similarly, Wallace and Davis (1974) also reinforced conversation of chronic psychiatric patients. Interestingly, instructions about the desired behavior and about the topics that should be discussed as well as reading material to provide patients with something to talk about did not increase conversation. When patients could earn points (sound of tone which was tallied for each 10 seconds of conversation) exchangeable for consumable items and privileges, conversation increased markedly.

Bennett and Maley (1973) reinforced conversation of two schizophrenic patients with tokens while they spoke with each other in experimental sessions. Social interaction increased in the sessions and on the ward as well even though consequences were not programmed for interaction on the ward. Similarly, Doty (1975) provided psychiatric patients with money contingent upon their social behaviors on the ward (e.g., proximity to others and social interaction). Social behaviors increased on the ward as well as in group discussion sessions although reinforcement was not based upon performance in these sessions.

Conversation has been reinforced in a group situation. Tracey, Briddell, and Wilson (1974) reinforced psychiatric patients for positive statements either about the hospital setting or about other people. In a combined multiple-baseline and ABAB design, the particular statements (i.e., about the setting or people) increased when token reinforcement and praise were delivered. Interestingly, as verbal behavior changed in the experimental sessions, nonverbal behavior changed on the ward. Patients increased in their attendance to hospital activities when they received tokens for positive statements about the activities. However, reinforcing positive verbalizations about others in the sessions was unrelated to verbalizations about others on the ward.

Other applications of token reinforcement in group situations have been reported. Liberman (1972) trained four withdrawn and verbally inactive patients to converse in a group. Conversation increased when token reinforcement was administered either according to individual or group performance. Similarly, Horn and Black (1973) reinforced verbalizations of chronic patients in a group situation. Group participation increased during the period in which tokens were administered.

Other Behaviors

Aside from focusing upon responses adaptive to the hospital environment, specific symptoms associated with diagnosed psychiatric impairment, and social skills, select behaviors have been focused upon in various programs with psychiatric patients. For example, Parrino, George, and Daniels (1971) used the withdrawal of tokens to reduce pill-taking. Only PRN (*pro re nata*) medication, i.e., those pills delivered "as the occasion arises" or "as needed" were included in the contingency rather than medication essential for the patients' health. Over a 20-week period, there was a substantial reduction in the use of medication. Upper and Newton (1971) used token reinforcement as part of a weight-reduction program. Two psychiatric patients received tokens, off-ward privileges, and approval for meeting a three-pound weight-loss criterion each week. The patients lost sufficient weight to reach their initial treatment goals over a period averaging slightly more than 6 months. Other programs have shown that token loss for weight reduction is effective with overweight psychiatric patients (Harmatz & Lapuc, 1968). Other behaviors focused upon token programs with psychiatric patients include violation of ward rules, performance of academic tasks, helping the staff, making eye contact with a therapist, and others (e.g., Calhoun, 1974; Linscheid, Malosky, & Zimmerman, 1974; Stahl, Thomson, Leitenberg, & Hasazi, 1974; Upper, 1973).

One area that has received attention in operant programs is developing in patients a more active role in their treatment than ordinarily is the case. Making suggestions for ward practices and management or decisions about treatment have been reinforced directly (O'Brien, Azrin, & Henson, 1969; Olson & Greenberg, 1972; Pomerleau, Bobrove, & Harris, 1972). For example, Greenberg *et al.* (1975)

placed psychiatric patients into decision-making groups and reinforced making treatment recommendations that would ensure the clients' progress through the hospital toward eventual discharge. Actual proposals had to be submitted by the staff and were differentially reinforced according to their feasibility and quality.

Other behaviors related to discharge have been reinforced. Henderson and his colleagues have established a token economy in a community-based facility for psychotic males (Henderson, 1969; Henderson & Scoles, 1970). Work, counter-symptom, and social behaviors were reinforced. For example, several jobs in the facility (e.g., kitchen steward, front-office clerk, janitor, messenger) earned tokens. Social behaviors included conversing superficially, interacting with others, and initiating conversation. Although most of the program focused on behaviors in the facility, many responses related to involvement in the community were also reinforced. For example, activities such as dances were scheduled in the community. Residents received tokens for interacting with others in a community setting. A given behavior (e.g., initiating conversation) paid more tokens when performed in a community setting than in the facility. Behaviors related to job procurement in the community were also reinforced. Securing job interviews, making phone calls to prospective employers, and obtaining employment were reinforced in the home facility. Interestingly, employers were involved in the token program by providing feedback so that consequences could be delivered for performance in the community (Kelley & Henderson, 1971).

THE MENTALLY RETARDED

Token economies with the mentally retarded have encompassed a variety of behaviors across clients of all levels of retardation (cf. Kazdin & Craighead, 1973; Thompson & Grabowski, 1972). The settings encompassed include institutions, day-care facilities, sheltered workshops, and special classrooms. For purposes of review, token economies will be considered according to ward and self-care behaviors, verbal behavior and language acquisition, social behaviors, and work-related behaviors. A large number of token programs with the retarded have been conducted in classroom settings. These programs will be included in a review of children and adolescents in classroom settings treated in the next section because classroom programs have focused on similar behaviors (e.g., attending to the lesson, academic proficiency) independently of the specific population.

Ward and Self-Care Behaviors

As with psychiatric patients, token economies with the retarded have devoted considerable attention to basic self-care behaviors. Of course, self-care responses represent a particularly important focus with programs for the retarded because fundamental skills (e.g., feeding, toileting) frequently are unlearned or performed inconsistently. Self-care responses are often focused upon in general ward programs.

One of the earliest programs for the retarded that encompassed a variety of behaviors was conducted at Parsons State Hospital and Training Center for institutionalized females (including children, adolescents, and young adults) with IQs up to 55 (Girardeau & Spradlin, 1964; Lent, 1968; Spradlin & Girardeau, 1966). The program evolved over several years in terms of the specific behaviors that were reinforced and the back-up events that could be purchased. In one of the reports, the range of behaviors was extensive (Spradlin & Girardeau, 1966). Table 8 illustrates the specific behaviors and the tokens earned for a group of 28 adolescents. Tokens (bronze coins) were exchangeable for consumable items (e.g., candy, fruit), clothes, cosmetics, and the rental of special equipment (e.g., watches, record players, bicycles). A global evaluation of the program after several years suggested that changes were achieved in self-care behaviors (e.g., dressing, personal grooming and cleanliness, walking with appropriate posture, sitting correctly), verbal behaviors (e.g., screaming, talking to oneself, swearing), and social behaviors (e.g., aggressive acts toward others, touching others) (Lent, 1968). Some of the residents were successfully returned to the community.

In some programs, select self-care behaviors are identified and serve as the main focus. For example, Hunt, Fitzhugh, and Fitzhugh (1968) administered tokens to improve the personal appearance of 12 retardates (mean IQ = 73) who were considered most likely to graduate to the community. Personal appearance improved under continuous and intermittent token reinforcement for some of the

Table 8. A Sample of the Behaviors Which Are Rewarded and Approximate Reward Amounts (Spradlin & Girardeau, 1966)

Behavior	Approximate reward amount
Making up bed	1 token
Dressing for meal	2 tokens
Brushing teeth	2 tokens
Taking shower properly	2 tokens
Helping clean cottage	Quite variable
Setting hair	4 tokens
Straightening bed drawer	2 tokens
Trimming and filing nails	2 tokens
Combing hair	1 token
Washing hair	2 tokens
Group play (30 min)	5 tokens
Coloring	Quite variable
Work placement in institution	10 tokens/day
Cleaning goldfish bowl	5 tokens
Feeding goldfish	1 token
Cleaning bird cage	5 tokens
School readiness tasks	Quite variable
Being on time at work or speech therapy	4 tokens
Shining shoes	5 tokens
General proper use of leisure time (20-min period)	4 tokens

clients, although the absence of baseline data precluded unambiguously interpreting the effects.

Horner and Keilitz (1975) carefully evaluated a token program designed to develop toothbrushing in moderately or mildly retarded institutionalized residents (mean IQ = 43). Separate individual steps of toothbrushing (e.g., picking up and holding the brush, applying paste to the brush, brushing the outside, biting, and inside surfaces of the teeth, and others) were trained with the use of prompts (instructions, demonstration, and physical guidance) if these were required. Tokens were exchangeable for sugarless gum. The effects of training were evaluated in a multiple-baseline design. These effects, presented in Fig. 11, show that each individual's behavior changed only when treatment was introduced. Similar effects were achieved with other subjects who received praise rather than tokens.

Campbell (1974) devised a token program to increase the physical fitness of institutionalized retarded boys (IQs ranging from 20 to 84). Tokens were given for attempting and completing various exercises including sit-ups, push-ups, head and shoulder lifts, touching one's toes, and running in place. Tokens (exchangeable for a wide range of back-up reinforcers from balloons to wristwatches) were provided for improvements in the number of exercises completed. Exercise increased for a group receiving contingent tokens relative to a control group that was instructed to practice but received no tokens.

Aside from self-care behaviors, token economies with the mentally retarded are used to control problematic behaviors on the ward. For example, Peniston (1975) implemented a token economy on a closed ward of institutionalized severely and profoundly retarded males (aged from 14 to 34 years). The main purpose of the program was to reduce disruptive behaviors including physical aggression, talking irrationally, and rocking behavior. Tokens and praise were delivered for not engaging in these behaviors as well as for specific tasks such as working in a sheltered workshop in the institution. Token reinforcement was supplemented with time-out from reinforcement (5-minute isolation), response cost (token withdrawal), and overcorrection for specific behaviors. In an ABAB design the overall program was shown to control the disruptive behaviors on the ward.

In addition to those reviewed, a large number of other token programs for the retarded have focused on several general ward behaviors including self-care, completing small jobs on the ward, attending or participating in rehabilitative or recreational activities, and social interaction (e.g., Anderson, Morrow, & Schleisinger, 1967; Bath & Smith, 1974; Brierton, Garms, & Metzger, 1969; Miller, 1973; Musick & Luckey, 1970; Sewell, McCoy, & Sewell, 1973; Wehman, 1974). The extent of these reports suggests that the token economy has been used quite frequently as means to structure institutional programs.

Verbal Behavior and Language Acquisition

Token reinforcement has been used to alter a variety of forms of verbal behavior including language acquisition, topographical features of speech, amount

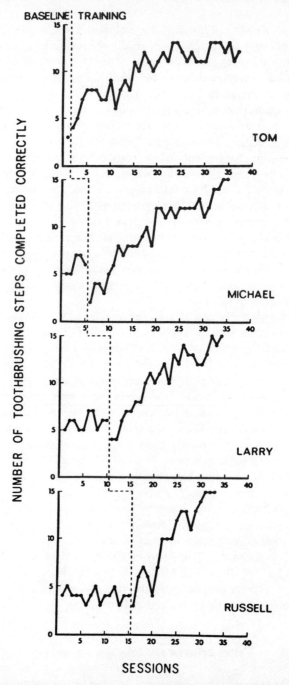

Fig. 11. The number of toothbrushing steps performed correctly. (The broken line through the individual graphs indicates termination of baseline and the beginning of training.) (From Horner & Keilitz, 1975; Society for the Experimental Analysis of Behavior.)

of speech about a particular content area, and the suppression of inappropriate speech.

Several studies have developed receptive and productive language and imitative speech of retarded clients, usually children, with social and token reinforcement (Baer & Guess, 1971, 1973; Brickes & Brickes, 1970; Guess, 1969; Guess & Baer, 1973; Guess, Sailor, Rutherford, & Baer, 1968; MacCubrey, 1971; Schumaker & Sherman, 1970). Diverse responses have been focused upon including use of verbs, plural nouns, past and present tenses, and others. Token reinforcement in language training is well illustrated in a program by Lutzker and Sherman (1974) in which the use of descriptive sentences involving plural and singular subject-verb agreement with moderately retarded residents (a child, an adolescent, and an adult) and normal toddlers was developed. In response to pictures, residents received praise and tokens (exchangeable for candy, soda, or toys) for correctly identifying the content of the pictures, action of the subject, and eventually for using sentences with subject and verb agreement. As more sentences were taught, correct untrained sentences were increasingly produced.

Wheeler and Sulzer (1970) developed the use of appropriate sentences with an 8-year-old boy who had been diagnosed as brain damaged, autistic, and retarded. The boy spoke in "telegraphic" English, leaving out most articles and auxiliary verbs when speaking. During training sessions, he received tokens (exchangeable for toys, food, being tickled) for correct sentences in response to pictures that he was instructed to describe. The responses generalized to novel stimuli not associated with reinforcement.

Twardosz and Baer (1973) trained two severely retarded institutionalized adolescents to recognize letters, numbers, and colors and to ask questions about particular stimuli that were not introduced into training. Tokens (exchangeable for candy) and praise were delivered for correct recognition of training items and for asking questions. The target responses increased markedly in the experimental sessions. Another investigation using food, praise, and tokens developed question answering in retarded residents (Clark & Sherman, 1975).

Aside from the development of language, token reinforcement has been used to alter topographical aspects of speech. In an interesting program, tokens were used to increase voice loudness of a mildly retarded adolescent girl whose speech was almost inaudible (Jackson & Wallace, 1974). During experimental sessions, the girl wore a lavalier microphone connected to a voice-operated relay. The relay was sensitive to preset levels of vocal intensity. Tokens (exchangeable for books, beauty aids, photo album, and other items) were automatically delivered for vocalizations that met the target level of intensity. The intensity requirements were made increasingly stringent as progressively louder vocalizations were shaped. To develop transfer from the experimental sessions to the classroom, common elements associated with these settings were maximized including the presence of other students in or near the training situation and the tasks on which vocalizations were reinforced.

Kazdin (1973a) used the withdrawal of tokens to suppress speech disfluencies in adult retardates (IQ range from 42 to 89). Not only were the punished disfluencies decreased, but nonpunished disfluencies decreased as well. The effects of treatment generalized to an interview not associated with the training procedure and were maintained up to one month after training had been terminated.

Griffiths and Craighead (1972) altered a misarticulation (incorrect production of the /l/ sound) using modeling and token reinforcement for correct articulations. Treatment effects did not generalize to extratherapy sessions in which the program was not conducted until the contingency was introduced into these sessions. Interestingly, reinforcing correct pronunciation in the extratherapy sessions in response to some of the stimuli used in therapy, led to the appearance of all responses focused upon in treatment.

While most token reinforcement procedures used to develop verbal behavior in retardates have focused upon the development of language, the content of language has received little attention. In one project, an attempt was made to develop specific language content of institutionalized retardates that reflected knowledge about the outside world. Keilitz *et al.* (1973) used tokens to increase verbalization about current events for three male institutionalized adolescent residents who were moderately or borderline retarded. Each boy individually viewed a brief televised newscast daily and verbalized the material after the telecast was over. The contingent delivery of tokens (exchangeable for pennies) markedly increased correct verbalizations about the telecasts.

Social Behaviors

Although operant programs have focused on social behaviors among the retarded (Hopkins, 1968; Stokes, Baer, & Jackson, 1974; Whitman, Mercurio, & Caponigri, 1970), few of these have been in the context of a token economy. In one program, Kazdin and Polster (1973) provided tokens to two withdrawn male adult retardates for talking with their peers in a sheltered workshop. During the three daily work breaks, the clients received tokens (exchangeable for consumable items in a "store" and special privileges) for each person they interacted with. An interaction was defined as a verbal exchange in which the client and a peer each made one statement to the other person that reflected some content area (e.g., the news, sports, television) rather than a general greeting. Peers were queried after each break to ensure that the clients' reports about their interactions were accurate. Discrepancies between peer and client reports about interacting resulted in the client not receiving a token for that interaction. Token reinforcement markedly increased the frequency of interactions.

Knapczyk and Yoppi (1975) used token reinforcement to increase the frequency of play behaviors of five institutionalized retarded children. During the token program, the children were told that they could earn points for playing nicely with each other. Points along with praise were administered for initiating either cooperative or competitive play. The points could be spent immediately after the play sessions for activities and games or after several tokens had been saved for larger and more delayed events (e.g., a field trip). During baseline and reversal phases of an ABAB design, the children engaged primarily in solitary and parallel play. During the token program, cooperative and competitive play increased markedly.

Token reinforcement has been used in experimental and quasi-experimental rather than applied programs to develop socially related behaviors. For example, Ross (1969) trained mentally retarded children in a special class to master syllogistic problems concerning the consequences of their behavior in social situations (e.g., "If I do this, then this results"). Stars (exchangeable for toys) were provided for correct answers to questions about consequences of behavior. Posttraining evaluation revealed that knowledge about correct situational behaviors increased relative to a control group that did not receive training. The actual social behaviors of the children were not studied. Similarly, Ribes-Inesta *et al.* (1973) increased physical contact of retarded and brain-damaged children by providing tokens when the children touched an adult in experimental sessions. The contact of children on the ward was not examined.

Work-Related Behaviors

Several programs have focused on aspects of work with the retarded in an attempt to develop community job placement. In one of the most elaborate programs toward this end, Welch and Gist (1974) provided trainable young retarded adults with tokens for job performance and other behaviors that would increase the likelihood of job placement. In a facility that the clients attended for 6 hours per day, tokens were administered for grooming, punctuality to work, academic, and homemaking behaviors.

Work behaviors were the main focus of the program. A large number of jobs were made available and grouped into levels. Lower level jobs included menial tasks (e.g., cleaning the toilet, mopping). Higher level jobs entailed greater responsibility (e.g., operating machines, checking absenteeism). The third and highest level included jobs of greater responsibility, decision making, and extreme proficiency (e.g., serving as an office courier, teacher aide, mail clerk). Progress through the levels was achieved by meeting a specific performance criterion. Although token earnings did not necessarily differ across levels, the jobs were considered to differ in prestige value. The goal of the leveled program was to place the individuals at the highest level into community work. The back-up reinforcers for the tokens included money, edible rewards, and privileges such as utilizing the phone, watching television, and others.

Several interesting features of the economy warrant mention in passing. First, the program included a banking system so that one could deposit and save tokens as well as write checks against their balance. Second, clients could earn sick leave (days of entitled absence) for working without absences for protracted periods. However, sick days at work which were not earned were associated with a large fine. This led to the third innovation. Clients could purchase the equivalent of workmen's compensation insurance. For a monthly premium (in tokens), clients were insured in the system against unexpected accident or illness that impeded their work. If a client could not work, his policy would pay some of the tokens due to the absence. Although the data for the program did not unambiguously demonstrate the effect of the contingencies in altering the target responses, several clients were successfully placed in the community.

In many programs, production rates have been altered with token reinforcement. Zimmerman *et al.* (1969) observed production rates for a piece-rate job for 16 young adult multiply handicapped and retarded clients. Before tokens were delivered, clients were given a "practice" period in which they received information as to how many tokens they would have earned had tokens been administered. In a subsequent phase, tokens actually were delivered. When individuals "practiced" earning tokens and only received feedback about their performance work improved over baseline rates. However, the addition of contingent tokens increased performance to a greater extent.

Hunt and Zimmerman (1969) increased productivity for institutionalized retardates in a simulated workshop by delivering tokens (exchangeable for canteen items) contingent upon improvement according to individualized criterion levels. Performance improved during the period in which tokens were delivered and during another period of the day in which no tokens were provided. Other studies have shown that productivity and accuracy of job completion can be altered with token reinforcement either alone or in conjunction with other reinforcers (e.g., Cotter, 1971; Jens & Shores, 1969; Karen, Eisner, & Enders, 1974; Logan, Kinsinger, Shelton, & Brown, 1971; Schroeder, 1972).

Other Behaviors

A major determiner of child placement in school is performance on various standardized psychological tests. Interestingly, some studies have shown that the conditions of test administration may dictate performance on these tests. In addition, training conditions such as experience with a program where academic responses are reinforced also contributes to performance on these tests.

Ayllon and Kelly (1972, Exp. 1) tested 12 trainable retardates (average IQ of approximately 47) with the Metropolitan Readiness Test under two conditions. Initially, the test was administered under standard testing procedures. On the same day, the test was readministered under token reinforcement. After each subtest, items were checked and tokens (back-up reinforcers unspecified) were administered for correct responding. Under conditions of token reinforcement, test scores increased. Increased performance on readiness and intelligence tests under conditions of token or food reinforcement also has been found for elementary school students of average intelligence (Ayllon & Kelly, 1972, Exp. 2; Edlund, 1972). In another experiment, Ayllon and Kelly (1972, Exp. 3) matched trainable retarded students on age, IQ, and Metropolitan Readiness Test performance and randomly assigned individuals to experimental or control groups. Experimental subjects received tokens for academic responses whereas control subjects remained in their original classes. The tokens were exchangeable for various items, activities, and special privileges (e.g., eating in a special dining area of the school lunchroom). The control group received their regular academic program in their original classrooms. After 6 weeks, the token reinforcement group showed superior performance on the Metropolitan Readiness Test than the control group when the test was administered under standard conditions. Both groups improved in a parallel fashion when the test was readministered under conditions of token reinforcement although the group previously exposed to a token program in the classroom maintained their superiority.

INDIVIDUALS IN CLASSROOM SETTINGS

Behavioral techniques have been applied more extensively to classroom settings than any other setting (cf. Kazdin, 1975b, 1975c). Often child behaviors are altered in the classroom by increasing the teacher's use of praise, attention, and approval as social reinforcers for appropriate behavior (e.g., Becker, Madsen, Arnold, & Thomas, 1967; Hall, Panyan, Rabon, & Broden, 1968; Kazdin & Klock, 1973; Madsen et al., 1968; Ward & Baker, 1968). However, for some children praise may not be sufficiently reinforcing to alter behavior or indeed may even be aversive. Token reinforcement is likely to be more effective than praise alone.

Token economies have been established in the classroom across diverse populations (e.g., "normal" children, retardates, delinquents), educational levels (e.g., preschool, high school, and college), and settings (e.g., classes in institutional settings, special education or adjustment classes) (see Drabman, 1976; Kazdin & Bootzin, 1972; McLaughlin, 1975; O'Leary, 1977; O'Leary & Drabman, 1971, for reviews). Typically, target behaviors include the reduction of disruptive behavior or inattentiveness to the lesson and improvement in academic responses. The present section will highlight programs across each of these areas and across different grade levels and populations.

The earliest report of a token economy that focused on attentive behavior in the classroom was a program for severely retarded children (IQs below 40) at Ranier School in Washington (Birnbrauer & Lawler, 1964). Behaviors that were reinforced included entering the classroom quietly, hanging up coats, sitting at the desk attentively, and working persistently on the task. Initially, candy was made contingent upon performance. Tokens were substituted and could be exchanged for candy and trinkets. The performance of 37 of 41 students improved on behavioral criteria, although a clear functional relation between behavior change and contingent tokens was not established.

The program at Ranier School also encompassed academic performance (Birnbrauer, Bijou, et al., 1965; Birnbrauer, Wolf, et al., 1965). The academic performance of retarded students (IQ ranging from 50 to 72) was altered by providing tokens on the basis of the number and accuracy of items completed on individual-tailored academic assignments and for cooperative behavior. Tokens could be exchanged at the end of class for edibles and small prizes. In an ABA design, which began with token-reinforcement phase, the effect of the program was demonstrated with 10 of 15 students. Because social reinforcement and time-out (10 minutes in a time-out area) were used throughout the study, it is unclear whether the failure of some subjects to show a reduction in performance during the reversal phase resulted from their unresponsiveness to the contingencies in general or whether changes effected with tokens were maintained by other programmed consequences in the situation. Since the early classroom programs at Ranier School, a large number of classroom programs have been implemented.

Disruptive Behavior and Inattentiveness

The majority of token programs in the classroom have focused on deportment. The goals usually are to decrease disruptive behavior and inattentiveness to the academic lesson. The disruptive or inappropriate behaviors usually focused upon have included getting out of one's seat or talking without permission, not attending to the teacher or one's lesson, playing with objects, performing aggressive behaviors, disturbing someone else's work, not complying with instructions, arguing or complaining, and other responses that are usually considered to be incompatible with academic performance. A large number of programs in preschool, elementary, and secondary classrooms for students identified "emotionally disturbed," retarded, and normal have shown the efficacy of token reinforcement in reducing disruptive behavior and increasing attentiveness (Ascare & Axelrod, 1973; Bucher & Hawkins, 1973; Bushell et al., 1968; Drabman et al., 1974; Evans, Howath, Sanders, & Dolan, 1974; Flowers, 1974; Herman & Tramontana, 1971; Kaufman & O'Leary, 1972; Kazdin, 1973g; Kuypers, Becker, & O'Leary, 1968; O'Leary & Becker, 1967; O'Leary, Becker, Evans, & Saudargas, 1969; Packard, 1970; Quay, Sprague, Werry, & McQueen, 1967; Ross, 1974b; Schmidt & Ulrich, 1969; Schwartz & Hawkins, 1970; Stedman, Peterson, & Cardarelle, 1971; Wagner & Guyer, 1971; Winett et al., 1971).

As an illustration, Broden, Hall, Dunlap, and Clark (1970) used token reinforcement to increase study behavior and to decrease disruptive behavior in a special education class of 13 junior high school students. The students were several years behind in academic areas and engaged in various problem behaviors (e.g., throwing things, refusing to obey the teacher). The students were in the same class for several periods of the day at which time a token program was implemented. Tokens were administered for study behavior including working quietly, completing assignments, and remaining in one's seat and were lost for disruptive behavior including talking or being out of one's seat without permission, arguing with the teacher, and others. Privileges such as leaving early for lunch, talking with others, going on a field trip, as well as snacks served as back-up reinforcers. The effect of the program was evaluated in an ABAB design (see Fig. 12). Although the experimenters had intended to withdraw the point system for a few days in a reversal phase, performance deteriorated on the first three periods of the first day. Thus, the program was reinstated on the first day of the reversal for the remaining periods. Reinstatement of the program increased study behavior. Subsequent data

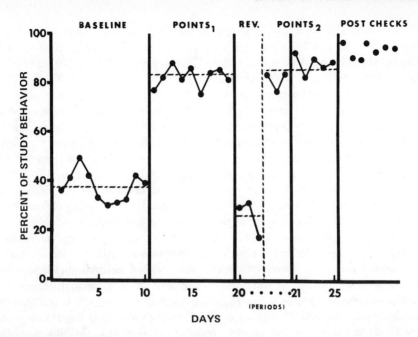

Fig. 12. A record of study behavior during the entire day under Baseline, Points$_1$, Reversal, Points$_2$, and Post Check conditions (from Broden, Hall, Dunlap, & Clark, 1970; Council for Exceptional Children).

over a 6-week period indicated that study behavior remained high as long as the program was in effect.

The majority of classroom token programs focusing on disruptive and attentive behavior have provided reinforcement on the basis of individual student performance. However, group contingencies have been commonly investigated in classroom settings. A number of classroom token reinforcement programs have used the "good behavior game" to alter behavior (Barrish *et al.*, 1969; Harris & Sherman, 1973; Maloney & Hopkins, 1973). The "good behavior game" consists of a group contingency where the class is divided into teams. Each team earns or loses points contingent upon behavior of the group. The team with the most points earned or least points lost, depending upon whether the system is based upon positive reinforcement or response cost, earns the reinforcing events. Outstanding performance by each team is reinforced by allowing each of them to receive the reinforcing event.

For example, Medland and Stachnik (1972) used the "good behavior game" in an elementary school class to reduce disruptive behavior (e.g., talking without permission, being out of seat, hitting others). The team with fewer marks given for disruptive behavior received extra recess. The game also included making explicit classroom rules and providing feedback for performance. Red and green lights controlled by student observers were used to signal how each group was doing. The program decreased disruptive behavior for each group.

The vast majority of token programs in the classroom have been conducted in a single classroom. Few large-scale applications have been reported. A notable exception was reported by Rollins, McCandless, Thompson, and Brassell (1974) in which several inner-city school classes encompassing grades 1 through 4 and 6 through 8 received a token reinforcement program. A comparison was made between 16 experimental classes that received a token program and 14 control classes that did not. Over 700 students participated in the study. Students in the experimental classes received tokens (exchangeable for candy, toys, school supplies, and activities) for attentive classroom behavior. Behavioral measures of disruptive and on-task behaviors over several months of the program revealed the superiority of the experimental classes. In addition, greater gains in measures of intelligence and reading and math achievement were found for the token program classes. These results, of course, are consistent with the existing literature on individual classrooms but are notable for the elaborate extension of programs across several classes.

The focus of many classroom token programs on attentive or disruptive behavior is based on the view that attentive behavior is a precondition for academic performance. Of course, to some extent this assumption is accurate because an abundance of disruption in a classroom can preclude presenting lessons. Many of the classroom programs that focus on disruptive behavior alter flagrant responses such as disruptive verbal and physical assaults and specific hyperactive behaviors rather than merely not paying attention to a lesson (e.g., Christensen, 1975; Christensen & Sprague, 1973; Kirschner & Levin, 1975). Other programs focus on extreme deficits in performing basic entry responses such as sitting in one's chair for more than a brief period (e.g., Bauermeister & Jemail, 1975; Twardosz & Sajwaj, 1972). In these cases, which may represent relatively extreme cases of inappropriate deportment, basic responses appear to warrant separate contingencies prior to or along with academic performance.

However, in several programs relatively mild levels of disruptive behavior and inattentiveness appear to be altered for their own sake without any clear relationship to academic performance. Of course, there are different reasons to increase attentive behavior and to reduce disruptive behavior in the classroom. Teachers in fact seem to be extremely concerned about disruptive behavior, perhaps even to a greater extent than academic performance *per se*. Also, some individuals hypothesize that it may be easier to alter disruptive behavior than academic achievement (Hamblin *et al.,* 1974). In any case, the importance of altering attentive behavior and minimizing disruption has become less clear. A major issue is that increasing attentive behavior and decreasing disruption does not necessarily increase academic performance.

In one investigation, a token economy was implemented with third-grade children of an inner-city school (Ferritor, Buckholdt, Hamblin, & Smith, 1972). The target behaviors were attending to the lesson (e.g., looking at or writing on one's paper, looking at the teacher when she was talking, not performing disruptive behaviors such as talking without permission or hitting others) and correctly completing arithmetic problems. Token reinforcement was introduced for attentive behavior and for correct academic performance at different points in time to determine the relationship between these responses. When tokens were delivered for attentive behavior, this behavior increased. However, academic performance was not affected. Similarly, when tokens were delivered for academic performance, this behavior increased. However, attentive behavior was not affected. Both attending and correct academic performance were increased only when tokens were concurrently delivered for each behavior. These results suggest that increasing attentive behavior may not affect academic performance and vice versa. Thus, the value of focusing on attentive behavior as an end in itself has been debated (O'Leary, 1972; Winett & Winkler, 1972). The issue is not entirely resolved. While some studies have shown that increasing attentive or reducing disruptive behavior has little or no effect on academic performance (Ferritor et al., 1972; Harris & Sherman, 1974), others have shown an effect (Iwata & Bailey, 1974).

Although increases in attentive behavior do not appear to usually affect academic performance, increases in academic performance have been shown in several studies to increase attentive and to reduce disruptive behavior (Ayllon et al., 1972; Kirby & Shields, 1972; Marholin, Steinman, McInnis, & Heads, 1975; McKenzie, Clark, Wolf, Kothera, & Benson, 1968; Sulzer et al., 1971; Winett & Roach, 1973). For example, Ayllon and Roberts (1974) provided tokens to disruptive fifth-grade students for completing reading assignments correctly. When academic performance increased, disruptive behavior (e.g., being out of one's seat, talking instead of working) decreased, although no specific contingencies were designed for disruptive behavior. These results suggest that disruptive behaviors and inattentiveness, areas of major concern to teachers, may be ameliorated by directly reinforcing academic performance. The importance of increasing academic performance *per se*, and its role in suppressing disruptive and inattentive behavior may explain why behavior modification in the classroom has increased its focus on academic behaviors in recent years (Kazdin, 1975b).

Academic Behaviors

Reinforcement has been used to alter a wide variety of academic skills including accuracy in reading, arithmetic, handwriting, and spelling assignments, completing in-class homework assignments, creative writing, and developing vocabulary (cf. Klein, Hapkiewicz, & Roden, 1973; O'Leary & O'Leary, 1972; Ulrich, Stachnik, & Mabry, 1974). Programs have focused on preschool through college students.

In an early application, Wolf, Giles, and Hall (1968) provided a remedial education program for low-achieving sixth-grade students who were at least 2 years below the norm for grade level in reading. Tokens, redeemable for candy, novelties, clothing, food, field trips, money, and other items were delivered for correctly completing classroom assignments. In separate experiments, the effect of magnitude of point rewards was demonstrated. Higher point values increased the amount of academic performance on a given task. Also, performance on specific tasks predictably followed shifts in point values to increases, decreases, and total cessation of points for particular responses.

Additional contingencies were described in this report including points for completing extra work assignments, attendance, grade averages, deportment, and cooperation. The effects of these contingencies were not systematically evaluated. At the end of a one-year period subjects who received the remedial program were significantly higher in achievement test performance than control subjects who did not receive the program. Also, remedial students showed higher public school grades than control subjects during the year of the program. Since experimental and control subjects had been matched on the basis of initial reading deficiency and assigned in blocks to groups prior to the program, the comparison suggests rather strongly that the gains resulted from differential treatment.

Another early application of token reinforcement was completed by Clark, Lachowicz, and Wolf (1968) who worked with school dropouts. Some dropouts received money for completing academic assignments each day over a 2-month period, while others received approximately the same amount of money for completing job assignments. For the academic group, money was differentially provided for completion of assignments in areas of academic deficiencies. A comparison of pre- and posttreatment achievement test scores indicated 1.3-year gain for the academic group and only a 0.2-year gain for the job group.

Since these early programs, token economies have been demonstrated to be effective in increasing the number of academic tasks completed in reading, arithmetic, language, and writing and the accuracy with which they are completed (Ayllon & Roberts, 1974; Brigham, Finfrock, Breunig, & Bushell, 1972; Chadwick & Day, 1971; Dalton, Rubino, & Hislop, 1973; Glynn, 1970; Hamblin et al., 1974; Haring & Hauck, 1969; Knapczyk & Livingston, 1973; Lahey & Drabman, 1974; Lovitt & Curtiss, 1969; Lovitt & Esveldt, 1970; McLaughlin & Malaby, 1972a, 1972b; Miller & Schneider, 1970; Rickard, Melvin, Creel, & Creel, 1973; Rosenfeld, 1972; Wilson & McReynolds, 1973).

Many of the programs focusing on academic behaviors have addressed relatively complex behaviors. For example, one area focused upon is writing compositions where judgments of appropriate and creative responses are somewhat difficult to specify. Maloney and Hopkins (1973) developed a token program for 14 elementary school students who voluntarily attended a remedial summer school session to increase writing skills. Points were delivered contingent upon the use of different adjectives, different action verbs, and for different sentence beginnings throughout a given story. The point contingency was introduced for these responses at different points in time. Points were delivered to each individual who was on one of two "teams" in class. The team that achieved a specified number of points earned early recess and candy. The results (which appear in Fig. 13) show that the compositional skill that was reinforced in a given phase (e.g., adjectives or action verbs) increased in that phase. Only when all of the responses were reinforced did they all increase (last phase). Interestingly, individuals unfamiliar with the project

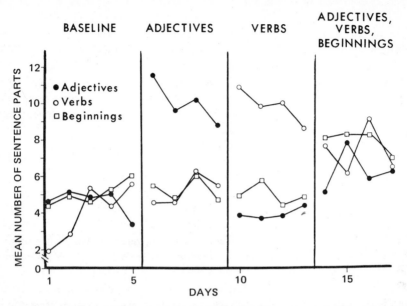

Fig. 13. Mean number of different adjectives, different action verbs, and different beginnings (from Maloney & Hopkins, 1973; Society for the Experimental Analysis of Behavior).

rated those stories written during the intervention phases as more creative than those written during the baseline phase.

Brigham, Graubard, and Stans (1972) also used token reinforcement to develop compositional skills in a fifth-grade class of behavior-problem students. Tokens (exchangeable for various privileges in the room) increased the number of words used although the effects were inconsistent for producing new words and different words in the compositions. Interestingly, rating of quality of the compositions increased when the contingencies for writing were in effect. Other investigations have shown the effects of reinforcement on composition writing (Van Houten, Morrison, Jarvis, & McDonald, 1974).

The above programs have shown the effects of token reinforcement on specific responses in classroom situations. Of interest are the effects of such contingencies on standardized test performance and measures of grade level. In many studies, the effects of the contingencies on achievement levels is difficult to infer because token reinforcement and nonreinforcement groups are not randomly composed or matched on variables which may be relevant (e.g., Rollins *et al.*, 1974). Also, the effect of token reinforcement cannot be separated from other components of the intervention (e.g., attending remedial sessions) (e.g., Wolf *et al.*, 1968). Nevertheless, the results of many studies show that students receiving token-reinforcement programs increase on achievement test scores (e.g., Stanford Achievement Test, Wide Range Achievement Test) relative to untreated control groups (Bushell, 1974; Clark *et al.*, 1968; Kaufman & O'Leary, 1972; Wolf *et al.*, 1968). Some studies have reported changes in tested intelligence (e.g., on the Wechsler Intelligence Scale for Children) as a function of token reinforcement in classroom settings, although methodological artifacts (e.g., differential regression among treatment and control groups) cannot be ruled out as a rival interpretation of the results (Dickinson, 1974; Mulligan, Kaplan, & Reppucci, 1973; Sachs, 1971). Superior academic performance associated with token reinforcement has been apparent immediately after the intervention and after a follow-up period up to a year after the intervention has been terminated (Dalton *et al.*, 1973).

A variation of token programs for academic performance has been extended to classes at the college and university level. In a sense, a token economy is used in any class where points are earned for performance and are "exchanged" for some back-up reinforcer (e.g., final course grade). However, operant principles have been extended explicitly to the college classroom in the method referred to as the Personalized System of Instruction (PSI) which has been developed by F. S. Keller (1966, 1968; Ryan, 1972). Personalized instruction refers to a way of structuring the course so that there are specific contingencies devised toward course completion. There are several distinguishing features of a course structured in this fashion such as allowing the student to complete the course requirements at his or her own pace, dividing the content of the course into units, providing exams over each unit, allowing repeated completion of exams until mastery is achieved, and determining the final grade based upon the number of units mastered. Students usually receive immediate feedback for test performance and may be encouraged to discuss areas with a proctor or tutor that need to be focused upon prior to retesting.

Self-paced courses at the college level have included a number of variations of the specific procedures espoused by Keller. Typically, in the Keller method, students are required to pass one unit before progressing to the next unit and their grades are largely determined by the number of units mastered by the end of the course. In some variations of the course, several options are added so that students can achieve final grades in diverse ways. For example, points are sometimes given for completing specific options (e.g., doing an experiment, reading extra material) and their value is applied toward the final grade (e.g., Born, Gledhill, & Davis, 1972; Fowler & Thomas, 1974). Essentially, the course is structured on a point economy with grades as the back-up event.

For example, Lloyd and Knutzen (1969) developed a self-paced instruction format to teach a psychology course on operant conditioning. At the beginning of the term, students received a list of ways to earn points toward their final grades and the point levels required to achieve specific grades. Points were given for a variety of behaviors including answering questions orally in class, answering written questions on text material, reporting on journal articles relevant to the course, reviewing films assigned for the course, conducting laboratory experiments, presenting a talk or a symposium in class, and others. For each assignment, a minimal level of points had to be achieved out of the total points possible to ensure some level of mastery. A number of options were available so that final point totals could be achieved through diverse academic and practical activities related to the course content. The course was self-paced so that students could cease working for the course when they achieved the point total and grade that they had wanted.

Personalized instruction has been used in a large number of classrooms and even in an experimental college where students enroll simultaneously in several courses that are programmed in this way (Malott, Hartlep, Keenan, & Michael, 1972). Several variations of individualized instruction have been investigated along with diverse parameters that contribute to its efficacy (e.g., Farmer, Lachter, Blaustein, & Cole, 1972; Johnston & O'Neill, 1973; Miller, Weaver, & Semb, 1974; Semb, 1974). Research has suggested that personalized instruction when compared to traditional lecture-course formats results in greater mastery and retention of course material (Alba & Pennypacker, 1972; Born et al., 1972; Cooper & Greiner, 1971; McMichael & Corey, 1969; Sheppard & MacDermot, 1970).

Other Behaviors

Classroom token programs have encompassed behaviors other than attentive and disruptive behavior, and academic performance. One area of interest to teachers has been developing instructional control over child behavior. For example, Baer, Rowbury, and Baer (1973) reinforced compliance with instructions in three preschool children who had multiple behavior problems. The children already were earning tokens (exchangeable for free time) by working on preacademic tasks (e.g., matching and manipulative tasks). To develop compliance, the teacher invited the students to perform various tasks (e.g., "Wouldn't you like to . . ." or "It's your turn to . . ."). Complying with these requests increased markedly when tokens were provided for performing these tasks in response to instructions. A time-out contingency (removing the child from work for 1 minute), added to the token-reinforcement contingency, increased compliance further. The effectiveness of token reinforcement in developing compliance to instructions has also been demonstrated with "emotionally disturbed" and retarded children (Fjellstedt & Sulzer-Azaroff, 1973; Zimmerman, Zimmerman, & Russell, 1969).

Speech also has been focused upon in the classroom. Johnston and Johnston (1972) used token reinforcement to develop correct articulation of consonant sounds in children with severe articulation problems. Correct verbalizations during activities including lessons and play in which the children normally conversed were reinforced with tokens (exchangeable for selection of a play activity). Failure to earn a predetermined number of tokens led to remaining inside while other children were at play for a 10-minute period. Incorrect speech sounds were verbally corrected by the teacher. In a multiple-baseline design across different time periods, token reinforcement was shown to increase the rate of correct speech sounds.

Social behaviors have been focused upon in the classroom (Sacks, Moxley, & Walls, 1975). Three socially withdrawn preschoolers in a nursery school class received cards for entering and remaining in a group and talking to adults or other children. The reinforcers were "mixie" cards. Pairs or triples of these cards could be put together in various combinations to form pictures of people or animals and had no other back-up value. Social interaction markedly increased with the contingent delivery of these cards.

In one classroom study, an attempt was made to evaluate the effects of token reinforcement on measures of self-concept (Parker, 1974). Disruptive elementary students in special classrooms who received tokens for attentive behavior increased in their self-concept (as measured by the Pictorial Self-Concept Scale for Children) after 8 weeks of the classroom management program relative to control subjects who did not receive the program. The inability to randomly assign subjects across groups interfered with unambiguously drawing conclusions about the effect of token reinforcement on self-evaluations.

McLaughlin and Malaby (1975a) delivered tokens to increase class participation in a science lesson. Tokens were provided to elementary school students for asking questions about a science film that was shown each day. As shown in a reversal design, tokens were associated with marked increases in question-asking.

The above studies illustrate the diversity of applications in the classroom. As noted earlier, the vast majority of programs have focused upon attentive behavior. Recently, academic responses have received increased attention, perhaps in response to the literature showing that improvement in attentive behavior does not necessarily improve academic performance.

CONCLUSION

Token economies with psychiatric patients, the mentally retarded, and individuals in classroom settings constitute the majority of applications. Research with each of these populations has firmly established the efficacy of token reinforcement in altering a wide range of responses.

With psychiatric patients, responses adaptive to the institution have been altered such as performing self-care behaviors, personal chores, and jobs in the hospital, attending and participating in activities, taking medication, and others. Alteration of these behaviors frequently has been associated with changes in symptomatic behaviors as well as increases in discharge and decreases in readmission rates. Many programs have focused directly on deviant and bizarre behaviors such as overt signs of hallucinations, delusions, and depression. An important focus has been on social behaviors including conversing and being with others. Select programs with psychiatric patients have altered such diverse behaviors as overeating, making recommendations for their own treatment, seeking jobs in the community, and working outside of the treatment facility.

With the mentally retarded, programs frequently have developed self-care behaviors such as dressing, grooming, personal hygiene, exercising, and others. Language skills have been developed including using diverse forms of receptive and productive language, asking questions, articulating correctly, and select topographical features of speech. Social behaviors have received somewhat less attention in token programs with the retarded relative to self-care and language. Yet, existing evidence suggests that social behaviors can readily be altered with retardates. Because the retarded frequently participate in sheltered workshop programs, token economies have been used to alter work-related behaviors. Production rates have received attention and have been readily amenable to token-reinforcement contingencies.

In classroom settings, token economies have been implemented from the preschool to the secondary level primarily for attentive, on-task, and disruptive behaviors. The focus on a deportment has decreased in recent years as it has become clear that improvements in attentive behaviors are not necessarily associated with improvements in academic productivity or accuracy. Thus, many recent programs have focused directly on academic performance, and interestingly, find that deportment improves as a concomitant effect. Increases in academic performance have been demonstrated on an extremely wide range of tasks including relatively complex responses related to creative writing. Academic gains are frequently reflected in improvements in achievement test performance. At the college or university level, point systems sometimes are used as part of courses programmed according to a Personalized System of Instruction. In these courses, point values are sometimes assigned to diverse responses that represent options for the student. The final course grade is the back-up event and, of course, depends upon the accumulated point values.

Overall, the capacity of the token economy to focus on diverse behaviors appears evident from the programs reviewed in the present chapter. The next chapter details greater diversity of the token economy in the populations focused upon.

REVIEW OF
TOKEN ECONOMIES II

5

The previous chapter detailed programs with psychiatric patients, the mentally retarded, and individuals in classroom settings. While that research illustrates the extensive literature on token economies, it does not convey the breadth of applications. This chapter reviews a number of other populations that have been exposed to token-reinforcement procedures including delinquents, adult offenders, drug addicts, and problem drinkers. In addition, diverse uses of token programs in outpatient treatment both with children and adults are reviewed. Finally, applications with select populations who have been infrequently treated with token programs and in select research areas are briefly cited.

PRE-DELINQUENTS AND DELINQUENTS

Behavioral programs for juveniles who have committed antisocial behaviors have proliferated (see Braukmann & Fixsen, 1975; Braukmann, Fixsen, Phillips, & Wolf, 1975; Davidson & Seidman, 1974; Stumphauzer, 1973, for reviews). Although programs often focus on eliminating specific antisocial behaviors, the view generally adopted is that these individuals do not know how to function adequately in their environment. Thus, token reinforcement has been used to develop particular competences as in academic performance, vocational skills, and social interaction. The objectives and types of programs vary as a function of the age of the individual, the setting to which he or she is assigned, and the behaviors that are focused upon. Before reviewing the behaviors focused upon in token programs for delinquents, a sample of major behavioral programs will be outlined. The select programs described include street-corner research with delinquents, a special education program for institutionalized offenders, and Achievement Place, a home-style facility for pre-delinquents.

From a historical standpoint, an important application of behavioral techniques with delinquents was made in naturalistic settings (Schwitzgebel, 1964; Schwitzgebel & Kolb, 1964; Slack, 1960). To establish therapeutic contact with delinquents, adolescent and young adult males who had been arrested or previously incarcerated were solicited on the street (e.g., street corner, pool halls) and invited to participate in a research project.

The individuals were told they could be hired for a "job" that entailed coming to the laboratory (a converted store front) and talking about their own experiences

113

into a tape recorder. Money was given for a few visits each week. Cigarettes, candy, or small amounts of money were used to shape prompt attendance and the use of positive statements (e.g., statements of concern or comments elevating another person's status). The youths were also paid for odd jobs at the meeting place, and, in some cases, eventually obtained outside jobs. A 3-year follow-up of 20 boys after termination of the project indicated that these youths showed fewer subsequent arrests and months of incarceration than control subjects previously matched on variables such as age, type of offense, length of incarceration, and others (Schwitzgebel, 1964). There was no significant difference in the number of individuals from each group who returned to reformatory or prison. Although token reinforcement (e.g., money) was used in this project, this was not of course a token economy in the usual sense. Also, delinquent behavior was not focused upon directly. Rather the investigators attempted to establish therapy contact and to develop interpersonal relationships.

One of the most ambitious projects with delinquents was conducted at the National Training School for Boys in Washington, D.C. Two projects referred to as CASE (Contingencies Applicable for Special Education) I and II established educational programs for delinquent youths who had committed serious offenses (e.g., homicide, rape). The initial demonstration project (CASE I) established a token program for 16 youths in a special education day program in the institution to raise the level of performance in specific academic areas and to alter attitudes of the youths toward public schools (Cohen, Filipczak, & Bis, 1968). Points, exchangeable for money, edibles, rental of a private office, lounge time, items from a mail-order catalog, and other events were earned for studying and achieving high levels of accuracy on programmed academic material. After several months, academic gains and increases in grade level were shown on standardized test performance.

The program was expanded (CASE II) to include a 24-hour token economy (Cohen & Filipczak, 1971; Filipczak & Cohen, 1972). Individuals were paid for completion of various academic tasks and for select social behaviors. The points (each worth one penny) could purchase a large number of events including rental of private living quarters, room furnishings, noninstitutional clothes, special foods, home visits, and others. Gains in the program were reflected on standardized achievement tests, as with CASE I. After release from the facility, inmates leaving the CASE II project showed lower recidivism rates (i.e., return to incarceration) and fewer warrants issued against them than other federal juvenile parole releases. However, trends in the data indicated that by 3 years after release, CASE II graduates approached the recidivism rate of the national norms.

Aside from work with already convicted delinquents, CASE programs have been applied as a preventive measure for junior high school students whose behavioral patterns (e.g., drop in grade levels, truancy, runaway from home, withdrawal from normal social contacts, misbehavior at home or in the community, and other criteria) suggest a high probability of later delinquent activities (Cohen, 1972; Filipczak & Cohen, 1972). In this program, referred to as PICA (Programming Interpersonal Curricula for Adolescents), the student attends a special half-day program during the morning and continues regular public school in the afternoons. In the special program, curricula covering academic and interpersonal topics are provided. Academic performance on self-instructional materials in language, reading, and mathematics as well as social behaviors including appropriate interaction with peers and teachers are reinforced with points exchangeable for money or recreational opportunities.

The effect of the program has been suggested in gains in attendance to school, grades in English and math, and overall IQ, and decreases in suspensions from school (Cohen, 1972). In addition to within-group improvements, academic test performance and grades at school were higher and instances of discipline lower for subjects who completed the program compared with randomly assigned control subjects not in the program (Filipczak & Cohen, 1972).

By far, the most extensively evaluated token program that focuses on a wide range of behaviors has been at a home-style facility referred to as Achievement Place. As described in Chapter 3, Achievement Place is a facility for pre-delinquent youths under 16 years of age who have been adjudicated and are sent to the facility for rehabilitation. There are separate facilities for males and females. The basic program is a token economy that incorporates several innovative procedures. Diverse behaviors have been developed or reduced with various contingency manipulations (Bailey et al., 1971; Bailey, Wolf, & Phillips, 1970; Fixsen, Phillips, Phillips, & Wolf, 1976; Fixsen, Phillips, & Wolf, 1972, 1973; Phillips, 1968; Phillips et al., 1971; Phillips, Phillips, Fixsen, & Wolf, 1973; Phillips et al., 1973). Some of the behaviors developed with token reinforcement include room cleaning, watching daily newscasts, saving money, articulating correctly, studying school assignments, and engaging in conversation. Some of the behaviors decreased include aggressive statements or threats to one's peers, use of poor grammar, tardiness in returning home from school or errands, and going to bed late.

There are several interesting features of the program at Achievement Place. One procedure is the managerial system of administering tokens. A boy in the facility serves as a manager and is allowed to assign jobs to be done and administer tokens or fines according to the performance of his peers (Phillips, 1968; Phillips et al., 1973). The privilege of being a manager either has been purchased by the highest bidder or obtained by being elected by one's peers. Although the manager can assign points to his peers, he himself receives points from the teaching-parents on the basis of how well the task is completed. Experiments indicated that managers can readily increase performance of the target behaviors in their peers, that points must be delivered for manager behavior if he is to continue to effectively administer the contingency with his peers, and that reinforcement by the manager tends to be as effective as contingencies administered on an individual basis by the teaching-parents. Interestingly, the opportunity to serve as a manager appears to be a reinforcer for the residents, particularly if the manager has the authority to provide reinforcing and punishing consequences (Phillips et al., 1973). There are several other innovative features of Achievement Place. As discussed earlier, a semi-self-government system has been developed to judge rule infractions and to prescribe penalties and self- and peer administration of the reinforcement contingencies were incorporated into the program.

At Achievement Place, a large number of investigations have evaluated the effects of specific techniques on overt target behaviors within the setting. However, additional studies have looked at more global changes and the broad range effects of the program. Eitzen (1975) evaluated the effect of participation in the Achievement Place program on the attitudes of pre-delinquents and compared these attitudes with boys who were the same age but had no record of delinquency. Questionnaire assessment of attitudes when the boys first entered the program, at various intervals throughout their stay, and after completion of their stay at the facility yielded interesting results. Achievement Place participants increased in their achievement orientation, their personal feelings of mastery over their fate, and favorable feelings toward themselves (i.e., self-concept). The boys did not change on a scale measuring Machiavellian attitudes and, thus, did not become more manipulative and deceptive in their own social relationships. (This latter scale was included in response to the perennial but unsubstantiated concern that behavior modification may teach individuals how to manipulate others.) Overall, the results showed that positive attitudinal changes were made as a function of the program. The pre-delinquent boys tended to be below their normal age peers in specific attitudes prior to treatment and either approached or surpassed this level after participation in the program.

Additional evidence has been gathered addressing the overall efficacy of the program by comparing the relative effects of three different "treatments" to which delinquents were assigned up to 2 years after individuals were released from treatment, namely, Achievement Place, an institutional boys' school, and probation (Fixsen et al., 1976; Phillips et al., 1973). Although all individuals had originally been processed by the same Juvenile Court, they were not randomly assigned to these three conditions. Thus, comparisons of follow-up must be made cautiously. Comparisons were made on measures of recidivism, police and court contacts after release from the program, grades and attendance at school, and school drop-out rates. After treatment was terminated, graduates of Achievement Place showed fewer contacts with the police and courts, fewer delinquent acts which resulted in their being readjudicated by the court, a markedly higher percentage of attendance to public school, and slightly higher grades than individuals who attended institutional treatment or were placed on probation. These results suggest that the effects of the program after treatment was terminated were vastly superior to traditional methods of treating delinquents. Because of the success of Achievement Place, other facilities developed based on the same general model (Liberman, Ferris, Salgado, & Salgado, 1975; Mahoney & Mahoney, 1973; Tsosie & Giles, 1973; Wasik, 1972; Williams & Harris, 1973). Outcome data on recidivism and academic performance from programs at other facilities have supported the results of the Achievement Place model (Liberman & Ferris, 1974).

In addition to the street-corner research, the CASE programs and their derivatives, and Achievement Place, a large number of other programs have been reported for delinquents in institutional settings (Fineman, 1968; Gambrill, 1974; Jesness, 1975; Jesness & DeRisi, 1973; Karacki & Levinson, 1970; Rice, 1970; Seymour & Stokes, 1976). These programs usually develop self-care, participation in activities, and vocational and academic behaviors. Tokens are exchangeable for such diverse events as money, rental of private quarters, the privilege of wearing noninstitutional attire, home furloughs, and others. Some token programs for delinquents and behavior-problem youths have been conducted in the home and at school (e.g., Carpenter & Carom, 1968; Rose, Sundel, Delange, Corwin, & Palumbo, 1970; Tharp & Wetzel, 1969). To describe the accomplishments of token programs for delinquents, the next section reviews the behaviors focused upon including antisocial behavior, social skills, and classroom and academic behaviors.

Antisocial Behavior

Because the delinquent or pre-delinquent has been selected on the basis of antisocial behavior, most programs might be expected to emphasize suppression of these behaviors. Yet, this is not always the case for several reasons. First, many of the problematic behaviors are of a relatively low frequency (e.g., homicide, car theft) and cannot be easily followed with immediate contingent consequences. Second, elimination of a particular inappropriate behavior does not ensure that appropriate behaviors will be performed in its place. Thus, most programs have tried to develop prosocial behaviors rather than merely eliminating problematic behaviors.

A few programs have focused on antisocial behaviors. Burchard and Tyler (1965) treated an institutionalized 13-year-old boy who was committed because of his destructive and disruptive behavior. In the institution, the boy was placed in isolation for a wide range of behaviors (e.g., breaking and entering, glue-sniffing, property damage). Tokens, backed with special privileges and canteen items, were provided for intervals in which he remained out of time-out. Although the combined or separate effects of token reinforcement and time-out were not experimentally evaluated, a decrease in antisocial behavior followed the intervention.

As part of a general program designed to develop personal, social, recreational, educational, and vocational skills, Burchard (1967) suppressed aggressive behaviors with institutionalized delinquent adolescents who were mildly retarded (IQ ranging from 50 to 70). Time-out from reinforcement was combined with token withdrawal to suppress behaviors including fighting, property damage, physical and verbal assault, cheating, lying, and others. The effect of the contingency was relatively small. In subsequent research, Burchard and Barrera (1972) found that the reduction of undesirable behaviors depended upon the duration of time-out and the number of tokens lost for undesirable behavior. Short durations or small fines when administered separately rather than in combination were not effective whereas relatively large durations or fines were. Other research has not consistently shown that long time-out durations or large fines are necessary to suppress behavior (Kazdin, 1975a).

Classroom and Academic Behaviors

Although token programs in classroom settings were reviewed earlier, it is useful to review applications with delinquents briefly here. Applications with delinquents are somewhat distinct because the programs often are conducted in special classes in residential settings. Also, events used as reinforcers may extend beyond the usual items used in most public school classrooms.

Several programs with delinquents have focused on academic behaviors. Tyler (1967) used token reinforcement to develop academic behaviors in an institutionalized delinquent who exhibited severe problem behaviors in the institution (e.g., running away, stealing). Tokens were delivered contingent upon daily and weekly evaluations of academic effort and actual test performance made in classes at the institution. The tokens could be exchanged for rental of a mattress for a 24-hour period, the right to wear noninstitutional clothes, canteen items, and money. The client's grade point average increased during the period in which tokens were delivered.

Tyler and Brown (1968) developed academic responses of 15 court-committed adolescents who resided in a state training school. Individuals received tokens in a classroom setting in the institution for their test performance based upon the television newscast of the previous evening. Tokens were exchangeable for canteen items and privileges in the cottage. As expected, the subjects responded with superior test performance when tokens were delivered contingently upon correct answers than when they were delivered noncontingently. Phillips et al. (1971) also showed that performance of pre-delinquents on quizzes covering daily newscasts as well as the number of individuals watching the news were increased with token reinforcement.

Bednar, Zelhart, Greathouse, and Weinberg (1970) provided delinquents with money for reading proficiency and word comprehension on programmed reading materials. Performance on reading proficiency tests was reinforced on a graduated scale so that increasingly higher performance resulted in higher amounts of money. Although both a reinforcement and nonreinforcement group improved in reading achievement, the reinforcement group was superior on measures of reading and word comprehension. Teacher ratings showed greater persistence, attentiveness, sociability, cooperation, and more favorable attitudes toward school for the group that received tokens.

Meichenbaum et al. (1968) altered inappropriate classroom behaviors of institutionalized delinquent females. Behaviors were categorized as inappropriate (unrelated to the task assigned by the teacher) or appropriate (related to class activity). The subjects received feedback notes from observers in the classroom when their behavior was appropriate. The notes could be exchanged for money. Appropriate behavior increased across morning and afternoon periods as token reinforcement was introduced.

At Achievement Place, point consequences have been used effectively to increase the completion of homework assignments and in-class performance at school (Bailey et al., 1970; Phillips, 1968). The program, designed to alter behavior at school, is particularly noteworthy because tokens were earned at the home for performance at school (Bailey et al., 1970). The boys carried daily report cards with them to school that were marked by the teacher. The teacher noted whether each boy obeyed the classroom rules (e.g., did not leave his seat or talk without permission, did not disturb others, etc.) and worked the whole period across two math periods. When the youths returned to Achievement Place, points exchangeable for special privileges for the remainder of the day (e.g., snacks, permission to go outdoors) were delivered on the basis of the evaluation. The results indicated that study behavior increased and rule violation decreased when points and back-up events were contingent upon classroom performance. Study behavior remained high even though the report card was utilized intermittently rather than daily.

The home-based reinforcement procedure has been shown to be effective in suppressing serious disruptive classroom behavior (e.g., physical aggression) and in increasing completion of academic assignments (Harris, Finfrock, Giles, Hart, & Tsosie, 1976; Kirigin, Phillips, Fixsen, & Wolf, 1972). Anecdotal reports have suggested that a home-based program for delinquents improved grade point averages and reduced tardiness and absenteeism (Marholin, Plienis, Harris, & Marholin, 1975).

Social Skills

Token reinforcement programs have focused upon developing positive social interaction involving several interpersonal skills. As noted earlier, developing social behavior can involve the suppression of antisocial behavior, although they are not necessarily the same. One can alter negative patterns of interacting without developing the necessary skills in dealing with others.

At Achievement Place, delinquent girls were trained how to accept criticism and negative feedback from others without reacting aggressively or arguing (Timbers, Timbers, Fixsen, Phillips, & Wolf, 1973). Positive responses were trained such as establishing eye contact, acknowledging feedback, and omitting behaviors such as throwing things, pouting, frowning, and stamping one's feet. Instructions and role-playing were used to develop basic skills. Tokens and praise were used to reinforce appropriate responses and fines were used to suppress inappropriate reactions. Appropriate reactions increased from a rate of 7% during baseline to over 83% after training.

Also at Achievement Place, conversational skills were taught to pre-delinquent girls (Minkin, Braukmann, Minkin, Timbers, Timbers, Fixsen, Phillips, & Wolf, 1976). In a multiple-baseline design across behaviors, the girls were taught to ask questions as part of conversation and to provide positive feedback (agreement and approval). Instructions, modeling, practice, and feedback were used to develop responses as the girls engaged in conversations with a number of adults. Tokens (i.e., money) were used to reinforce appropriate responses. An interesting feature of this demonstration is that the effect of training was validated by showing that conversational ratings after treatment were similar to levels of females (female college students and junior high school students) who were normally conversant.

ADULT OFFENDERS

Recent programs in prison settings for adult offenders have utilized principles of operant conditioning (cf. Ayllon et al., 1977; Kennedy, 1976; Milan & McKee, 1974; Schnelle & Frank, 1974). A number of programs have been controversial in part because of the reliance upon aversive conditions and techniques that infringe upon the rights of prisoners (Subcommittee on Constitutional Rights, 1974). In the main, token programs and other behavioral techniques have focused upon developing behaviors adaptive in the prison rather than preparing the individual for postprison adjustment.

One prison program that has received widespread attention was referred to as START (Special Treatment and Rehabilitative Training) and was conducted in Missouri by the Federal Bureau of Prisons (1972). The program focused on severely problematic prisoners who were aggressive, resisted authority, and consistently were placed in solitary confinement because of their behavior (1% of the Federal prisoners). The program was designed to segregate these individuals in a special prison, and, through behavior modification, to develop their behavior so that they could return to the normal prison environment. Thus, the program did not attempt to rehabilitate individuals for community release but rather to return them to their penal institutions for completion of their sentences.

The specific contingencies of the START program evolved over the few years in which it was in effect (see Subcommittee on Constitutional Rights, 1974). At its most complex state, the program consisted of a multilevel system that relied heavily upon a token economy. At the initial level, inmates were not on the economy. They earned minimal privileges and were confined to their cell continuously barring a few scheduled activities such as showering or exercising twice a week. Individuals could advance to the next level by cooperating with the rules, maintaining neat personal appearance, following their schedule, refraining from threats to staff or peers, not fighting, and similar behaviors. These behaviors earned them the right to shave, shower, and exercise twice weekly, and to possess small amounts of tobacco, a bible, and select personal items (e.g., comb, soap). At a higher level, individuals were placed on a point economy where a large number of behaviors (e.g., personal hygiene, work performance, taking self-improvement courses) were rewarded with points on a daily basis. The points could be exchanged for back-up events including commissary items, renting personal or institutional equipment (e.g., musical instrument, radio), making calls outside of the institution, taking time off from work in the prison with pay, and others. For rule infractions in the program, the inmates could be assigned to their cells for amounts of time considered to be necessary by the staff. During isolation, an inmate could not earn or spend points.

Higher levels of the system increased the demands and emphasized different behaviors (e.g., self-improvement rather than personal hygiene) and attempted to fade tokens. Thus, after target behaviors were performed consistently, the individual received tokens but these were not exchangeable for back-up events. All privileges were automatically provided. Finally, tokens were no longer earned at all. At the highest level, the inmate was performing his behavioral requirements for self-improvement on a contingency contract basis. After satisfactory performance at this level, the individual was considered for transfer to a conventional prison.

The effects of the program were evaluated on 20 men who were referred to START at the inception of the program. The outcome data over a 2-year period did not suggest that the original goals were attained, i.e., return individuals to their original open prison environment. Only one of 20 individuals at the end of a 2-year period completed all phases and was returned to conventional prison life. The others were either still in the program, transferred out of the program because they had failed to progress, released to the hospital or psychiatric units affiliated with the program, or completed their sentence.

The program was wrought with controversy and litigation because of the infringement of the Constitutional rights of the inmates (cf. Steinman, 1973; Subcommittee on Constitutional Rights, 1974). Prisoners could not earn anything that would not already be available to them in conventional programs since the purpose of START was to return prisoners to normal prison conditions. An exception to this was that inmates could earn a return of statutory good time (time off their sentences for good behavior) which had been lost previously for recalcitrant behavior.

The START program relied upon involuntary assignment of prisoners, deprivation of privileges normally provided in prisons, earning back lost privileges, submissive behavior on the part of the prisoners, and aversive conditions in general (e.g., isolation and alleged unplanned abuse such as shackling prisoners to their beds) (cf. Cohen, 1974; DeRisi, 1974; Saunders, Milstein, & Roseman, 1974). After 2 years, the START program, under litigation, was voluntarily terminated. (Ethical and legal issues raised by behavior modification programs including issues raised by START are discussed in Chapter 10.)

Another prison program that utilized a token economy and other behavioral techniques was conducted at Draper Correctional Center in Alabama, a maximum security state institution for young offenders serving for their first or second felony conviction. In the early development of the program, the major emphasis was on individualized programmed instruction (McKee, 1970, 1971, 1974). Completion of academic behaviors was rewarded with money or the selection of a variety of privileges and activities such as free time (Clements & McKee, 1968). Both the quantity and quality of academic work performed in the classroom improved. Also, a large percentage of subjects earned high school equivalencies (McKee & Clements, 1971).

As the program evolved, a token economy was established in an experimental cellblock to increase performance of behaviors such as getting up on time, making one's bed, working on academic activities, and performing maintenance jobs (e.g., sweeping the hall, emptying trash) (Milan & McKee, 1974; Milan, Wood, Williams, Rogers, Hampton, & McKee, 1974). Back-up reinforcers included events and privileges such as access to a lounge, a television viewing room, and poolroom, time to visit others or attend recreational activities, and the purchase of commodities at a canteen. Also, individuals could purchase items from department store catalogs with tokens. Fines were used to decrease entering a "reinforcing area" without first paying tokens, for leaving the cellblock for another place in the institution without punching out, and for overdrawing their token banking accounts.

Several experiments were conducted to evaluate the effects of the program (Milan et al., 1974). Token reinforcement was shown to be more effective in increasing self-care behaviors than was praise and feedback for performance, coercive and aversive control procedures that are commonly relied upon in prisons (e.g., intimidation, threats, ultimatums, extra work), and the noncontingent delivery of tokens. Also, the magnitude of reinforcement was related to performance in the economy. Up to a limit, increasing the number of tokens delivered increased inmate attendance to activities (i.e., attending a television news program). Other experiments demonstrated that participation in an education program during leisure time could be increased by making the exchange of tokens for back-up reinforcers contingent upon academic performance and that an undesirable behavior (leaving the cellblock without punching out) could be suppressed by reinforcing an incompatible response (reinforcing those with time off the cellblock for marking their time of departure) (Milan *et al.,* 1974).

The educational program illustrates the careful evaluation of contingencies which influenced inmate behavior in the token-economy cellblock (Milan *et al.,* 1974). In this program, an attempt was made to increase inmate completion of programmed instructional materials in areas where inmates had specific educational deficiencies. After baseline observation of inmate participation in the program once the study materials were made available, token reinforcement was used to increase performance. Points were delivered contingent upon successfully passing mastery tests for the programmed materials. The number of points earned was based upon the estimated amount of time that was required to complete the material upon which the test was based. After the token economy was in effect, it was modified so that a "license" was required by the inmates to exchange the points they earned for back-up events in the cellblock and at the canteen. The license could only be earned through participation in the remedial education program. To obtain a license, the inmate had to participate in the program and earn points toward the purchase of the license. The license usually could be earned in less than 10 hours of study per week. Essentially, the license itself was a new back-up event (and discriminative stimulus) that was required to utilize other reinforcing events.

The effect of the token economy with and without the license contingency can be seen in Fig. 14 which presents several phases over a protracted ABAB design. The data for both the percentage of inmates participating in the program and the minutes that inmates spent studying increased as a function of the token program. Adding the license contingency markedly increased performance over the level achieved with the original token program. (Figure 14 also shows that announcing the changes in interventions did not systematically alter behavior.) Overall, the results show the utility of the token program in enhancing educational performance of the inmates.

Aside from the demonstration of the contingencies on behavior in the cellblock, follow-up data were reported for one of the programs at Draper (Jenkins, Witherspoon, DeVine, deValera, Muller, Barton, & McKee, 1974). A comparison was made between individuals receiving token reinforcement versus traditional prison treatment (e.g., either the usual prison program or special vocational training). Follow-up data did not indicate generalized treatment effects. Immediately after release, participants on the token program tended to show fewer arrests and a

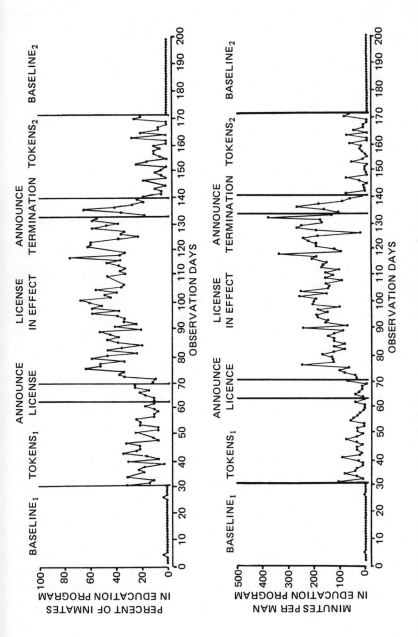

Fig. 14. Daily percentage of inmates participating in the education program (upper portion) and daily average number of minutes spent studying academic material (lower portion) (from Milan, Wood, Williams, Rogers, Hampton, & McKee, 1974; Rehabilitation Research Foundation).

lower recidivism rate than groups receiving traditional treatment although these differences were not statistically significant.

There are several interesting features of the programs conducted at Draper Correctional facility that warrant mention, particularly in light of the controversy generated over the rights of inmates in prison behavior modification programs. One feature was that individuals could leave if they were dissatisfied with the conditions. Although inmates were assigned or transferred to the token-economy cell-block, they could submit a request to leave and be transferred within a few days (Milan *et al.,* 1974). A second feature is that the reinforcing events did not rely upon deprivation of reinforcers that would normally be available. While many events normally used as reinforcers were used to back-up tokens, other events ordinarily unavailable were also included over the course of the program such as visiting a women's prison, renting television, having the opportunity to go fishing, having an interview with a parole board member, and others (McKee, 1971). Other features of note were that inmates had a role in establishing contingency contracts, the focus on positive behaviors rather than mere compliance with rules, and the careful evaluation of the effects of the program.

Another token economy with several innovative features was reported by Ayllon *et al.* (1977) which was implemented at a medium security prison. Inmates who applied for participation in the program could earn points for academic and vocational performance. Points were provided for completing academic assignments and mastery tests in mathematics, English, and areas of vocational skills. Aside from test performance, inmates received points for general ratings of work performance and attitudes toward others, as reflected in social interactions. In several experiments, the investigators demonstrated that the contingent delivery of points altered academic and vocational test performance and work attitudes.

A few commendable features of this program warrant mention in passing. Initially, the points earned in the prison could be spent for a large number of reinforcing events including diverse canteen items and special privileges. Some of the special privileges included making more frequently available events already provided in the setting. For example, inmates normally could make one 10-minute phone call every three weeks before the program. Yet, with points inmates could earn up to over six calls in a three-week period. Similarly, prior to the program, inmates could receive and write a limited number of letters from and to their immediate family. During the program inmates could earn unlimited correspondence to anyone for points.

Aside from expansion of normally presented privileges, new reinforcers were introduced into the system. For example, one event was a system in which inmates could redeem their points for trading stamps to be used by someone outside of prison. Letters were sent to individuals on the outside who selected gifts from a trading stamp catalog. The inmate could exchange points earned in prison for trading stamps that were automatically mailed to the recipient.

A second noteworthy feature is that the opinions of the inmates were solicited to evaluate the program. Questionnaire responses of the inmates indicated that the program was perceived as resulting in positive changes in work behavior and was rated as generally excellent. In addition, during the program, the majority of inmates felt better about being in prison and about education. Overall, 88% of the inmates were pleased with the program. Given the aversive qualities of some prison programs and the concern over inmate rights, evaluation of the opinions of the clients represents an important innovation.

Another program for adult offenders was reported by Bassett, Blanchard, and Koshland (1975) who implemented a token economy on a minimal security prison farm. Self-care, work, and participation of rehabilitative activities earned points that could be spent for a variety of consumable items and privileges. In one experiment, the effect of the token reinforcement was demonstrated in keeping the inmates up on current events, defined as responding to quizzes based upon daily television newscasts. In another experiment, token reinforcement was shown to control the amount of time that inmates attended a learning center during their free time where they worked on programmed educational materials. In an ABAB design, the amount of time that inmates attended the center was shown to be a function of the number of tokens delivered for hours of studying. The greater the number of tokens provided, the greater the time spent at the center.

Not all token programs for adult offenders have been conducted in prisons. In some instances, individuals whose behavior is regarded as incorrigible are placed in a psychiatric facility and treated separately from other prisoners. Lawson, Greene, Richardson, McClure, and Padina (1971) implemented a token economy in a correctional hospital for men who were unable to adjust to prison life and who exhibited a variety of incapacitating behaviors. The majority of men had been convicted for murder, manslaughter, or robbery and were sent to the unit because they were "troublemakers," were apathetic in carrying out simple aspects of their functioning (e.g., eating, dressing, toileting), and engaged in minimal activities. To overcome the general inactivity, a token program was implemented in which attending activities (e.g., gym, music, occupational therapy), arising, washing, dressing on time, and volunteering for jobs were reinforced. Commissary items and access to recreational facilities served as back-up reinforcers. Over several months of the project, token earnings increased suggesting an increase in participation in activities. However, the role of the contingencies was not experimentally validated. Other programs for adult offenders conducted in hospital wards have included sex offenders and psychiatrically impaired criminals (Laws, 1971, 1974).

The token economy has been extended to "delinquent soldiers" in a treatment program designed to return the soldiers back to active duty (Boren & Colman, 1970; Colman & Boren, 1969; Ellsworth & Colman, 1970). These soldiers (average age 20 years) were diagnosed as character and behavior disorders (Colman & Baker, 1969). They had violated rules of the military (e.g., repeated absence without leave) and had a history of difficulty with the law, school, and their families in civilian life. The program, conducted at Walter Reed General Hospital, was implemented in a psychiatric ward housing up to 18 soldiers. Tokens were delivered for diverse behaviors and exchangeable for privileges such as semiprivate rooms, access to television or a pool room, weekend passes, and other events. In several experiments, token reinforcement was shown to increase the number of individuals who engaged in exercise, attended group meetings, spoke up at a group meeting, and described current personal problems (Boren & Colman, 1970).

Aside from demonstrating behavior change in the hospital ward, follow-up evidence suggested that the effects of the program extended beyond treatment (Colman & Baker, 1969; Stayer & Jones, 1969). Follow-up assessment at 3, 6, and 9 months revealed that 69.5% of the treated soldiers completed their tour of duty in the service or were serving in good standing (i.e., "successes"), whereas 30.5% were administratively discharged, were AWOL, or in a stockade (i.e., "failures"). A comparison group that did not undergo treatment showed only 28.3% success and 71.7% failures by the same criteria.

DRUG ADDICTS AND PROBLEM DRINKERS ("ALCOHOLICS")

Behavior modification techniques have been applied to individuals who are drug addicts and problem drinkers (see Gotestam, Melin, & Ost, 1976; Miller & Eisler, 1976, for reviews). Generally, treatment techniques for addictive and maladaptive approach behaviors have relied upon procedures classified under the rubric of aversion therapy (Rachman & Teasdale, 1969). Token reinforcement has been applied in relatively few situations.

Applications with Drug Addicts

O'Brien et al. (1971) reported a program with drug addicts based upon the Premack principle. Patients admitted for addiction but with diverse psychiatric diagnoses, received consequences for general ward behaviors including self-care (e.g., arising on time, dressing, making one's bed), compliance with rules (e.g., not bringing others to their rooms, remaining out of bed during the day), and attending activities (e.g., work). Whether the patient completed various behaviors was indicated daily on a publicly displayed chart. The number of "yes" ratings (indicating completion of a target behavior) on the chart were exchangeable for various privileges including access to radio, television, or a recreation room, passes, visitors, and others. In an AB design with 150 patients, performance improved during the 35 weeks in which the program was in effect.

Glicksman, Ottomanelli, and Cutler (1971) reported a token system with hospitalized male heroin addicts, over half of whom had been previously incarcerated. Patients received credits for various ward behaviors such as participating in group therapy and educational programs in the hospital. Global behavior ratings (e.g., how the individual relates to himself and others) and select specific responses (e.g., punctuality, cleanliness) observed by ward personnel served as a basis for awarding credits. The only back-up reinforcer for the credits was applying for discharge once a patient accumulated a large number of points. Yet, earning the specified number did not necessarily result in discharge as it has in some studies (Linscheid et al., 1974), because staff exercised their own discretion in making the final decision. The success of the program was suggested by the short period of patient hospitalization (4 months) relative to patients assigned to other wards at the same time (7.5 months). However, an analysis of the role of the contingencies on carefully specified target behaviors and of the role of the program in effecting rapid discharge was not made.

One of the most systematic token programs for drug addicts has been conducted at a psychiatric hospital in Sweden (Eriksson, Gotestam, Melin, & Ost, 1975; Gotestam & Melin, 1973; Gotestam, Melin, & Dockens, 1975; Melin & Gotestam, 1973; Melin, Andersson, & Gotestam, 1975). The program treated addicts who had a protracted history of amphetamine or opiate use. The program focused upon behaviors considered adaptive to the hospital such as self-care, attendance to activities, and work. Initially, performance of specific behaviors were exchanged directly for privileges although a token program was developed to mediate the exchange (Melin & Gotestam, 1973).

The token program was organized into three phases as a leveled system consisting of detoxification, treatment, and rehabilitation phases. Detoxification was a brief phase traversed by the patients as soon as their daily urine samples revealed no trace of drugs. After detoxification, the patient began a token program where diverse behaviors earned tokens including self-care (dressing, cleaning one's room, exercising), engaging in hospital activities (e.g., attending a morning conference where they planned the day), and performing jobs (e.g., serving meals, serving as a host or hostess for visitors). Tokens purchased privileges such as leaving the ward for brief periods or for days at a time, consumables, reading material, and tickets to special events (e.g., cinema or football).

In one of the reports, the effect of the reinforcement contingencies was carefully evaluated on the adaptive target behaviors in an ABAB design (Eriksson et al., 1975). After baseline, the token-reinforcement program, described above, was implemented, withdrawn, and eventually reinstated. In the final phase, the tokens were delivered noncontingently. The effect of the token program on total activity of the patients (i.e., the percentage of activities that were performed of those possible) appears in Fig. 15. The program clearly established functional control over behavior.

An important measure of success of the program was the percentage of patients who were intoxicated (i.e., had injested drugs during the program, as measured by urine samples). The number of patients with positive urine samples decreased while the token program was in effect. At different points in this program, follow-up data have been reported (Melin & Gotestam, 1973). Evaluation up to one year after release from the program indicated that the percentage of drug-free patients was higher than for nonrandomly assigned control patients and for patients who had attended the hospital prior to the inception of the behavioral program. While the comparisons at follow-up are equivocal because of the comparison groups, the results suggest durable effects.

There have been programs for drug addicts other than those reviewed above (e.g., Clayton, 1973; Coghlan, Dohrenwend, Gold, & Zimmerman, 1973). However, in these programs token reinforcement often is combined with other procedures

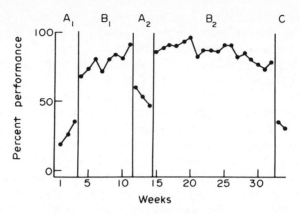

Fig. 15. Percentage of activities performed by the patients during Baseline (A), Token Reinforcement (B), and Noncontingent Token Reinforcement (C) Conditions (from Eriksson, Gotestam, Melin, & Ost, 1975; Pergamon Press).

such as group, family, and activity therapy or integrated with psychodynamic principles. More importantly, the effect of treatment has not been systematically evaluated in these reports.

Applications with Alcoholics

The earliest systematic extension of the token economy to alcoholics was made by Narrol (1967). In this program, 17 hospitalized chronic alcoholics received tokens for work on or off the ward. Tokens could be exchanged for room and board, ground privileges, recreation, and clothing maintenance, all of which had to be purchased by the patients. Additional back-up events which did not have to be purchased included tickets exchangeable for canteen items, passes, attendance to a meeting of Alcoholics Anonymous, individual therapy with a therapist of one's choice, and other privileges. The program was divided into five levels in which increasingly greater freedom (e.g., from a closed to an open ward), status (e.g., from on-ward to off-ward work), and privileges (e.g., increasingly greater ground privileges) were provided. Although the program was not systematically evaluated, the patients in the program were reported as working more hours per day than patients not in the program.

Cohen, Liebson, Faillace, and Speers (1971) used token reinforcement with four chronic male alcoholics living in a six-bed alcohol behavior research ward. Patients earned money for grooming, ward maintenance, and abstinence on the ward. Money could be spent on outside shopping, movies, ballgames, and consumable items. The magnitude of reward varied according to whether the individuals had been abstinent for consecutive days or had consumed the alcohol made available in the unit. Delay of reward was associated with an increase in alcohol consumption although abstinence could be increased by providing higher magnitudes of reward. If individuals were provided with a small amount of alcohol and received tokens for not drinking beyond that small amount, further consumption still tended to occur. However, this too could be altered as a function of the amount of reward provided for abstinence.

In another report of this program, different contingencies were systematically evaluated in an attempt to develop moderate drinking in a male alcoholic (Cohen, Liebson, & Faillace, 1971). To control drinking, the patient received an "enriched" environment if his drinking did not exceed a criterion level (5 ounces per day). Drinking above that level led to an "impoverished" environment. The enriched environment included the opportunity to work in the ward to earn money, use of a phone and a recreation room (including television, games, pool table), attendance to group therapy, regular diet, reading material, and a bedside chair. The impoverished environment consisted of the absence of each of these (e.g., pureed food rather than regular diet and the loss of all privileges for a 24-hour period). Access to a particular environment was determined by consumption on a given day. In separate experiments using ABAB designs, the effects of earning the living conditions was clearly demonstrated. Alcohol consumption was low and usually remained within the allowable limit when consumption earned more desirable living conditions than when the conditions were provided on a noncontingent basis.

Miller, Hersen, and Eisler (1974) examined the effect of contingency contracts in reducing drinking of chronic hospitalized alcoholic patients. The patients participated in a token economy where self-help and work activities could earn ward privileges and where they could receive alcohol. To limit drinking, patients were exposed to different experimental conditions including verbal or written instructions (signed and agreed to in contract form) to limit alcohol consumption. Half of the patients in these conditions received tokens for limiting drinking and lost them for exceeding a specified criterion, whereas the other half received no consequences. The tokens could be spent in the ward economy on items such as cigarettes and clothes. The main finding was that token-reinforcement contingencies led to a larger number of patients limiting their drinking relative to patients who received no consequences. The effect of tokens was not differentially influenced by the manner in which instructions were provided.

As evident from the brief review, token programs focusing on addictive behaviors are relatively sparse. There are a few applications of token reinforcement that are used in outpatient therapy that have focused on addictive behavior. These applications do not constitute general token programs in institutional settings and will be treated in the discussion of outpatient programs reviewed next.

OUTPATIENT APPLICATIONS

Token economies discussed in previous sections have covered programs in hospital, institutional, and educational settings. However, token reinforcement has been extended to outpatient treatment where the individual is not in a special setting. For purposes of discussion, three outpatient applications of token reinforcement may be distinguished, namely, outpatient behavior therapy, interventions in the natural environment with children and adolescents, and interventions in the natural environment with adults.

In outpatient behavior therapy, individuals come to therapy for some treatment where therapy interventions are designed, described, and implemented. In outpatient behavior therapy, as distinguished here, the therapist plays a major role in actually administering the reinforcement contingencies. With interventions conducted in the natural environment, the therapeutic program is designed or agreed to in therapy but is carried out elsewhere (e.g., the home). Therapy may consist of describing aspects of behavior therapy or helping individuals plan the therapy program that is to be implemented in the natural environment. The program often is designed in the form of a contingency contract where the conditions in the natural environment that constitute the intervention are made explicit in written form and are agreed to by all participants.[1]

Outpatient Therapy

The first application of token reinforcement is outpatient behavior therapy in which tokens are used to treat behaviors focused upon in therapy sessions. The individual client attends therapy and receives or loses tokens based upon his performance. In these applications, money usually is used as the reinforcer and may be delivered or withdrawn contingent upon client behavior. For example, Mahoney, Moura, and Wade (1973) instructed obese individuals to self-reward, self-punish, or to both reward and punish with money contingent upon weight loss or gain at weigh-ins. The money which was rewarded or fined was taken from an initial deposit provided by the subjects at the inception of treatment. Subjects who self-rewarded for weight loss lost more weight than those who self-punished for weight gain.

Ross (1974a) developed a contingency contract for a woman who was a chronic nail-biter. The contract specified that her nails had to remain at an acceptable length and increase in length when measured (every 9 days) to avoid loss of money. The fine consisted of sending precompleted money orders to an organization that she regarded as repugnant. If nail length increased, one of the money orders was returned to her rather than to the organization. Nail length increased during this program and was maintained after 6 months. Several other

---

[1] The distinction of outpatient applications of token reinforcement is designed to facilitate organization of the review of token programs. The distinction itself breaks down upon finer analysis because many outpatient programs referred to here as outpatient behavior therapy include therapist-administered contingencies in therapy sessions as well as client-administered contingencies in the natural environment. The applications referred to as interventions in the natural environment are not entirely distinct from outpatient behavior therapy that uses token reinforcement. Yet, it is meaningful to discuss programs that differ somewhat on the location of therapy (e.g., in distinct therapy sessions) and the individuals who manage the contingencies (e.g., a professional therapist or a parent or spouse).

investigations and case reports have utilized the contingent delivery or withdrawal of money to alter overeating, cigarette smoking, alcohol consumption, and drug abuse (e.g., Abrahms & Allen, 1974; Boudin, 1972; Elliott & Tighe, 1968; Jeffrey, 1974; Miller, 1972).

In most applications of token reinforcement in outpatient behavior therapy, the therapist delivers consequences when the client attends select therapy sessions. Thus, the therapist does not actually provide immediate consequences for behaviors as they occur in the natural environment. Occasionally, the therapist may directly deliver immediate consequences for performance in the natural environment. For example, Miller, Hersen, Eisler, and Watts (1974) used token reinforcement to alter the drinking of a 49-year-old male who was being treated as an outpatient. Alcohol consumption was assessed through blood/alcohol concentration (measured by breath samples and analyzed by gas chromatography). The times of assessment were randomly determined twice each week and the individual was phoned immediately prior to assessment and visited by someone to obtain a breath sample. After baseline, the individual was given $3.00 in coupon booklets contingent upon the absence of traces of alcohol in his blood. These coupons could be exchanged for items (e.g., cigarettes, meals, clothing) at the hospital commissary. After the contingent phase, the coupons were delivered independently of blood/alcohol concentration. In the final phase, the coupons were again contingent upon behavior. Fig. 16 shows that blood/alcohol concentrations were lower during the treatment phases than during baseline or noncontingent reinforcement phases. Interestingly, the program appears to have controlled drinking in the natural environment

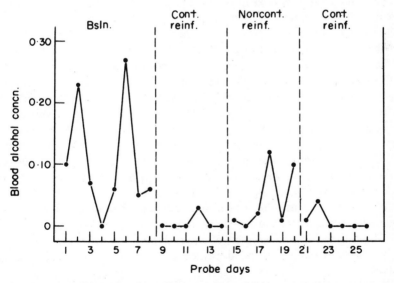

Fig. 16. Biweekly blood/alcohol concentrations across Baseline, Contingent Reinforcement, Noncontingent Reinforcement, and Contingent Reinforcement Phases (from Miller, Hersen, Eisler, & Watts, 1974; Pergamon Press).

rather than in an inpatient-treatment facility, at least while the contingencies were in effect.

In some applications of token reinforcement, the therapist provides consequences for behavior during the treatment session. Biofeedback is an example where reinforcing or punishing consequences are delivered for behavior during the session (see Blanchard & Young, 1974). Data for a continuously monitored physiological response are displayed or "fed back" to the client visually (on a dial or meter) or auditorially (by a tone). Although most studies use feedback alone to increase or decrease particular responses, occasionally token reinforcement in the form of money is used. For example, money was used to increase the heart rate of a college student whose rate was below normal (Scott, Peters, Gillespie, Blanchard, Edmunson, & Young, 1973; Exp. 2). To demonstrate the effect of reinforcement (rather than to achieve some protracted clinical change), the individual was given money (1¢) for every 10 seconds of emitting a heart rate above the criterion. The criterion for earning money was gradually increased so that a higher rate was progressively shaped. Contingent money increased heart rate relatively markedly, as demonstrated in an ABAB design. Other investigators have shown the effects of monetary reinforcement on heart rate (e.g., Sirota, Schwartz, & Shapiro, 1974).

Interventions in the Natural Environment with Children and Adolescents

Token reinforcement often is used to alter behavior in the natural environment. Unlike outpatient behavior therapy discussed earlier, the contingencies are implemented in naturalistic settings rather than in therapy sessions and are managed by someone who can control the contingencies of reinforcement directly. Most of the applications of token reinforcement in the natural environment have been conducted in the home to control child behavior identified by the parents as problematic, although some extensions have been made to the community at large.

Christopherson, Arnold, Hill, and Quilitch (1972) helped two families implement a token program to control child behavior. In one of the families, three children (aged 5, 8, and 9 years) received points for completing individually designated chores (e.g., making beds, hanging up clothes, feeding a pet, cleaning the bathroom) and were fined for not completing chores or for undesirable social behaviors (e.g., bickering, teasing, social interaction after bedtime, and whining). Points could be spent on basic privileges around the house or special events such as going to a movie, a camping trip, or a picnic. The parents consulted with the therapist who taught them how to use the system in a few therapy sessions, a home visit, and phone calls. The parents were completely responsible for implementing the program and collecting the data to assess its efficacy. Overall, the token program increased the performance of chores and decreased performance of undesirable social responses.

Frazier and Williams (1973) helped parents design and implement a token program in the home of a 6-year-old boy. The boy had engaged in self-stimulatory behavior ("rocking and bumping") since he had been 6 months old. After baseline observations by the mother, tokens (punched holes on tickets personalized with his name) were delivered for each hour without rocking. The tokens were delivered by the mother and exchanged for pennies provided at the end of the day by the father. The money could be accumulated for a weekly shopping trip with the father for playthings. Treatment also included restricting rocking to particular places (in a chair facing a corner) and time-out (5-minute isolation for rocking). After approximately 3 weeks of treatment, all procedures were gradually withdrawn and rocking was completely eliminated. Additional observations including a 6-month and a 1-year follow-up indicated no recurrence of self-stimulatory behavior.

A programmatic series of investigations of treatment in the natural environment has been completed by Patterson and his colleagues who have designed treatment programs for families of children with severe conduct problems at home and at school (e.g., stealing, running away from home, setting fires, fighting) (Patterson, 1974; Patterson, Cobb, & Ray, 1972, 1973; Patterson & Reid, 1973; Patterson et al., 1969). The families were referred for treatment by community agencies such as the Juvenile Court, schools, and mental hygiene clinics. The parents were trained to conduct behavior modification programs in their homes. Training included reading programmed materials on behavior modification, practicing the identification of target behaviors, collecting data, and actually implementing techniques. The parents also received feedback in the home and worked with others in group situations to develop their skills. Treatment also was carried out at school which involved school personnel, experimenters, and parents. Token systems were developed by the parents in the form of contingency contracts with children to control individually tailored aggressive and disruptive behavior. At school, attending lessons, talking appropriately to the teacher or to peers, complying with instructions, and similar behaviors were reinforced.

Generally, the training program has been shown to markedly decrease deviant behaviors at home and at school to the level of peers whose behaviors are not problematic. Illustrative results of a program to alter the problem behavior of 27 boys in their homes are presented in Fig. 17 (Patterson, 1974). The data scored by observers in the home, show the rate of deviant behavior (e.g., noncompliance, negativism, whining, yelling, physical aggression, and others) in baseline, during the intervention (or parent training) phase, and up to 12 months after treatment termination. The shaded portion of the figure shows the level of deviant behaviors of children who were not identified as behavior problems but were matched on several variables (e.g., socioeconomic status, age of "target" child, number of children in the family). The deviant behavior of this latter group suggests the level of acceptable disruptive behavior for the age group. As evident from the figure, after the training of parents was completed, deviant child behavior fell within the "normative" range. Interestingly, treatment gains also appear to have been maintained several months after the parent training program was terminated.

Token economies conducted in the home have been successful in altering a wide range of behaviors related to interaction patterns with siblings, compliance with adult demands, school attendance, performance of various jobs and chores, idiosyncratic bizarre behaviors, and acts of destruction (Alvord, 1971; Arnett & Ulrich,

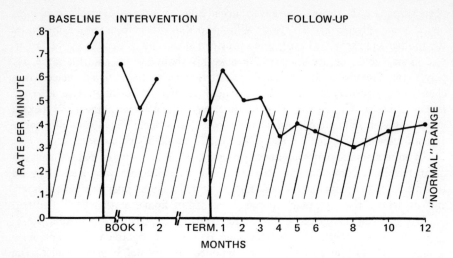

Fig. 17. Total deviant behavior in the home (from Patterson, 1974; American Psychological Association).

1975; Ayllon, Smith, & Rogers, 1970; Doleys & Slapion, 1975; O'Leary, O'Leary, & Becker, 1967; Tharp & Wetzel, 1969; Thomas & Walter, 1973; Wiltz & Gordon, 1974).

In addition to applications in the home, token reinforcement has been used in community-based treatment. Ayllon and Skuban (1973) treated a 9-year-old autistic child who failed to comply with adults' requests and engaged in a high rate of temper tantrums. Treatment took place 5 days per week and from 5 to 7 hours per day for 35 days. Observation of behavior and treatment was carried out in everyday settings such as the zoo, museums, parks, movies, swimming pools, shopping centers, markets, and so on. These settings were focused upon to expose the child to situations where normal conduct was expected (i.e., compliance with requests and nontantrums). To develop compliance with adult requests, money (pennies and nickels) was delivered if he complied within a 10-second time limit after the request. The money could be used to operate pinball and vending machines and for privileges such as swimming, playing catch with and being tickled by the therapist, access to play equipment, and others. Tantrums were followed by a reprimand ("no") and time-out from reinforcement (a period in which no money could be earned). Eventually, money was provided for performing academic work. Late in training, money was replaced with praise and periodic treats to provide a transition to "natural contingencies." The treatment procedures appeared to be extremely effective in reducing tantrums in response to instructions and in increasing compliance. Tantrums were almost completely eliminated and compliance was almost perfect at the end of treatment.

Fo and O'Donnell (1974) developed a large-scale reinforcement program in the community for youths (between the ages of 11 and 17) who were having various behavior and academic problems. Adults were recruited from the community (between the ages of 17 and 65) to work in a "buddy system" in which they conducted behavior modification individually with the youths. The adults were trained via modeling, role playing, monetary reinforcement, instructions, prompts, and praise for performing behaviors in training sessions and behaviors in relation to the youths with whom they were working. The adults established a relationship with the youths, engaged in a variety of activities (e.g., arts and crafts, fishing, camping), and implemented and evaluated a behavior change program. Different interventions were used to alter behaviors of the youths such as truancy, fighting, not doing chores at home, not completing homework, and staying out late. When the adults provided social reinforcement (a warm positive response) in response to appropriate behavior of the youths or provided social reinforcement and money, the youths' behaviors were altered over an intervention period that averaged 6 weeks. In the case of truancy, a behavior problem of many of the youths, the intervention effects were marked. This investigation is of particular interest because community members were recruited, received a nominal fee, and effectively altered behavior without professionals having contact with the youths. The adults attended several training sessions prior to the intervention and continued in the sessions throughout the duration of the project.

In a subsequent report of the efficacy of the buddy system, Fo and O'Donnell (1975) examined the court records of individuals who participated in the program during the year of treatment. The data indicated that youngsters who had committed major offenses (e.g., auto theft, burglary) prior to the program committed fewer major offenses that led to commitment during the program than similar subjects who received no treatment. Yet, subjects who had not committed major offenses prior to participating in the buddy system *increased* in their rate of major offenses during treatment relative to similar subjects who received no treatment. These results essentially showed that youths who committed more serious acts profited from treatment whereas those who committed less serious acts somehow became worse from treatment. The study appeared to control for regression artifacts that might otherwise account for such a pattern in the data. It is unclear why the youths who committed less serious crimes did better in no treatment than in the buddy program.

Interventions in the Natural Environment with Adults

Token reinforcement has been used with adults in the natural environment relatively infrequently. An extremely interesting application of token reinforcement was developed by Stuart (1969a, 1969b) to treat marital discord. Based on the assumption that marriage depends upon reciprocal positive reinforcement exchanged by spouses, a reinforcement program was developed in the homes of couples with severe marital conflict. Although couples identified various target behaviors, only two were focused upon, viz., conversation and sexual interaction. For each of several couples, the wife identified as an important target behavior increased conversation with the husband. To increase conversation, defined somewhat individually across couples, the wife provided tokens to the husband for conversing when at home if a criterion amount of conversation had been met. If the husband failed to converse appropriately within a given time, he was prompted (offered constructive suggestions by the wife). Tokens earned for conversation by the husband could be exchanged for physical affection, which was selected by the husband as a class of behaviors to be increased. A list was constructed individually for each couple but included behaviors such as kissing, light petting, heavy petting, and intercourse. Each couple received several sessions (seven sessions over a 10-week period).

Over the course of treatment, weekly hours of conversation and frequency of sexual intercourse increased, as measured by data obtained by the spouses. In addition, a global self-report measure indicated that ratings of marital satisfaction increased over treatment. Follow-up data obtained by phone or mail indicated that the effects of treatment were maintained up to a year after treatment (Stuart, 1969a, 1969b). While this initial application of token reinforcement may appear to oversimplify marital conflict, the couples in the program considered treatment as a last resort prior to obtaining divorce. Although treatment focused on well-defined behaviors (e.g., conversation and sexual interaction), the likelihood of generalized effects is plausible. In any case, this report stands out as a unique application of token reinforcement in outpatient treatment.

In another program conducted in the natural environment, Miller (1972) used a contingency contract with a husband and wife to control excessive drinking of the husband. The couple had been referred to treatment because the husband's excessive drinking appeared to be a major source of marital discord. By mutual consent in a written contract, the couple agreed that up to three drinks per day was an acceptable limit for the husband's drinking. The drinks were to be consumed before dinner in the presence of the wife. Any drinking outside of this resulted in a monetary fine of $20.00 which was to be paid to his wife and to be spent as frivolously as possible. To reduce the wife's critical and disapproving comments for drinking, a $20.00 cost was imposed upon her for negative verbal behavior related to drinking. Drinking gradually dropped below the agreed-upon level of drinking with some days of complete abstinence. A 6-month follow-up revealed maintenance of drinking within the acceptable level.

OTHER POPULATIONS AND APPLICATIONS

The token economy has been extended to other populations not included in the above review. These extensions represent areas with only select studies.

Geriatric Residents

One area of interest that has received little attention is the treatment of geriatric residents. Sachs (1975) employed social and token reinforcement (back-up events unspecified) for diverse behaviors with residents of a nursing home. Select residents received consequences for behaviors such as going for walks in the home rather than remaining confined to a chair or for brushing their teeth. Although most of the demonstrations were AB designs and precluded drawing causal inferences about the intervention, the contingent delivery of consequences appeared to be associated with improvements in behavior.

In one demonstration, praise rather than tokens was used to increase social interactions of a 91-year-old male confined to a wheelchair. Staff praised the resident's initiation of conversation and verbal responses to others. The effects of the program, evaluated in an ABAB design, are presented in the cumulative record in Fig. 18. The results indicate increased contacts during the reinforcement phase (steep slope) and virtual cessation of social contacts during baseline and reversal phases (horizontal slope).

In another program with geriatric patients, Libb and Clements (1969) used token reinforcement to develop exercise in four patients (late 60s and early 70s) who carried diagnoses of chronic brain syndrome. Tokens (marbles exchangeable for cigarettes, candy, gum, and peanuts) were delivered automatically as the patients pedaled a stationary bicycle. Marbles were delivered on the basis of the number of wheel revolutions completed. For three or four patients, the delivery of tokens increased exercise.

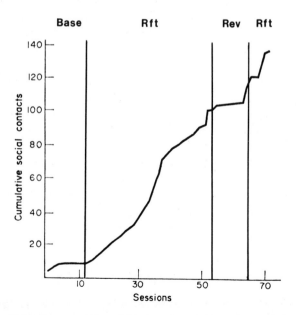

Fig. 18. Cumulative time spent in social interaction across Baseline, Reinforcement, Reversal, and Reinforcement Phases (from Sachs, 1975; Pergamon Press).

Stutterers

The token economy has been extended to the treatment of stuttering (Andrews & Ingham, 1972; Ingham & Andrews, 1973; Ingham, Andrews, & Winkler, 1972). These investigators have designed a token program with a large number of chronic stutterers who participated in a day-hospital program. Tokens were delivered contingent upon decreases in stuttering (percentage of syllables stuttered) and lost for increases in stuttering in ordinary conversation at scheduled sessions. Tokens were exchangeable for necessities and luxuries including meals, tea, coffee, magazines, cigarettes, beer, and others. In separate experiments, specific variations of the contingencies were evaluated. Token reinforcement was found to be less effective than token reinforcement combined with response cost. The degree of reduction of stuttering was found to be a function of the magnitude of reward (i.e., tokens). Noncontingent tokens were less effective than contingent tokens in decreasing stuttering. Finally, token reinforcement was found to be a useful adjunct with other procedures used for the treatment of stuttering (cf. Ingham & Andrews, 1973; Ingham *et al.,* 1972). Other investigators have used tokens to decrease stuttering (Lanyon & Barocas, 1975; Leach, 1969; Rickard & Mundy, 1965).

Aphasics

Aphasia is regarded as an interference with language processes as a result of brain injury. The focus of treatment has been to attempt to communicate with aphasics and to stimulate language processes as much as possible. Ayers *et al.* (1975) used token reinforcement to alter language processes of four adult aphasics (46–78 years old) whose aphasia was attributed to cerebral vascular accidents. Individualized programs were devised to focus on the particular deficits that the patients evinced. The aphasics, who resided in a convalescent center, received individual treatment within the center in which they received marks contingent upon performing the select target responses. The marks were tallied and progress was charted over the sessions. However, the marks were not redeemable for back-up events.

Interestingly, the reinforcement program effectively altered diverse behaviors including grammatical closure drills (requiring the client to supply the missing word in a series of sentences), misarticulation, and identifying objects. Subsequent to treatment, patients improved in general areas on a test used to isolate response deficits in aphasics. That is, their deficits were categorized as less severe.

Social Extensions

A major application of token reinforcement has been to social and community problems such as pollution, energy conservation, employment, job performance, and several others. Social applications of reinforcement techniques have increased in recent years (Kazdin, 1975b, 1977a). Because of the diversity of research in this area and the recent interest in social problems, token economies in social settings will be reviewed in a separate chapter.

CONCLUSION

The token economy has been applied to delinquents, adult offenders, drug addicts, problem drinkers, children and adults in outpatient treatment, geriatric residents, stutterers, and aphasics across an extremely wide range of behaviors.

A relatively large amount of research has been completed with programs for delinquents. Programs have altered classroom and academic behaviors, antisocial acts, and social skills. While many of the changes reflect specific behaviors within the treatment setting, more global changes within the setting and the community have also been demonstrated. For example, the Achievement Place program has been shown to improve attitudes of the participants as well as many overt behaviors. Also, suggestive data indicate that adjustment in the community after treatment is superior after participation in the Achievement Place program than in a traditional institutional program or even probation.

Programs with adult offenders have focused primarily on behaviors adaptive within the setting including self-care behaviors and participation on jobs. However, programs have attempted to incorporate token reinforcement into an educational program where performance on academic tasks is reinforced. Prison programs in general have received increased public attention because select programs have utilized procedures that threaten or sometimes blatantly violate the rights of inmates. Although these programs have been the exceptions, some of the litigation they have spawned has had important implications for programs with institutionalized and involuntarily confined populations.

Relatively few token economies have been implemented with drug addicts and problem drinkers. These programs usually focus on adaptive behaviors within the hospital including self-care, job performance, attending activities, and so on. Programs for problem drinkers occasionally have attempted to directly control alcohol consumption by providing alcohol in the treatment setting and reinforcing abstinence or moderate drinking.

Outpatient applications of token reinforcement have included diverse types of programs. In outpatient therapy, tokens, usually in the form of money, are provided by the therapist in the therapy session. This procedure has been used with behaviors such as overeating, cigarette smoking, alcohol consumption, and others. Outpatient applications also include interventions in the natural environment. In these cases, contingencies are implemented entirely in the natural setting (e.g., the home, at school, the community). The therapist serves primarily as a consultant to facilitate in the design of the program. Usually, programs in the natural environment focus on problem behaviors of children and the administration of reinforcement contingencies by the parents. Occasionally, token-reinforcement contingencies are designed for adults in the home so that points, perhaps backed by money, are delivered for changes in select problem behaviors or interaction with one's spouse.

Generally, the use of token reinforcement in outpatient treatment dramatically extends the focus of early programs in the field. Evidence strongly supports the notion that treatment programs can be effectively conducted in the natural environment. Possibly, the increased use of reinforcement programs in extratreatment settings will emphasize the prevention as well as amelioration of behavior problems.

The use of token reinforcement has been extended to other populations and areas of study than those reviewed above. Select applications with geriatric residents, stutterers, and aphasics were discussed. Overall, the token economy appears to be readily adapted to diverse treatment populations and settings.

The review of token economies has emphasized the accomplishments of diverse programs. In subsequent chapters, various problems and issues related to the efficacy of the token economy will be discussed. Thus, a more critical evaluation of the results will be made. The evaluation begins with Chapter 6 which discusses important obstacles and problems that arise in a token economy. In Chapter 7, the generality of behavior changes effected with token reinforcement is discussed. Finally, the evaluation is culminated in Chapter 8 where specific limitations of contemporary research on the token economy are elaborated.

MAJOR ISSUES AND OBSTACLES
IN ENSURING EFFECTIVE PROGRAMS

6

Implementing effective token economies requires more than merely selecting target behaviors, a medium of exchange to serve as the token, back-up events, and providing rules to specify the contingencies. There are many issues and obstacles that need to be addressed which underlie the efficacy of a given program. This chapter discusses major issues which pertain directly to the efficacy of the token economy, namely, training staff or other behavior-change agents to implement the contingencies, altering the contingencies for clients who do not respond to the initial token program, and utilizing economic variables to sustain or augment high levels of performance on the part of the clients.

TRAINING STAFF AND OTHER BEHAVIOR-CHANGE AGENTS

A major consideration in establishing an effective behavior modification program is training agents to administer the contingencies correctly (Atthowe, 1973; Berkowitz & Graziano, 1972; Hall & Baker, 1973; Kazdin, 1976a; Kazdin & Moyer, 1976; O'Dell, 1974; Patterson, 1976; Yen & McIntire, 1976). These agents, referred to here collectively as staff, include attendants, aides, teachers, parents, spouses, peers, and others who, in a given program, may be responsible for administering the contingencies. In behavior modification programs, perhaps more than any other treatment, nonprofessional staff are responsible for the effect of treatment. The staff are responsible for what behaviors are reinforced, punished, and extinguished and, hence, determine the efficacy of the contingencies. If a program is to be effective, the contingencies must be applied to the client in a systematic and consistent fashion. If a consequence is delivered inconsistently by a staff member (i.e., contingently on some occasions and noncontingently on other occasions), behavior is not likely to improve as it would if the consequence were delivered contingently all of the time (Redd, 1969; Redd & Birnbrauer, 1969).

Training individuals in frequent contact with the clients is especially important because their behavior often contributes directly to deviant client behavior. Aides, parents, teachers, and others inadvertently reinforce the deviant behaviors they would like to suppress (Ayllon & Michael, 1959; Buehler et al., 1966; Gelfand, Gelfand, & Dobson, 1967; Wahler, 1972). Thus, staff have to be trained to carefully control the consequences they administer.

STAFF TRAINING TECHNIQUES

Generally, some specific training procedures must be provided to ensure that staff can effectively administer a behavioral program. Occasionally, successful programs have been managed by individuals who have received minimal training. For example, in a token program conducted in the home, parents successfully implemented the contingencies after only a few hours of contact with a therapist (Christophersen *et al.*, 1972). Ideally, if the target behaviors are well specified, few judgments of the staff are required, and the contingencies can be administered without extensive training. Yet, in many settings staff must monitor a large number of behaviors, clients, and contingencies, and training is essential to provide diverse skills in handling a variety of exigencies.

A large number of techniques have been used to train staff. Many of the techniques include the same behavior modification techniques that are used to alter client behaviors. The major techniques used for staff training include instructional methods, feedback, self-monitoring, modeling, role-playing, and social or token reinforcement.

Instructional Methods

Instructional methods such as lectures, discussions, inservice training, workshops, and course work frequently are used to train behavior modifiers. In most cases, the effect of instructional techniques on behavior is not evaluated. Yet, there is sufficient evidence that instructional methods alone, at least as ordinarily practiced, are not very effective in developing behavior modification skills.

Instructional techniques have taken many forms. In some situations, staff members are merely told to behave in a particular way in relation to the clients (i.e., reinforce certain behaviors). This type of instruction either has transient effects or results in virtually no change in the behavior of staff (e.g., Gelfand, Elton, & Harman, 1972; Katz, Johnson, & Gelfand, 1972; Kazdin, 1974a; Pommer & Streedbeck, 1974). Even if staff are given verbal or written reminders to use reinforcement, their behavior does not change reliably (Katz *et al.*, 1972; Quilitch, 1975).

In other forms, instructional methods consist of planned lectures and instructions. Didactic procedures appear to be effective, if supplemented with training in the actual situation in which the techniques will be employed (Nay, 1975; Paul, McInnis, & Mariotto, 1973). Instructional training without accompanying practical training does not appear to be sufficient to train behavior modifiers (Gardner, 1972). Instructional methods do provide an individual with knowledge of the principles of behavior modification but provide little or no proficiency in actually carrying out the procedures (Gardner, 1972; Nay, 1975).

In one program, instruction has been successful in training teachers and parents. The method is referred to as the Responsive Teaching Model (Hall, 1972; Hall & Copeland, 1972). For example, to train teachers, a college-level course is provided where the basic principles of behavior modification and techniques for classroom management are taught. All of the procedures required to implement and evaluate a behavior modification project (e.g., selection and measurement of target behaviors, experimental design) are trained so that an individual can both administer programs and scientifically validate their efficacy. Individuals who have previously completed the course discuss their own projects. Each student in the course executes a behavior modification project. In doing this, the student carefully defines and measures behavior in the classroom situation, assesses reliability of the recording procedure, applies experimental procedures systematically, and scientifically verifies whether the procedures caused behavior change. The Responsive Teaching Model appears to successfully combine academic and practical training. Thus, it goes beyond mere instructional methods which characterize the majority of staff training programs.

As a general statement, instructional methods as usually practiced (workshops, discussions, inservice training) have little demonstrated efficacy in changing staff behavior. The procedure referred to as Responsive Teaching is a special case in which instructional methods (lectures and discussions) are integrated with supervised practical experience (implementing and evaluating a program).

Feedback

Feedback refers to providing information about the adequacy of performance. Feedback is inherent in virtually all forms of response consequation including delivery of praise, tokens, reprimands, and even providing no consequences at all. Yet, feedback can be used independently of other reinforcers by simply providing the individual with information about what he or she is doing.

Typically, in training staff, feedback consists of verbal or written reports of behavior. For example, Panyan, Boozer, and Morris (1970) trained attendants to teach self-help skills in retarded residents. Staff were responsible for conducting sessions with the residents. Initially, no consequences were associated with conducting or failing to conduct the sessions with the residents. A feedback procedure was implemented which consisted of providing staff with a "feedback sheet" which told the percentage of sessions they conducted out of all those possible. With feedback, staff performance was at a substantially higher level.

Cooper, Thomson, and Baer (1970) used feedback to train two preschool teachers to attend to appropriate child behavior. In-class feedback (administered verbally to one teacher and in written notes to the other teacher) included the number of times the teacher attended to appropriate child behavior (e.g., studying, following directions). Feedback also was provided after class on the percentage of time the teacher attended and failed to attend to appropriate behavior and the specific child behaviors that occurred. Feedback was markedly effective in increasing the attention of one teacher to appropriate child behavior. The teacher for whom training was only moderately effective did not read the feedback notes given to her until the end of the day. Thus, feedback was delayed.

Gelfand *et al.* (1972, Exp. 1) provided daily videotape feedback to staff in a day-care facility for disturbed, brain-injured, and retarded children. Staff viewed taped replays of their individual training sessions with a child. During the early part of training, experimenters praised appropriate staff behavior on the tape (administering events contingently), ignored inappropriate behavior (negative attention), and gave instructions. Videotape feedback and verbal consequences increased staff delivery of contingent consequences, decreased negative attention, and was associated with changes in child behavior. Withdrawal of feedback did not result in a return of behavior to baseline levels in a period of several days.

McNamara (1971) used a telemetric signaling system to provide immediate feedback to one elementary school teacher. The signals consisted of mild (nonpainful) electric shock which informed the teacher whether he attended to appropriate or inappropriate child behavior (using a different number of shocks to signal which category of child behaviors had occurred). Teacher attention to appropriate behavior (hand-raising) increased whereas teacher attention to inappropriate behavior (calling out) decreased. Child behavior also changed in response to changes in teacher behavior.

Although feedback has been effective in the studies reviewed above as well as others (Barnard, Christophersen, & Wolf, 1974; Parsonson, Baer, & Baer, 1974; Quilitch, 1975), it has produced little or no effect in many studies (e.g., Breyer & Allen, 1975; Cossairt, Hall, & Hopkins, 1973; Rule, 1972). Thus interventions with more consistent effects on staff behaviors have been sought.

Self-Monitoring

Self-monitoring refers to collecting data on one's own behavior. Collecting data on one's own behavior sometimes is used as a technique to help individuals perform particular target responses either in its own right or as an adjunct to other staff training techniques. With self-monitoring, the staff members collect data on responses that need to be increased (e.g., amount of contingent approval delivered) as well as decreased (e.g., verbal reprimands or threats). Self-monitoring might be viewed as a special case of feedback because an individual receives feedback for his or her own performance based upon the data collected and presumably alters his or her own behavior in response to this feedback. However, self-monitoring has been distinguished in behavior modification research as a therapeutic technique in its own right and is not always interpretable from the standpoint of feedback (Kazdin, 1974d).

A few applications of self-monitoring in the area of staff training suggest the utility of the technique. Herbert and Baer (1972) used self-monitoring to train two mothers to alter the behavior of their children in the home. The two children performed a variety of deviant behaviors including breaking, tearing, and throwing things, ritualistic behaviors, aggressive acts, and others. The mothers were instructed to record their attention (e.g., comment, praise, instruction, suggestion) to appropriate child behavior. When the mothers self-monitored appropriate attention, their attention increased. As expected, appropriate child behavior increased in those phases in which contingent maternal attention increased. Similarly, Thomas (1972) found that having teachers self-monitor their own behavior led to systematic behavior change. Teachers viewed videotapes of their classroom activity and recorded the frequency of praise, token delivery, and instructions, all of which increased during subsequent classroom sessions.

Not all of the evidence for self-monitoring as a staff training technique is favorable. Van Houten and Sullivan (1975) instructed teachers to count each time they praised a child for his work, to calculate their praise rate in responses per minute, and to chart this rate after each session. Self-monitoring had a negligible effect on the behavior of one teacher and no effect on the behavior of another. In general, the effects of self-monitoring as a behavior-change technique, have been weak, transient, and often nonexistent in a plethora of self-monitoring studies (Kazdin, 1974b). Thus, self-monitoring when used alone is unlikely to be an effective intervention for staff behavior.

Modeling and Role-Playing

Modeling has been used infrequently as a staff training technique. With modeling, the staff member must observe someone else perform the target behaviors usually in the situation in which staff perform. The value of modeling in teacher training has been suggested in one report (Ringer, 1973). In a fourth-grade class, an investigator (model) administered verbal and token reinforcement to decrease inappropriate child behavior including getting out of seat, talking, and fighting. The investigator circulated in the class in a "random fashion" and administered praise and tokens (by initialing a student's special card) for appropriate behavior. Eventually, both the teacher and investigator administered praise and tokens. Each day the teacher was given responsibility for administering an increasingly greater proportion of the tokens. In a short time, the investigator was completely out of the classroom. Inappropriate behavior was maintained at a rate lower than baseline when the teacher had complete responsibility for the program. Child behavior was slightly better when the experimenter had been in the room. Yet, the teacher was able to maintain substantial control after viewing a model administer tokens. The modeling procedure did not alter the teacher's use of contingent attention.

Modeling also has been used to develop staff-patient interaction in a behavior modification program. Wallace, Davis, Liberman, and Baker (1973) evaluated different techniques designed to increase the staff's attendance to a social interaction hour held daily for patients and staff. Instructions to attend the activity or ensuring that the nursing staff had no competing duties during the hour did not affect staff performance. However, when professional staff (i.e., psychologist, research assistant responsible for the program, or the nursing supervisor) attended the sessions, both staff and patient attendance were high, as demonstrated in an ABAB design.

Modeling has been used in other programs designed to develop the use of behavior modification skills in parents and siblings (Cash & Evans, 1975; Engelin, Knutson, Laughy, & Garlington, 1968; Nay, 1975).

Role-playing has been used as a training technique in which individuals rehearse the behaviors that are going to be used in administering a reinforcement program. Role-playing usually includes modeling to demonstrate the desired behaviors and is followed by practice of the modeled behaviors in simulated situations. For example, Jones and Eimers (1975) trained elementary school teachers after school by modeling diverse behaviors (e.g., use of praise, disapproval, instructions, time-out) and having the teachers practice the modeled skills. Participants in the sessions alternated playing the role of the teacher and a "good" or "bad" student. Prompts, feedback, and praise by the experimenter supplemented enactment of the appropriate behaviors. Training was associated with marked changes in inappropriate student behaviors in the teacher's original classrooms. Other studies have shown the efficacy of role-playing as a staff training technique (Gardner, 1972; Jones & Miller, 1974).

Social Reinforcement

Praise, approval, and attention have been used to alter staff behavior, particularly teacher behavior. For example, Cossairt *et al.* (1973) carefully evaluated the effect of comments by a consultant on teachers' behavior. An experimenter met individually with two teachers after each class to discuss their behavior. In these sessions, the experimenter provided feedback by telling the teachers how often the students paid attention and the amount of praise given. This did not produce systematic changes in teacher or student behavior. To introduce social reinforcement into the meetings, the experimenter praised the teacher for using praise in the classroom. During the sessions, the experimenter would say things such as, "You certainly have the ability to hold their (the students') attention with your praise." Positive comments by the experimenter markedly increased both teachers' use of praise in their classrooms. In addition, the students in each class showed dramatic increases in attending behavior when the teacher increased her use of praise. Thus, the experimenter's contingent praise notably affected the behavior of the students. Other studies have shown that teacher behavior can be altered by praise from a consultant (Brown, Montgomery, & Barclay, 1969; McDonald, 1973).

Perhaps the most creative use of social consequences has been using the clients themselves to alter staff behavior. Graubard and his colleagues have trained special education students who were considered to have behavior problems to alter the behaviors of their teachers (Graubard, Rosenberg, & Miller, 1971, 1974; Gray, Graubard, & Rosenberg, 1974). Special education students (ages from 12 to 15) received instruction and practice in behavior modification including role-playing and studying videotapes. To increase positive teacher-student contacts, the students were taught to reinforce appropriate and to extinguish or mildly punish inappropriate teacher behavior. Students reinforced teacher responses by smiling, making eye contact, sitting up straight, and making comments such as "I work so much better when you praise me." To discourage negative teacher contacts, statements were made such as "It's hard for me to do good work when you're cross with me" (Graubard *et al.*, 1974; Gray *et al.*, 1974). The effect of contingent student social consequences on teacher behavior was evaluated in an ABA design. The results of several students across a total of 14 teachers appear in Fig. 19. These results reveal that during the intervention phase, positive teacher comments increased and negative comments decreased relative to baseline and reverted in the direction of baseline when social consequences were withdrawn. In a similar demonstration, Seymour and Stokes (1976) trained institutionalized delinquent girls to solicit reinforcement from the staff. When the girls' work appeared good and when staff were nearby, the girls called the staff's attention to their work which increased staff delivery of reinforcement.

Token Reinforcement

Tangible conditioned reinforcers have been used effectively in a number of instances of training staff. For example, McNamara (1971) evaluated the effect of token reinforcement and response cost (withdrawal of tokens) in altering teacher attention to appropriate and inappropriate behavior. Two teachers received points for attending to appropriate child behavior; another teacher lost points (from an

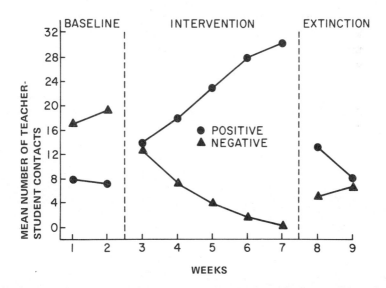

Fig. 1. Mean number of positive and negative teacher–student contacts during Baseline, Intervention, and Extinction Phases (from Graubard, Rosenberg, & Miller, 1974; Scott, Foresman, and Company).

initial sum delivered noncontingently). Points were exchangeable for cans of beer. In some cases, extra beer was given for particularly good teacher performance (i.e., no instances of attention to inappropriate student behavior in a given session). The contingent delivery or removal of points was effective in altering teacher behavior. Performance gains were especially evident when the bonus contingency (extra beer) was in effect.

In an institutional setting, Bricker, Morgan, and Grabowski (1972) provided video feedback to attendants for their performance on the ward of mentally retarded residents. While staff viewed themselves on tape, they received verbal praise and trading stamps from the project director. The reinforcers were contingent upon staff interacting with the residents, as previously recorded on videotape. The reinforcement system increased staff-resident interaction on the ward by 700% relative to baseline. The changes appeared to result from the reinforcement because removal of the videotape feedback did not result in a loss of the target behavior.

Monetary incentives for staff behavior have been provided in several programs (Ayllon & Azrin, 1968b; Cohen & Filipczak, 1971; Katz et al., 1972; Peine & Munro, 1973; Pommer & Streedbeck, 1974; Pomerleau et al., 1972; Pomerleau, Bobrove, & Smith, 1973; Watson, 1976; Wolf et al., 1968). For example, Pomerleau et al. (1973) evaluated the effect of feedback, noncontingent $20 cash awards, and contingent $10, $20, or $30 awards for staff on the behavior of patients participating in a token economy in a psychiatric hospital. Contingent events were delivered for improvement shown by specific patients to which the staff members had been assigned. If a patient showed the greatest improvement, as measured by biweekly inventory ratings of appearance, verbal behavior, and adaptation to ward routine, the staff member responsible for that patient earned the award. Feedback and noncontingent delivery of money, respectively, had little or no effect on behavior. However, contingent cash awards markedly affected patient behavior. When cash awards were discontinued for the staff, inappropriate patient behavior increased markedly.

Katz et al. (1972) compared separate interventions to increase the frequency that psychiatric aides reinforced patient appropriate behavior. Instructions to reinforce behavior and verbal prompts had virtually no effect on staff performance. However, contingent monetary reinforcement (a $15 cash bonus for increasing their reinforcing contacts with the patients) markedly altered staff behavior.

Investigations relying upon token reinforcement in the form of money, trading stamps, or points redeemable for back-up events have shown consistent effects on behavior. The effects have been shown to result both in changes in staff and client behavior.

Combined Techniques

Individually discussing staff training techniques implies that the procedures usually are used in isolation. In many studies, techniques have been used individually so that their relative impact on staff behavior can be determined. However, in many programs techniques are combined. For example, Gladstone and Sherman (1975) trained high school students to teach profoundly retarded children to follow simple instructions. Training consisted of showing each trainee a videotape in which a model worked with a retarded child and demonstrated the techniques to be used during the intervention phase. In addition to viewing a model, trainees rehearsed the procedure using an experimenter as a subject and practiced delivering instructions, reinforcement, using and fading physical prompts, and ignoring disruptive behavior, all of which had been demonstrated on the videotape. Finally, the trainees practiced on a child and received corrective feedback (e.g., Your prompts . . . came too quickly") and praise (e.g., "Your enthusiasm is great . . .") from the experimenter. After each of these training techniques were completed, the trainee worked with a target subject. The overall training package was effective in training the students to develop behaviors with profoundly retarded subjects. Of course, the effect of the individual components used in training including modeling, rehearsal, feedback, and praise, could not be determined. Yet, the investigation illustrates the combined use of several procedures to develop behavioral skills.

As another example of multiple training techniques, Fo and O'Donnell (1974) recruited individuals from the community to serve as "buddies" and behavior-change agents for youths who had behavior problems in the home and at school. The buddies received training sessions that included modeling, role-playing, instructions, prompts, and praise for learning how to identify problem behaviors, collect data, deliver consequences contingently, and to design and execute intervention programs. Monetary reinforcement also was provided for specific behaviors such as contacting the youths, submitting weekly behavioral data, completing the assignments with the youth and attending training sessions which continued throughout the duration of the project. The buddies were able to alter specific problem behaviors of the youths including truancy, fighting, not doing home chores, and others by administering social and token (monetary) reinforcement contingent upon desired behaviors.

Another training program with multiple procedures was designed to train hospital attendants to complete various assigned tasks on the ward including observing and recording patient self-care (e.g., grooming, bedmaking) and work behaviors, distributing lunch tickets to patients who performed select target behaviors, graphing patient performance, implementing individualized programs, and others (Hollander, & Plutchik, 1972; Hollander, Plutchik, & Horner, 1973). A 6-week course was provided to discuss and to demonstrate the principles of operant conditioning. After the course, staff participated in question and answer sessions, observed modeled demonstrations, and role-played the target behaviors. Finally, tokens in the form of a nationally known brand of trading stamps redeemable at a stamp redemption center were delivered for completion of assigned tasks. The effects of the stamp contingency are illustrated in Fig. 20. As evident in the figure, implementation and withdrawal of the stamp contingency controlled the percent of tasks that staff completed on the ward. Several other investigators have combined

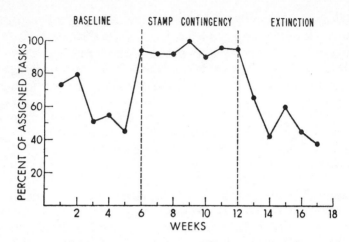

Fig. 20. Percent of assigned tasks completed by attendants during Baseline, Stamp Contingency, and Extinction Phases (from Hollander & Plutchik, 1972; Pergamon Press).

diverse procedures as part of staff training including classroom instruction, role-playing, modeling, feedback, and others (cf. Gardner, 1972, 1973; Paul & McInnis, 1974).

Training Techniques: General Comments

As usually conducted, staff training consists of providing information about behavior-change techniques in the form of didactic workshops or instructional methods. As noted earlier, instructional methods of training have not been consistently associated with behavior change. The failure of workshops to clearly develop the behaviors for which they are designed has led some authors to question the ethics of even conducting them (Stein, 1975). It seems clear that merely disseminating information about the desired staff behaviors is not adequate to alter behavior.

Altering behavior of the staff perhaps can best be viewed as a behavior modification program separate from the token economy relied upon to alter client behavior. The methods and principles required to change client behavior are needed to alter behavior of the staff. Target behaviors need to be identified for the staff, data need to be collected to determine levels of performance of the target behaviors, an intervention needs to be implemented and experimentally evaluated, and data need to be collected continuously to ensure that behavior is maintained or, if not, that some means are taken to maintain staff behavior.

In any staff training program, data need to be collected to evaluate the procedures. Evaluation needs to include all, or at least most, of the behaviors that training is designed to alter. It cannot be assumed that altering one or a few behaviors will generalize to other responses. Evidence suggests that only those specific behaviors trained are altered and not related responses which would make the staff more effective behavioral engineers (Cooper *et al.*, 1970). Moreover, if a response is trained to be executed only a few times during the day, it may not generalize to nontrained times during the same days (Fielding, Errickson, & Bettin, 1971). If the goals of training include changes in several staff behaviors, which are performed at several times, or in various situations, training has to include the range of circumstances.

After the desired target behaviors have been established, it is important to ensure that the behaviors will continue to be performed when training is terminated. The procedures to establish and maintain behavior may be different. A number of programs have shown that merely training staff does not guarantee that these behaviors are maintained. Indeed, as a general rule as soon as extrinsic consequences for attendant and teacher behavior are withdrawn, behavior reverts to pretraining levels (Brown *et al.*, 1969; Cooper *et al.*, 1970; Katz *et al.*, 1972; Kazdin, 1974a; Martin, 1972; Panyan *et al.*, 1970; Pomerleau *et al.*, 1973), although there are exceptions (Cossairt *et al.*, 1973; Parsonson *et al.*, 1974; Van Houten & Sullivan, 1975). Thus, it appears that specific contingencies or maintenance strategies need to be invoked to maintain staff behavior after training is terminated. Currently there is little research addressing itself to long-term maintenance of staff behavior. Some procedures are needed to cue staff members to utilize the behaviors they have developed and to provide consequences contingent upon this performance (cf. Van Houten & Sullivan, 1975).

At the very least, staff behavior needs to be assessed periodically to determine whether performance has deteriorated over time and to evaluate the efficacy of maintenance techniques. The evaluation of staff behavior need not introduce insurmountable problems related to the collection of data on staff. Some training programs have evaluated staff behavior by looking at the behavior change of their clients (Pomerleau *et al.*, 1972, 1973; Walker, Hops, & Johnson, 1975). Of course, there is a potential problem with indirectly evaluating staff performance in that temporary client behavior change can result from staff performing undesirable behaviors that were not developed during a staff training program (e.g., heavy reliance upon aversive contingencies such as corporal punishment, frequent fines, and isolation). Yet, data on client behavior usually are more easily obtained than data on staff behavior. Thus, data on client behavior provide a readily accessible resource to reflect staff performance.

THE EFFECTS OF TOKEN ECONOMIES ON STAFF BEHAVIOR

Various programs have reported that implementing a token economy has a general effect on the staff including an increase in morale or even decreased absenteeism (Atthowe & Krasner, 1968; Winkler, 1970). In many programs, the report of staff changes are anecdotal. Even where an attempt is made to assess inadvertent changes in staff, the results may be ambiguous. Usually, staff are not randomly assigned to treatment and control groups (e.g., wards) and staff who participate in a token program may differ from control staff along a wide variety of dimensions including age, salary, years of education, and other subject or demographic variables. In addition, prior to administering a token program, staff often participate in a training program which can alter various attitudes and behaviors (Paul & McInnis, 1974). Despite these differences, a number of studies have provided suggestive evidence pertaining to the effects of token programs on staff behavior.

The effect of a token economy on staff behavior was demonstrated in a psychiatric hospital where a comparison was made between the unprogrammed reinforcement rates of aides participating in a token economy ward with aides participating in a traditional care ward (Trudel, Boisvert, Maruca, & Leroux, 1974). Aides on the token economy ward were found to provide over six times more positive attention (praise, smiles, approval) for appropriate patient behavior (e.g., patients greeting each other or a staff member, performing tasks on the ward, responding to the requests of others) than aides on the control ward. In addition, aides on the control ward ignored approximately twice as much appropriate patient behavior as aides on the token economy ward. Punishment was infrequent and did not differ across wards.

Several studies have been made within subject comparisons to show that various staff behaviors increase during the administration of a token program. Chadwick and Day (1971) found that having teacher and classroom aides administer tokens for academic behavior increased their use of approval and decreased their use of disapproval during the phases in which tokens were delivered. Similarly, Breyer and Allen (1975) unsuccessfully attempted to increase a teacher's use of praise and to decrease her use of reprimands by providing praise and feedback for teacher behavior. However, when a token economy was implemented for the children, teacher behaviors were altered in the desired direction. In a detention setting for adjudicated males, Gambrill (1974) found that implementing token economies in various units in the facility was associated with a decrease in the amount of discipline (sending an inmate to his room) administered by the staff. Although most programs show an increase in approval and a decrease in disapproval of staff who administer token programs, there are exceptions. For example, Drabman (1973) found that two teachers decreased their use of praise and reprimands when they implemented a token program in the classroom.

The effects of implementing a token program on staff behavior was carefully demonstrated in a kindergarten class (Mandelker, Brigham, & Bushell, 1970). Six children were divided into groups (Group A and Group B). In different phases over the course of the project, the teacher delivered tokens contingently (for correct handwriting) to one of the groups while the other group received tokens noncontingently. The tokens were exchanged for a variety of activities such as gym, games, walks, stories, and a snack. The effects of delivering tokens contingent upon student behavior was evaluated on teacher contact with the children (e.g., talking to, facing, touching or looking at the materials of a child). The results of the program are displayed in Fig. 21. As evident from the figure, the number of teacher contacts per child increased from baseline rates when the teacher delivered tokens contingently. As the contingencies changed from contingent to noncontingent delivery across phases, so did teacher contacts with her students. These results

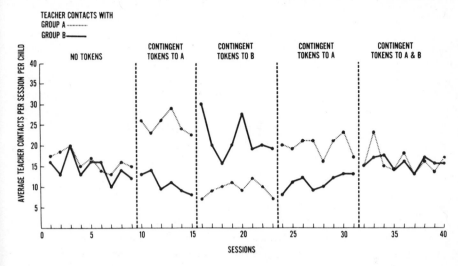

Fig. 21. Mean teacher contacts per session per child for Group A and Group B (from Mandelker, Brigham, & Bushell, 1970; Society for the Experimental Analysis of Behavior).

suggest rather clearly that social interaction may increase as a result of contingent token delivery.

Aside from effects on specific behaviors, research suggests that the token economy may alter staff evaluations of the patients. McReynolds and Coleman (1972) evaluated attitude changes made by staff working on a token economy ward for psychiatric patients. After one year, staff considered more patients responsive to staff members and capable of extra-hospital adjustment. Attitudes of staff on the token ward were more favorable regarding the potential benefits of treatment than were the attitudes of staff on a similar ward. Further, staff on a token ward rated a larger percentage of patients as aware of themselves and their behavior and as experiencing more normal emotions than did staff not on a token economy ward. Other investigators have found that staff participating in a token economy ward have more favorable attitudes about the program and problems of the patients than do staff participating in a ward with conventional treatment (Milby, Pendergrass, & Clarke, 1975).

UNRESPONSIVE CLIENTS

An extremely important issue in token economies, and indeed in any treatment intervention, is the failure of some individuals to respond to treatment (Kazdin, 1972a, 1973d). In many token economies, some individuals do not respond at all to the reinforcement contingencies. Individuals probably are differentially responsive to any given contingency included in treatment. To illustrate this, one program categorized psychiatric patients according to their responsiveness to a contingency designed to increase their attendance to occupational therapy in the hospital (Allen

& Magaro, 1971). Some patients attended therapy without receiving tokens; others only when tokens were delivered. A final group did not attend at all even when tokens were provided for attendance. This latter group simply did not respond to the contingency as originally programmed.

The failure of clients to respond to token reinforcement has been reported with programs across diverse treatment populations including psychotic patients, retardates, delinquents, and children in elementary and secondary school classrooms (Atthowe & Krasner, 1968; Ayllon & Azrin, 1965; Betancourt & Zeiler, 1971; Hunt et al., 1968; Hunt & Zimmerman, 1969; Liberman, 1968; Lloyd & Garlington, 1968; Logan et al., 1971; Tyler & Brown, 1968; Zimmerman et al., 1969). Although a large number of programs have reported individuals who fail to respond, it is difficult to provide a percentage to represent the number of nonresponders. Because some programs have focused on a small number of clients, a few non-responders would spuriously inflate this percentage (cf. Kazdin, 1973d).

The failure of individuals to respond in the predicted manner should come as no surprise. Laboratory research has shown somewhat different response characteristics of populations who are focused upon in treatment (e.g., psychiatric patients, retardates, and autistic children). For example, in his study of free-operant responses in a laboratory task (lever-pulling), Lindsley found that psychotics responded inconsistently, with low rates, and pauses or interruptions in responding which were not as apparent in the responses of nonpsychiatric patients (Lindsley, 1956, 1960, 1963). Similarly, laboratory research has suggested that response rates of retardates sometimes are more variable from session to session, show more frequent pauses, and are less well controlled by the schedule of reinforcement than the rates of nonretardates (Barrett & Lindsley, 1962; Ellis, 1962; Ellis, Barnett, & Pryer, 1960; Spradlin, 1962). With autistic children, some research has shown that unresponsiveness to complex reinforcement schedules, rates of learning, and the development of superstitious behavior is more likely than with normals (Ferster & DeMyer, 1961, 1962; Metz, 1965). Thus, laboratory findings have shown complexities and idiosyncrasies in the responding of various treatment populations. These studies do not suggest that there are intractable response deficits or behavioral characteristics that necessarily impede performance but only that contingencies which usually lead to particular response characteristics with normals might have somewhat different effects with specific populations.

EXPLANATIONS OF UNRESPONSIVENESS

There are several possible explanations for unresponsiveness.[1] First, when subjects do not respond to the reinforcement contingencies, this might reflect the

---

[1] The term "unresponsiveness" here is used as a convenient term to denote that individuals may not respond to a given contingency. The term is used merely to describe performance in relation to a particular contingency and is not intended to reflect a general trait or characteristic of the individual per se. Indeed, as the discussion indicates later, alteration of the contingency in any of several different ways can increase responsiveness to the program.

lack of appropriate or sufficiently powerful reinforcers which serve to back up the tokens (Ayllon & Azrin, 1965). Indeed, when tokens have no actual back-up value, behavior usually does not change (Bushell *et al.,* 1968; Schaefer & Martin, 1966). Thus, when patients do not respond to a token system, it may be that the back-up reinforcers are not adequate incentives for performance. Second, clients may not respond because the reinforced responses are not in their repertoires. For example, although tokens might be delivered for dressing and grooming oneself, the component behaviors (e.g., putting on clothes, buttoning a shirt) may not be in the client's repertoire. Unless the contingencies are individualized and terminal responses are shaped, individuals may fail to respond to contingencies which only reinforce terminal performance.

A final explanation for unresponsiveness is a failure of the clients to understand the relationship between performance and token reinforcement. In most programs, clients are told what the contingencies are. Despite these instructions and experience with the contingencies either directly or vicariously through the performance of others, individuals may not comprehend the contingency. Some studies might be cited in support of this explanation because they suggest that delivering contingent consequences does not alter behavior until instructions about the reinforcement contingency are made explicit. For example, in a study of psychiatric patients, individuals received candy, cigarettes, or a beverage if they selected utensils with which to eat their meal rather than eat food with their hands without utensils (Ayllon & Azrin, 1964). The contingency alone did not increase the percentage of patients who selected utensils. However, when instructions were added explaining the relation between the target response and the reinforcing consequences, selecting utensils increased markedly. Not all studies have shown that instructions stating the contingency facilitate performance (Kazdin, 1973g; Resnick, Forehand, & Peed, 1974).

There are a number of other interpretations that might be advanced to explain why some clients fail to respond to the contingency. For example, for some clients the delay between performance and token reinforcement or the delay between token reinforcement and the exchange of tokens for back-up reinforcers may be too great to affect responding.

TECHNIQUES TO ENHANCE RESPONSIVENESS

As indicated above, several token-reinforcement programs have reported clients who fail to respond to the program. Such a consistent finding has been considered to imply that there are clients for whom token programs are not likely to be effective (Allen & Magaro, 1971). While this might well be true, current evidence does not support such a claim. If an individual does not respond to a particular token reinforcement program, this does not mean that he would fail to respond to any token program. Individuals who do not respond to the initial token program may readily respond when the contingencies are altered. There are several procedures that can be used to augment the responsiveness of clients to reinforcement contingencies including response priming, reinforcer sampling, selecting increasingly potent reinforcers, modeling of target behaviors and reinforcer utilization, and adding individualized contingencies or group administration of consequences to the existing program.

Response Priming

Frequently, tokens cannot be delivered because the client never performs the target response. In some cases, the response is not in the individual's repertoire and has to be shaped; in other cases the response is in the repertoire but simply is not performed. The response can be primed to enhance performance. Response priming consists of a procedure to initiate early steps in a response sequence. Prompts such as gestures and instructions serve a response priming function. However, response priming refers to a particular method to prime behavior. Specifically, the method involves requiring an individual to engage in the initial components of the sequence of behaviors (chain) that leads to completion of the response. By engaging in responses that are early in the sequence of responses, the probability of performing the final behavior is increased.

The procedure is based upon the notion of chaining. Any response can be divided into a sequence of smaller responses. The responses are maintained by reinforcement at the end of the response sequence. The influence of a terminal reinforcer on earlier responses in the chain is much weaker than is its influence on responses later in the chain. For responses early in the sequence the final reinforcer is delayed whereas for responses late in the sequence the reinforcer is more immediate. Also, conditioned reinforcers that control performance of each response in the sequence are less potent early in the chain than are the conditioned reinforcers late in the chain. Thus, to increase the likelihood that the response sequence will be completed and that behavior will be reinforced, the individual can be advanced to responses further along in the chain toward the final reinforcer.

Response priming has been used to initiate responses that otherwise have an exceedingly low frequency. For example, O'Brien et al. (1969) used response priming to increase the frequency that chronic psychiatric patients made suggestions for managing the ward. The goal of the program was to involve patients more directly in their treatment by having them make specific suggestions regarding treatment procedures. Meetings were held in which patients were to suggest anything they wished to have changed or improved on the ward. During baseline, staff sounded chimes on the ward and announced that patients could come to the meeting and make suggestions. After baseline rates of suggestions were gathered, the priming procedure was implemented. Patients were *required* to attend the meetings. If patients attended the meeting (an initial response in the chain of behaviors leading to a suggestion), suggestions should increase. Suggestions made by patients increased markedly when attendance was required. When priming was withdrawn and attendance was not required, the number of suggestions decreased. Later in the study, one group leader was told to follow the suggestions whenever patients made them at the meeting, whereas another group leader was told not to follow the suggestions. When suggestions were followed (reinforced) they increased, whereas when they were ignored they decreased.

In token economies, response priming may be useful to initiate responses of individuals who have not responded to the contingencies. If the response can be initiated, the terminal response is more likely to be reinforced. Through repeated instances of priming the response and reinforcement of the primed response, the contingency may take over and continue to influence behavior without the priming procedure. In this way, individuals who have failed to respond to the contingency may increasingly perform the behavior.

Reinforcer Sampling

Some individuals may not respond to the contingencies because the events provided to them as reinforcers are not sufficiently reinforcing. The clients may utilize the reinforcers infrequently or in many cases not at all (Atthowe & Krasner, 1968; Ayllon & Azrin, 1968a; Curran, Lentz, & Paul, 1973). If the events were provided noncontingently, they might be utilized. However, when the events are contingent upon behavior, they may not be worth the effort. Thus, some patients may not respond to the reinforcers provided. The task for the program is to increase the reinforcing value of those events used to back up the tokens so that the clients are more likely to earn tokens to purchase back-up events.

Reinforcer sampling refers to a procedure to increase the utilization of reinforcing events in the setting. It is a special case of response priming where the responses that are primed involve utilization of reinforcing events. Utilization of the reinforcing events, as any behavior, can be divided into a sequence of responses. If the initial responses in the sequence can be primed, the likelihood of completing the sequence is increased, as described earlier in the discussion of response priming.

Reinforcer sampling initiates utilization of a reinforcer by providing a sample of the event. The procedure is similar to providing small free samples in sales and advertising when companies promote a new product. By providing a free sample, the companies attempt to increase utilization of the product once the sample is used. In token economies, reinforcer sampling resembles the promotional procedures with which most people are familiar.

Ayllon and Azrin (1968b) used reinforcer sampling to increase the frequency that psychiatric patients engaged in various activities that served as back-up reinforcers for tokens. A particularly interesting application was to increase the attendance of patients to religious services. The services in the hospital lasted 15 minutes and were conducted by a hospital chaplain. Ordinarily, to determine who wished to attend the service, an attendant announced that the service was scheduled and the chaplain walked through the ward, greeted each patient, and personally invited each one to attend. Patients attended the services by depositing a token into a turnstile at the entrance of the room. After baseline observations, a sampling procedure was invoked in which the patients were required to enter a room for 5 minutes where they could hear and see the service. After 5 minutes, the patients had the opportunity to enter the service or return to the ward. Providing a brief sample of the service increased utilization of this event. During a reversal phase in which the sample was discontinued, the frequency of attendance decreased. Yet, some patients continued to attend although they had not attended during the original baseline.

Reinforcer sampling has been used to increase utilization of a number of events such as attending diverse activities (e.g., social events, movies, exercise opportunities, snack bar, pool room, and others), taking passes in the community, taking a walk on the hospital grounds, and eating meals (Ayllon & Azrin, 1968a, 1968b; Curran *et al.*, 1973; McInnis, Himelstein, Doty, & Paul, 1974; Sobell, Schaefer, Sobell, & Kremer, 1970). These investigations indicate that providing a small sample of a reinforcer increases its utilization. An important feature of the procedure is to only provide a sample of the event rather than the entire event. In one study, when individuals were given an entire meal instead of a sample, purchase of the meal with tokens decreased (Sobell *et al.*, 1970). Obviously, if the event is delivered in its entirety without cost, it is unlikely that there will be any incentive to earn and spend tokens for the event.

An important condition is that individuals have sufficient tokens to utilize the reinforcing event (cf. McInnis *et al.*, 1974). If individuals participate in a program where most of the tokens earned must be used to purchase reinforcers required for basic living conditions, increased utilization of nonessential reinforcers is not likely to occur. Finally, evidence suggests that over protracted periods, utilization of the reinforcing event may decrease even after the sampling procedure has been implemented. Possibly, "booster" sessions in which individuals periodically are exposed to the sampling procedure would be valuable to maintain utilization of the events (McInnis *et al.*, 1974). In any case, reinforcer sampling provides a technique that might be useful for clients who fail to respond to the contingency. Sampling the reinforcer could help establish the incentive value of events available in the setting.

Selecting Increasingly Potent Reinforcers

As indicated above, some patients may be unresponsive to the contingencies because the reinforcers are not sufficiently potent to provide an incentive for behavior change. Reinforcer sampling can enhance the use of the reinforcer to overcome this problem. Another solution is to select reinforcers that are likely to be extremely potent. Obviously, there are restrictions on the type of reinforcers that might be selected from the standpoint of ethical considerations so that food, water, sleeping quarters, and similar events are not withheld to develop performance. Yet, there may be behaviors in an individual's repertoire that could be used such as high probability behaviors that would be positive reinforcers in their own right or serve to back up the tokens.

For example, in a study of chronic schizophrenics, two patients were identified who did not respond to food and cigarettes as reinforcers to maintain work behavior on the ward (Mitchell & Stoffelmayr, 1973). Reinforcer sampling apparently was not effective to increase utilization of these events. To develop work behavior (stripping wire from a coil), a high probability behavior was selected as a consequence for performance. For the two inactive patients, sitting appeared to be very frequent and was therefore selected as the reinforcer. During work, the patients were asked to stand. While working on the task was being shaped, the patients were allowed to sit only after completing a small portion of work. Gradually, the task requirements were increased before the patients were allowed to sit. Eventually, sitting was allowed while completing work if the rate of behavior maintained a predetermined level. Contingent access to sitting dramatically increased the time the previously unresponsive patients spent working. The encouraging feature of this study is that it suggests that selection of alternative reinforcers might be useful in cases when clients do not initially respond to events ordinarily available as part of the program.

Similarly, Ayllon, Garber, and Pisor (1975) implemented a reinforcement system in a disruptive third-grade class where students earned tokens for completion of academic work. Although disruptive behavior declined, it returned to high levels after only a few days. Delivering tokens for nondisruptive behavior and fining tokens for disruptive behavior also had only a transient effect. Because the usual reinforcement system was not working, the authors altered the contingency. Specifically, a procedure was used in which the teacher sent a "good behavior letter" home to the parents of each child daily if the child had performed well that day. The parents were informed to expect a letter each day unless the child had been a severe behavior problem. The students earned the "good behavior letter" on the basis of their point losses for disruptive behavior. The parents also were instructed to express their approval and disappointment if the letter was or was not sent home, respectively, and developed other contingencies at their own discretion. The contingent delivery of the letter dramatically reduced disruptive behavior whereas the noncontingent delivery of the letter did not. These results show that initial failure of the program to produce consistent and durable changes was readily altered by modifying the consequences for behavior.

Modeling Influences

Responsiveness to contingencies can be increased by exposing clients to other individuals (i.e., models). Observing others appears to have vicarious effects both in the behaviors that are reinforced and in the utilization of reinforcers. To increase performance of a response, an individual can be exposed to others whose performance of the response is reinforced. Both laboratory and applied evidence has shown that individuals who observe models receive reinforcing consequences for engaging in certain behaviors are more likely to engage in those behaviors. Alternatively, individuals who observe models receive punishing consequences are less likely to engage in the punished behaviors (Bandura, 1969, 1970). These two processes are referred to as vicarious reinforcement and vicarious punishment, respectively.

A number of investigations have shown the effect of vicarious reinforcement in educational, and to a lesser extent, in psychiatric and rehabilitation settings (Broden, Bruce, Mitchell, Carter, & Hall, 1970; Brown & Pearce, 1970; Christy, 1975; Drabman & Lahey, 1974; Flowers, 1974; Kazdin, 1973b, 1973c, 1977c; Kazdin, Silverman, & Sittler, 1975; Tracey, Briddell, & Wilson, 1974). For example, in the classroom studies, providing praise or feedback to one student for attending to the lesson and working on the assignment increases these behaviors in adjacent individuals who are not directly praised. It is not entirely clear why the behavior of adjacent individuals changes even though they receive no direct consequences. The behavior that the model performs when the reinforcer is delivered does not always dictate the behavior of the observer. For example, praising an individual for *not* paying attention sometimes increases the attentive behavior of adjacent individuals (Kazdin, 1973b, 1977c). This suggests that delivery of praise may serve as a cue to others that appropriate behavior may be reinforced and increases the likelihood of appropriate behavior. Although vicarious punishment has been less well investigated than vicarious reinforcement, research in the classroom has shown its effectiveness. When a teacher punishes one student, other students may behave more appropriately as well (Kounin, 1970; Kounin & Gump, 1958).

Although research clearly shows that vicarious reinforcement and punishment can alter behavior, it has not been systematically applied in token economies to alter the behavior of patients who are initially unresponsive. Thus, its effectiveness remains to be demonstrated. On the one hand, it may seem that individuals in a token program are constantly experiencing vicarious reinforcement (by seeing others receive tokens) and vicarious punishment (by seeing others lose tokens). Yet for clients who are initially unresponsive, the vicarious consequences might have to be programmed more systematically and frequently than ordinarily is the case.

Aside from increasing or decreasing target behaviors through vicarious processes, observation of others can increase utilization of reinforcing events. Ayllon and Azrin (1968b) suggested that to increase utilization of reinforcing events, individuals can be exposed to others who are using the events. The procedure is referred to as *reinforcer exposure* (Ayllon & Azrin, 1968b). Exposure to others engaging in reinforcing events resembles reinforcer sampling. The main difference is that reinforcer sampling involves participating in the event directly (e.g., watching part of a movie, eating part of the ice cream cone), whereas reinforcer exposure involves merely observing others participate in the event (e.g., watching others view a movie or eat an ice cream cone). Presumably, in most cases of reinforcer sampling, reinforcer exposure also is operative because some people are using the event.

Ayllon and Azrin (1968b) utilized reinforcer exposure to increase the likelihood that a psychiatric patient would seek discharge from the hospital. The patient seemed resistant to return to the community and indeed had been rehospitalized within 10 days of a previous discharge. To enhance the likelihood of discharge to a half-way house, the patient along with an attendant visited the half-way house for short periods. During the visit, she observed former patients engaging in variety of activities. On some of the visits the patient engaged in the activities there such as eating with the other residents (i.e., reinforcer sampling). Eventually, after increased observation and participation of activities, the patient requested to stay at the half-way house. After being discharged, she remained there for over one year.

One of the best examples of exposure and modeling influences on performance in a token economy was obtained in studies of reinforcer sampling (Curran *et al.*, 1973; McInnis *et al.*, 1974). In this token economy, patients who had unpaid fines could not participate in reinforcing activities until the fines were paid. This procedure, of course, led to relatively low utilization of the reinforcers. For all patients, a reinforcer sampling procedure was used to increase the likelihood that patients who were eligible (i.e., had no fines) would utilize the events. The sampling procedure did not affect performance for the people who were eligible to utilize the events. When the fines no longer restricted reinforcer utilization, the sampling procedure was reinstituted. This increased utilization of reinforcing events by individuals who previously were not eligible. Interestingly, there was an increase in the utilization of reinforcers by patients who had been eligible to use the events all of the time. This suggests that modeling effects were operative. Clients may have been influenced in their utilization of the reinforcers by observing others utilize these events, i.e., reinforcer exposure (cf. Doty, McInnis, & Paul, 1974; McInnis *et al.*, 1974).

There are several ways in which vicarious effects can be incorporated into a token economy including the conspicuous delivery of reinforcing or punishing consequences in the presence of individuals who have not responded well to particular contingencies or by exposure of individuals to others who are partaking of the reinforcing consequences. The precise control that these procedures exert over performance, particularly for individuals who initially have not responded to direct contingencies, remains relatively unexplored as an area of research.

Individualized Contingencies

In most token economies, the behaviors that earn reinforcers and the reinforcers for which tokens can be spent are the same for all of the clients. Occasionally, individuals may be exposed to a few additional contingencies that focus on idiosyncratic behaviors. However, the main contingencies are standardized for most of the clients. In these cases, the lack of responsiveness to the contingencies frequently can be altered by merely adding individualized contingencies.

For example, in one program a mentally retarded resident had failed to respond to a token reinforcement contingency designed to develop attendance to occupational therapy (Winkler, 1971a). The resident, who had been hospitalized for 22 years, spent much of her day lying around the ward instead of going to therapy. Because the usual contingency did not increase attendance, an avoidance contingency was added to the program. The resident was told that until she went to occupational therapy, she would have to pay double the usual prices whenever she spent her tokens. As soon as this schedule was introduced, she began to attend occupational therapy regularly.

Similarly, in a token program in an elementary school classroom, children received tokens for completing daily academic assignments, remaining in their seats, and other appropriate classroom behaviors and were fined for not working and for disruptive behaviors (McLaughlin & Malaby, 1974c). One girl had not responded adequately to the contingencies to which the others were exposed so an individualized contingency was added. Specifically, the girl was told that she had to complete all of her work in order to earn the privileges that were available in class on the next day. She was removed from the class and placed on this individualized contingency. In a reversal design, the individualized contingency was shown to consistently improve performance in language and spelling.

In light of the research currently available, it appears that individualized contingencies are effective in increasing responsiveness to the program. Thus, when individuals are not initially responsive, the contingencies can be altered so that a special individualized contingency is implemented. The standardized contingency to which all individuals in a program are exposed might serve as a screening device to determine the individuals for whom individualized attention is required.

Group Administration of Consequences

In the majority of token economies, clients receive consequences on the basis of their own performance. The consequences may be administered on an individual basis whether an individualized, standardized, or group contingency is used. In some programs, the group administration of consequences has been used to increase the responsiveness of a given client.

Earning tokens for one's peers is frequently used to enhance performance in a token economy. Tokens are earned by other clients in the setting on the basis of the performance of a particular client. For example, in a token economy for a ward of chronic female psychotic patients, a 50-year-old patient diagnosed as schizophrenic and brain damaged had not responded well to the program, as evidenced by low rates of responding, earning just enough tokens to survive, engaging in little social interaction, inappropriate verbal behavior, irregularly completing work, and unkempt appearance (Feingold & Migler, 1972). To enhance responsiveness, an arrangement was made whereby two other patients on the ward received the number of tokens earned by the target patient. The contingency was somewhat stringent for the two other patients because the only way they could earn tokens was through the target patient's behavior. (To maintain high levels of performance of the two patients, they had to earn their previous sum of tokens to be able to spend the tokens earned by the target patient.) In an ABA design, the target patient's job performance and self-care behaviors as well as the total number of tokens earned were shown to increase as a function of the special contingency.

Interestingly, the social interaction of the patient with one of the peers increased and a friendship appeared to develop during the contingency.

A similar contingency was used to increase the responsiveness of an elementary school student who was frequently out of her seat (Wolf *et al.*, 1970). The girl was given tokens at the beginning of class and lost them if she was out of her seat when a timer sounded (VI: 10 schedule). Points remaining at the end of the day could be exchanged for candy, snacks, clothes, and other events. Although this procedure led to some change, she was still frequently out of her seat. To enhance performance, a group contingency was invoked in which all points remaining at the end of the day would be divided among herself and four peers. Earning points for her peers improved her behavior drastically.

UNRESPONSIVENESS: SUMMARY

In light of the above procedures and the research bearing upon their efficacy, it appears that the initial unresponsiveness of a client to token reinforcement contingencies can be altered with some modification of the program. At present at least, it is unclear whether a given client or type of client is unresponsive *per se*. The main issue raised when a client fails to respond is what alternative contingencies or additional techniques can be applied to effect change. The lack of responsiveness is a stimulus for further programming rather than a sign that the client is unsuited to the manipulation of contingencies in general. Of course, there might well be clients who, after exposure to several alternative procedures, fail to respond. However, the limiting conditions of token reinforcement contingencies and their many variations remain to be determined.

ECONOMIC VARIABLES IN TOKEN ECONOMIES

The initial token economies explained client behavior completely in terms of operant principles. Client behavior change was attributed to the contingent delivery of generalized conditioned reinforcers and the lack of change to the failure to identify effective reinforcers or to implement the contingencies in accord with well-established parameters of reinforcement (e.g., considerations based on delay, magnitude, quality, and schedules of reinforcement).

Advances in token economy research have established that token programs can be viewed from the standpoint of economic systems. The similarity between token economies and national economies can be more readily seen by applying the terminology of economics to concepts usually discussed in operant terms. In a token economy, target or token-earning behaviors represent work *output*. The delivery of tokens for behavior can be viewed as *income* or *wages*. Also, the amount of tokens spent on back-up events constitutes *expenditures*. The amount of tokens accumulated by a client can be viewed as *savings*, already a notion readily referred to in token economies.

Winkler and his colleagues have been instrumental in showing the relevance of economic laws and concepts to token economies (Battalio, Kagel, Winkler, Fisher, Basmann, & Krasner, 1973, 1974; Kagel & Winkler, 1972; Winkler, 1971a, 1971b, 1972, 1973a, 1973b; Winkler, Kagel, Battalio, Fisher, Basmann, & Krasner, 1973; Winkler & Krasner, 1971). Two interrelated lines of research have stemmed from the rapproachement of economics and operant psychology. First, the token economy provides a situation where many predictions from economic theory can be tested directly. The token economy provides a unique situation where many features not manipulated in national economies (e.g., price changes, wages, savings) can be readily altered. Second, various economic features of a token program have direct implications for client behavior and the therapeutic effects achieved by the program. Interrelationships between client income, expenditures, and savings, all contribute to work output or token-earning behaviors. The present discussion provides an overview of the application of economic concepts and research that appear to be particularly relevant in developing an effective token program and have applied significance.

INCOME AND SAVINGS

An important determiner of performance in a token economy is the level of savings. Savings in a token economy usually is simply the discrepancy between income (the amount of tokens earned) and the amount spent (or fined). At some point, it might be expected that as savings accrue, behaviors which earn tokens will decline. Winkler (1972) has hypothesized that there is a *critical range of savings* which refers to the level of accumulated tokens beyond which token-earning will decline. Tokens will be accumulated until this level is reached. While savings is below the level, token-earning behaviors are performed at a relatively high rate.

The critical savings level for a given client is likely to be a complex function of a variety of variables such as the range and number of back-up reinforcers available, the ways in which tokens can be earned, the attractiveness of the back-up reinforcers, and other features. In a given token program, the critical level can be determined by examining the level of savings at which performance of token-earning behaviors declines and removing or negating the tokens saved to see whether performance increases until that level of savings is achieved again (Winkler, 1972).

From a practical standpoint, it is not essential to determine the critical level of savings for individual patients to profit from the relationship between savings and performance. The applied significance of this relationship was demonstrated by Winkler (1973a) in a token economy with psychiatric patients. Patients could earn tokens for performing a variety of self-care behaviors (e.g., getting up, dressing, bedmaking, having a neat appearance) and spend tokens for back-up reinforcers (e.g., two meals, tea, watching television, items at a ward store). To examine the role of savings, Winkler substituted a new token in the system of an ongoing token economy. The new token was earned for the target behaviors and was the only medium of exchange allowed to purchase back-up reinforcers. This procedure effectively eliminated all savings the patients had previously accumulated. In one experiment, the price of back-up reinforcers was decreased and in another the price remained the same, when the new tokens were introduced. Independently of the price manipulations, when savings were eliminated, token-earning behaviors increased (Winkler, 1973a, Exp. 1 & 2). Generally, token-earning behaviors increased when savings were low and decreased when savings again became high. As token earnings were accumulated and surpassed levels of savings previously maintained by the patients, token-earning behaviors decreased. That is, performance deteriorated as a function of surplus savings.

One implication from the research on savings is that there should be some means to delimit accumulations of tokens so that performance gains continue. It is difficult to specify the level of savings that is associated with stable performance of token-earning behaviors. The level may vary across individuals so that a given level may be sufficient to impede performance of some individuals but not have any effect on others. For example, individuals who hoard tokens might require an exorbitant number of tokens to reach a critical level of savings, if this level could be reached at all.

An important factor that influences the effect of savings on performance is the utility of the savings. If there are many goods to purchase with tokens and the price of some of them is relatively high, the level of savings at which performance will be impeded will be higher than if there are few goods with relatively lower prices. Also, if consequences are provided for saving such as interest rates or direct reinforcement for depositing as in some token programs (Phillips et al., 1971; Liberman et al., 1975), the level of savings may have to be higher to directly influence token-earning behaviors.

The level of client savings is important because it influences the effects of other variables that might be manipulated to enhance client performance. An important variable sometimes altered to enhance performance in a token economy is the magnitude of reinforcement (i.e., the number of tokens) (e.g., Bassett et al., 1975; Bigelow & Liebson, 1972; Wolf et al., 1968). Generally, increases in the number of tokens for a response increases performance of that response. However, the effect of an increase in magnitude is a function of the level of savings. If accumulated savings are high, an increase in magnitude may have little or no effect on performance. If savings are low or have been maintained at a stable level, increases in magnitude can markedly affect performance (Winkler, 1973a).

EXPENDITURES AND BACK-UP EVENTS

Related to savings is the level of expenditures or consumption in an economic system. Expenditures vary as a function of different incomes. The relationship that describes how much individuals or groups spend as a function of different incomes is referred to as a *consumption schedule*. The general relationship is that individuals with greater income spend more. Generally, the amount of spending falls within 10% of the amount of income. However, this percentage does not hold for individuals with extreme incomes. Individuals with extremely low incomes spend a greater percentage and individuals with high incomes spend a much lower percentage.[2]

Consumption schedule is an important notion in predicting behavior in token economies, particularly in relation to savings. An individual's income and level of consumption (expenditures in a token economy) determine the amount of savings. Performance of token-earning behaviors may decrease, as noted earlier, if savings are relatively high and surpass a critical level for a given client. A decrease in performance of token-earning behaviors can be averted by reducing savings. An alternative to reducing savings directly is to alter the consumption level of the client so that more tokens are spent (Winkler, 1972).

One procedure to increase consumption is to expand the range and attractiveness of items that can be purchased. With more tokens being spent, the same output (token-earning behaviors) produces a lower level of excess income (savings). Thus, performance increases to keep the level of savings at a critical level and to maintain a higher level of consumption. A second procedure to increase consumption is to "force" spending. Tokens can be dated (marked with the date of administration) and allowed to be negotiable for only a given time period (e.g., one month) (cf. Atthowe & Krasner, 1968). If tokens are not spent within the time period, they lose their value. With forced spending, consumption of back-up reinforcers increases. (A potential disadvantage of forced spending is that it eliminates savings altogether and the advantages that savings might provide such as allowing individuals to earn toward delayed reinforcers.) A third procedure is to alter the price of items periodically to stimulate spending. The efficacy of this procedure may depend on the type of items that are sold, as discussed below. Prices determine the amount of income that is expended and, thus, have implications for consumption and savings.

The effect of prices on back-up events is in part a function of the types of back-up events that are employed. An important economic notion which helps classify back-up events is the *elasticity of a demand curve* which refers to a way of measuring the responsiveness of consumer demand for an event based upon price changes (Winkler, 1971b). Elasticity of a demand curve is defined as the percentage change in demand resulting from a 1% change in price (Winkler, 1971b). For some items, a price drop leads to a relatively larger percentage of demand than the actual drop. These items have *elastic* demands. The percentage of sales of an item with an elastic demand increases to a greater extent than the percentage reduction in price. For other items, a decrease in price of a given percentage will result in an increase in demand that is smaller than the actual percentage price drop. These items have *inelastic* demands. Generally, items that are "necessities" in an economy and for which there are no close substitutes are inelastic whereas luxury items usually are elastic.

---

[2] At a given point in time, expenditures can exceed income because some savings may exist. If expenditures consistently exceed income, then of course, expenditures must eventually fall below income.

The importance of separating elastic from inelastic items is that price changes in a token economy may have different effects on behavior depending upon the elasticity of the demand curve of the items for which the prices are altered. Winkler (1971b) examined the elasticity of cigarettes which was a back-up reinforcer for tokens in a token economy for psychiatric patients. The price of cigarettes was lowered by 66.67% (from three to one token) to determine the effect of price changes on consumption. Sales increased 228.5% after the price change. This suggests that the item showed an elastic demand curve because a given percentage drop resulted in a greater percentage of consumer demand.[3] The elasticity of another reinforcer, tea breaks in the morning, afternoon, and evening, was examined by increasing the price of tea 100% (from one to two tokens). This percentage increase was larger than the reduction in consumer demand. This indicated that tea had an inelastic demand curve (which may be a function, in part, of the established value of tea breaks in Australia where this program was investigated).

Elasticity of a demand curve is not an academic issue. Knowledge of elasticity can be important if the manipulation of prices of back-up events is used to stimulate spending. The elasticity of the items for which price increases or decreases are made can determine the effect on expenditures. Increasing prices of items with inelastic demand (i.e., necessities) is more likely to stimulate an increase in spending. In contrast, increasing the prices of items with elastic demands (i.e., luxuries) may decrease spending. With nonessential items, individuals may no longer continue to purchase the item at the same level when there is a price increase and thus decrease their expenditures (cf. Hayden, Osborne, Hall, & Hall, 1974). Also, the individuals may merely shift from spending tokens on one luxury item to another one that is a close substitute but less expensive. Because it is important to keep spending high, insofar as this attenuates high levels of savings, increasing the prices of items with inelastic rather than elastic demand curves is likely to be more effective.

The type of expenditures made in an economic system may change as a function of income so that it is essential to provide both items with elastic and items with inelastic demand curves. Generally, as income increases, the proportion of income spent on necessities decreases whereas the proportion spent on luxury and semiluxury items increases. This relationship, described by Ernst Engel in the nineteenth century, has been found in analyses of national economies and has been extended to token economies. In a token economy, one would expect that as token earnings increase, the proportion spent on nonessential items (i.e., luxuries) would increase.

Winkler (1971b) examined spending of high- and low-income patients in a token economy, mentioned earlier. He studied token expenditures on meals (assumed to function as a necessity) and canteen items (assumed to function as luxuries). As predicted from the relationship described by Engel, Winkler found

---

[3] The formula for computing elasticity of demand actually entails more than a comparison of simple percentages between price change and consumer demand. See Winkler (1971b).

that at higher incomes, a greater proportion of expenditures went to canteen items and a lower proportion of expenditures went to meals than at lower incomes.

ECONOMIC VARIABLES: GENERAL COMMENTS

Economic variables can directly influence the performance of token-earning behaviors and, thus, bear upon treatment outcome. Of course, some considerations independent of economic variables determine whether token-earning behaviors are therapeutic. The behaviors that earn tokens have been assumed in the present section to have therapeutic relevance. Also, improvement in client behavior depends upon altering the requirements of the specific behaviors that earn tokens so that performance requirements are increased and progressively more complex behaviors are developed.

Although economic variables are now known to be extremely relevant for designing and maintaining effective token economies, research has only begun in this area. At present, the range of variables and the extent of their impact has yet to be determined. While the present discussion has mentioned select economic variables and their individual effects on behavior, token-earning behavior is likely to be a complex function of several economic variables operating simultaneously (Fethke, 1972, 1973). Nevertheless, at this early stage, general recommendations can be made about maintaining effective token programs. Many of these were mentioned in the discussion of savings, income, expenditures, and back-up events. It is useful to reiterate general statements here.

As a common denominator of many of the points raised earlier, client consumption of back-up events needs to be maintained at a relatively high level. In the usual token program, consumption levels are not altered on a systematic basis which might account for some of the attentuated performance levels. Programs need to develop means to sustain high levels of expenditure. Introducing special reinforcing events that are highly attractive and expensive and utilizing items that need to be continually rented rather than purchased once in an all-or-none fashion are likely to stimulate expenditures. The use of periodic auctions of special events or privileges also appears to be a highly valued back-up event that stimulates spending.

To maintain high levels of consumption, price changes might be more systematically manipulated than has been the case in most programs. Frequently, a program undergoes inflation in an attempt to maintain token-earning behaviors. As wealth increases, prices for back-up reinforcers are increased. As evident from the previous discussion of expenditures and back-up reinforcers, increasing prices requires understanding the different types of back-up reinforcers and the effect of price changes with each. Correcting high earnings that might impede performance requires more than merely increasing prices of back-up events.

Finally, some means needs to be provided in a token economy to protect against high levels of savings. While there is no need to eliminate savings from a token program, particularly since developing saving behavior as a target response may be desirable, wealth needs to be minimized. Savings can be minimized in a number of ways including increasing consumption level, devaluating tokens periodically, and providing a limit to the life of tokens once they are earned.

It is important to stress that a concern with economic variables and their alteration are not academic. Variations in economic variables have already been used in select token programs to enhance performance. For example, Winkler (1971a) varied the prices of back-up reinforcers contingent upon the behavior of psychiatric patients. To increase performance, contingencies were devised so that patients would have to pay double the usual price of back-up events if specific target behaviors were not performed. As noted earlier, one patient was told that if she missed occupational therapy she would have to pay double for the prices of back-up reinforcers. This contingency dramatically accelerated attendance. Also, a reduction in the price of back-up reinforcers was used as a positive reinforcer for another patient if she attended occupational therapy. Similarly, using a reversal design, Hersen *et al.* (1972) demonstrated that psychiatric patients consistently earned a greater number of tokens in separate phases when there was inflation (i.e., doubling of the price of back-up reinforcers). The above results suggest that manipulation of prices, and by implication wages, savings, and consumption all can be used to influence behavior for therapeutic ends.

On a practical level, specific limitations of the program may attenuate the influence of the economics of a token economy. First, the relationship between performance and economic variables depends in part upon the extent to which tokens are the sole means of obtaining back-up reinforcers. In many programs, for example, in psychiatric hospitals, patients have money to which they are entitled that is independent of their token earnings (Turton & Gathercole, 1972). Events that cannot be purchased with tokens might be purchased directly with money thereby circumventing the demands for increased token-earning behaviors (Hayden *et al.,* 1974).

Second, the relationship between performance and economic variables will not be maintained if tokens or back-up events can be obtained illegitimately. Occasionally, clients arrange their own exchange systems with staff and peers to obtain tokens or back-up events illicitly within the system (Hall & Baker, 1973; Liberman, 1968). In any case, it is important to acknowledge that the influence of economic variables in a token program depends upon controlling the ways in which tokens and back-up events can be obtained.

CONCLUSION

A major task in token economies is to ensure that individuals who are responsible for the program administer the contingencies consistently and correctly. The staff who administer the contingencies determine the effectiveness of the program. Hence, staff training plays an important part in the effects of a program. Several methods of training staff to administer reinforcement contingencies have been evaluated. Providing some form of instruction seems to be the most frequently used method of training. Regrettably, the evidence strongly suggests that instructional procedures are not very effective in altering staff behavior in relation to the client. Other methods such as providing feedback for performance, having the staff member self-monitor performance, and providing modeling and role-playing experiences have been more successful than instructional methods of training. However, providing social or token reinforcement for staff behavior have been the most effective methods of ensuring changes in staff behavior. Money or trading stamps contingent upon staff behavior have proven extremely potent as behavior-change techniques.

An interesting feature of token economies is that they sometimes are associated with changes in the staff who administer the contingencies. For example, with psychiatric patients, research has shown that staff participating in a token economy tend to view the patients more favorably and tend to provide more positive attention than staff participating on a ward conducted in a traditional manner. In other settings such as the classroom, token economies have been shown to alter the manner in which the teacher responds to the student. Overall, evidence suggests that the token economy may be useful because of the changes it effects in the behavior of the staff as well as the clients.

Token economies across a large number of populations have reported some small number of individuals who fail to respond to the contingencies. Some authors have suggested that individuals who do not respond initially to the contingencies may not be suitable for a treatment program based upon reinforcement techniques. Yet, research demonstrates that individuals who do not initially respond to the program may readily respond when the contingency is altered. Techniques are available to alter the contingency and to increase the likelihood that the individual performs the target behavior and encounters the reinforcing consequences. Response priming, reinforcer sampling, selecting increasingly potent reinforcers, modeling influences, individualized contingencies, and group administration of consequences provide methods of increasing the efficacy of treatment.

Viewing the token economy from its economic characteristics has revealed the importance of the interrelationships between performance of the target behaviors (work output), token earnings (income), spending (expenditures), and savings. Client performance of behaviors that earn tokens tends to decline as savings increase and, indeed, may reach a plateau when a critical level of savings has been accumulated. The extent to which high savings are accumulated has to be controlled to ensure that performance does not reach a plateau.

To sustain high levels of performance, the token economy must ensure that the tokens earned are spent and that excess wealth is not achieved by the clients. Spending can be stimulated by varying the range of back-up events that are available, by increasing the prices of some back-up events, by providing auctions, and similar techniques. Essentially, these techniques need to be considered in sustaining the effectiveness of the program. Occasionally, variation of specific economic features of the contingency has been used to increase the performance of clients who have not responded. For example, price changes in the back-up events can stimulate performance. Thus, manipulation of economic variables constitutes another means of increasing responsiveness to the contingencies. Although economic variables in a token economy have only received attention relatively recently, it is already apparent that these variables determine the effect of the program on performance.

Staff training, increasing client responsiveness to the contingencies, and the manipulation of economic variables refer to important considerations in changing client behavior in a token economy. Other extremely important issues are the extent to which treatment effects are maintained after the program is withdrawn and transfer to nontreatment settings after the client leaves the treatment setting. The next chapter is devoted to these issues and discusses the generalization of behavior changes in a token economy.

GENERALIZATION OF
BEHAVIOR CHANGE

7

There is little question that behavioral changes have been effected across diverse populations and settings as a result of token reinforcement. Establishing an effective method of behavior change leaves unanswered a major question, namely, to what extent are the changes maintained once the token economy is terminated and the client leaves the setting in which the program was in effect? This is a major question and applies to any area where behavior change is a goal. For example, in education, the question is to what extent do the behaviors developed in school (e.g., reading, composing, logical and creative thinking) remain after the contingencies of school (e.g., grades) are withdrawn and an individual is placed in nonschool settings. In psychotherapy, rehabilitation, and treatment in general the question of maintenance of behavior and extension of behavior to extratreatment settings is obviously important.

The issue raised by the above question is referred to as stimulus generalization or the extent to which changes in behavior extend across stimulus conditions. There are two types of stimulus generalization.[1] One type of stimulus generalization, referred to as *response maintenance* or *resistance to extinction,* is the maintenance of behavior in a given setting after the intervention (e.g., token reinforcement) has been terminated. Maintenance of behavior can be viewed as generalization across stimulus conditions because behavior change in one situation (e.g., a ward in which

[1] As mentioned in Chapter 1, the term "stimulus generalization" technically may be less appropriate than the alternative terms used here such as response maintenance or resistance to extinction and transfer of training. The notion of generalization includes not only a description of a spread of effects across stimulus conditions but also an implicit explanation of the process by which this occurs. Stimulus generalization refers to a change in behavior across stimulus conditions which bear resemblance to the stimulus conditions in which training was conducted. Behavior generalizes because of the similarity of the new stimuli to the original stimulus associated with training. And, the extent of generalization is a function of the degree of similarity (cf. Kimble, 1961). In behavior modification programs, the extent to which generalization of treatment effects is a function of similarity of the stimulus conditions is not usually evaluated. It is unclear that transfer occurs as a function of generalization across stimulus dimensions. Thus, it is more appropriate to use notions as resistance to extinction and transfer which merely denote that behavior change occurs across situations rather than suggest the reason for this effect. Incidentally, the same logic applies to the notion of response generalization which implies generalized changes as a function of similarity of the untrained responses to the response which has been trained. A descriptive term such as concomitant or current behavior change might be preferred in place of response generalization (Kazdin, 1973f).

a token program is in effect) may generalize to another situation (e.g., the same ward after the program is withdrawn). Maintenance of behavior perhaps is the most important issue after behavior change has been achieved. The goal of most programs is to ensure that changes are sustained after treatment has been eliminated.

The second type of stimulus generalization, referred to as *transfer of training,* is a transfer of behavior from the situation in which the program is conducted to another situation in which no program has been in effect. Transfer of training is of obvious import given that a concern of most treatment interventions is that behavior change achieved in one setting (e.g., hospital or classroom) carry over to other settings (e.g., the community and at home).

Although response maintenance and transfer of training can be distinguished, very often they go together. For example, in most token programs in institutional settings there is a two-fold goal, namely to ensure that behavior change is maintained when the individual is removed from the program (maintenance) and placed in a nontreatment setting (transfer).

Only recently has research in applied behavior analysis begun to systematically investigate response maintenance and transfer. The experimental designs frequently used in applied operant research in some way may have impeded the search for maintenance and transfer techniques (Hartmann & Atkinson, 1973; Kazdin & Kopel, 1975.[2] For example, the familiar **ABAB** design requires demonstrating transient effects of the intervention to show a functional relation between treatment and behavior change. Ideally, repeated implementation of the intervention is associated with behavior change while withdrawal of the intervention is associated with a return of behavior to baseline or near baseline levels. If behavior does not return to or approach baseline when the intervention is withdrawn, there is no clear demonstration of a functional relationship. Hence, the return of behavior to baseline during a reversal **(A)** phase is desired from the standpoint of the experimental evaluation (to show an effect of the intervention), although it is not desired from the standpoint of clinical effectiveness (where a permanent change is desired). The competing interest in clear experimental results and maintenance of behavior perhaps has led to sacrificing one for the other. For example, researchers sometimes recommend implementing intervention phases for only brief periods so that behavior is not likely to be maintained when the intervention is withdrawn (Bijou, Peterson, Harris, Allen, & Johnston, 1969). This recommendation, of course, emphasizes the importance of the return of behavior to baseline conditions.

The multiple-baseline designs, the next most commonly used evaluation technique in applied operant research, may present a problem in evaluating transfer of behavior. In these designs, the effect of an intervention is evaluated by showing that behavior change either across several different behaviors, individuals, or situations occurs only when the intervention is applied. The intervention is implemented at different points in time across the baselines to show specific changes in behavior.

---

[2] For a detailed description of experimental designs in applied behavior analysis, the reader may wish to consult any of several references (e.g., Baer *et al.,* 1968; Hersen & Barlow, 1976; Kazdin, 1977b).

In the multiple-baseline design across settings, implementation of the intervention in one setting might be associated with change in another setting. If there is transfer, this introduces ambiguity in the experimental evaluation. Obtaining a generalized treatment effect across settings interferes with a clear demonstration that the intervention was responsible for behavior change.

At first glance, the ABAB and multiple-baseline designs appear to be incompatible with the study of maintenance and transfer. However, this is not necessarily the case. For example, in an ABAB design, procedures to develop response maintenance can be implemented and evaluated after a functional relation has been demonstrated between the intervention and behavior (Kallman, Hersen, & O'Toole, 1975; Kazdin & Polster, 1973). Similarly, the multiple-baseline design across situations may be used to determine precisely what techniques promote transfer of behavior across situations and stimulus conditions (situations, therapists, or settings) (Stokes et al., 1974). Although the ABAB and multiple-baseline designs are not necessarily incompatible with investigating maintenance and transfer, there is some agreement that the designs do not usually direct an investigator's attention or priorities to these areas (Stokes & Baer, 1977).

MAINTENANCE OF BEHAVIOR AFTER WITHDRAWAL OF TOKENS

Although the majority of programs show that cessation of token delivery is associated with a return of behavior to baseline or near baseline levels, this is not always the case. In a number of programs, behavior change has been maintained after reinforcers are no longer delivered. For example, Surratt, Ulrich, and Hawkins (1969) provided token reinforcement for completion of academic assignments with elementary school students. After tokens were no longer provided, three or four students performed better 6 weeks after the program than during the initial baseline. Similarly, Thomson, Fraser, and McDougall (1974) developed speech in two withdrawn chronic schizophrenics. The delivery of praise and food contingent upon intelligible words and eventually complete sentences dramatically increased verbal behavior. A year later, data indicated that verbal behavior had maintained a level close to treatment even though the contingency had been completely withdrawn. In addition, Carroccio, Latham, and Carroccio (1976) reduced the frequency of a stereotyped head/face touching response of a schizophrenic patient by providing tokens for low rates of responding. The behavior remained at a low level 15 months after all contingencies had been withdrawn.

A large number of other programs in diverse treatment, rehabilitation, and educational settings have indicated that behaviors are maintained after the contingencies are withdrawn (Ayllon, 1963, 1965; Greenwood et al., 1974; Hall & Broden, 1967; Hamilton & Allen, 1967; Hewett, Taylor, & Artuso, 1969; Kazdin, 1971a; Keith & Lange, 1974; O'Leary, Drabman, & Kass, 1973; Rosenbaum et al., 1975; Whitman et al., 1970; Zimmerman, Overpeck, Eisenberg, & Garlick, 1969). Despite the above investigations, behaviors usually extinguish when a program is withdrawn.

When responses are maintained after consequences have been withdrawn, the reason usually is unclear. Various explanations of unplanned response maintenance can be offered. First, it is possible that behaviors developed through a reinforcement program may come under control of other reinforcers in the setting (Baer & Wolf, 1970; Baer *et al.,* 1968; Bijou *et al.,* 1969). Events associated with the delivery of a reinforcer acquire reinforcement value (Medland & Stachnik, 1972). Even though the programmed reinforcers are withdrawn, behavior may be maintained by events that have acquired reinforcing properties. For example, behaviors may be maintained in a classroom in which token reinforcers were previously used because the teacher has been consistently associated with token reinforcement. The teacher may be a more powerful reinforcer after the program and maintain performance of the students without using other reinforcers such as tokens (Chadwick & Day, 1971).

Second, and related to the above reason, after withdrawing extrinsic reinforcers, consequences that result directly from performing the target behavior may maintain that behavior. Many behaviors result in their own reinforcing consequences. For example, social skills may be maintained because once they are developed they are associated with responses from others.

In one report, a preschool child received praise and teacher attention for using outdoor play equipment in an attempt to develop her motor skills and social contact with other children (Buell, Stoddard, Harris, & Baer, 1968). Social interaction (e.g., talking with and touching other children) increased even though it was not directly reinforced. When social reinforcement was withdrawn for playing with equipment, this behavior decreased in frequency. However, the child continued to interact socially with others. Possibly, social contact of the child was maintained by the reinforcing aspects of interacting (Baer, 1968; Baer & Wolf, 1970).

A final explanation of unplanned response maintenance is that even though the program is terminated and reinforcers are withdrawn, the agents administering the program (parents, teachers, staff) have changed in their behavior in some permanent fashion. The agents may continue to use the principles of behavior modification even though a specific program has been withdrawn (Greenwood *et al.,* 1974). For example, if a contingency contract system in the home is withdrawn, a child's behavior may still be maintained at a high level. Although the contract is terminated, the parents may provide reinforcement (allowance, praise) and punishment (loss of privileges) more systematically than they had prior to the contract system. As noted in the previous chapter, there is little evidence showing that the behavior of those who administer a behavior modification program is permanently altered. As soon as a program is withdrawn, the agents who administer the program frequently revert to behaviors previously used to control client behavior.

Each of the above explanations of response maintenance is usually offered *post hoc.* After behavior fails to reverse when the program is withdrawn, an investigator may speculate about why this occurred. Any of the explanations may be correct in a given instance. Yet, maintenance of behavior is more clearly understood when it is *predicted* in advance on the basis of explicit procedures that are used to develop resistance to extinction rather than when extinction does not occur and has to be explained.

In spite of the above examples and explanations of response maintenance, removal of the contingencies usually results in a decline of performance to baseline or near baseline levels. A reversal of behavior to baseline levels has been shown after the program is withdrawn across a wide range of settings and clients (Baer & Stokes, 1976; Davison, 1969; Kazdin, 1975a; Kazdin & Bootzin, 1972; Kazdin & Craighead, 1973; Marholin, Siegel, & Phillips, 1976; O'Leary & Drabman, 1971; Stokes & Baer, 1977). If response maintenance is a goal of the program, it has to be systematically programmed into the setting rather than assumed to be an automatic consequence of the program (Baer *et al.*, 1968).

TRANSFER OF TRAINING ACROSS SITUATIONS

In most token economies, altering behavior in one situation does not result in a transfer of those changes to other situations either while the program is in effect or after it has been withdrawn (Kazdin & Bootzin, 1972). Indeed, the range of stimulus conditions controlling behavior often is quite narrow. Typically, behavior changes are restricted to the specific setting in which training has taken place and to the presence of those who administer the program (Lovaas & Simmons, 1969; Redd, 1969; Redd & Birnbrauer, 1969). However, examples of transfer of behavior across situations have been reported.

For example, in a psychiatric hospital, two patients received tokens for talking with each other during daily sessions conducted in a special treatment room (Bennett & Maley, 1973). Dramatic changes were made in talking during the sessions. In addition, social interaction on the ward increased even though it was not reinforced. When reinforcement was withdrawn in the individual sessions, performance on the ward was maintained.

In the classroom, token programs altering behavior during one part of the day such as the morning or afternoon period sometimes demonstrate behavior change at other periods of the day even though performance during these other periods is not reinforced (Kazdin, 1973g; Walker, Mattson, & Buckley, 1971). Additionally, programs for pre-delinquents have shown that behavior changes achieved in one situation in the treatment setting transfer to other situations in the setting or to other settings such as one's home (Bailey *et al.*, 1971; Kifer, Lewis, Green, & Phillips, 1974). Despite a few encouraging findings of transfer to training to settings in which the program has not been conducted, they are in the minority. As with response maintenance, transfer of training usually is not found (e.g., Broden *et al.*, 1970; Isaacs *et al.*, 1960; Kuypers *et al.*, 1968; Meichenbaum *et al.*, 1968; O'Leary *et al.*, 1969).

TECHNIQUES TO PROGRAM MAINTENANCE AND TRANSFER

The principles of operant conditioning would suggest that behavior would not be maintained after token reinforcement is terminated or would not extend to new situations where the contingencies are not in effect. If behavior is a function of its consequences, performance should adjust to the contingencies which operate in a given situation. Individuals would be expected to discriminate in their performance across situations as a function of different contingencies. If token reinforcement is withdrawn or is not available in a situation, performance of a target response should decline. This suggests that if behaviors are to be maintained or to transfer across situations, specific contingencies have to be in operation continuously and extended into the transfer situations. In some situations, discussed below, changes *are* made in the "natural environment" such as the home to support continued change. This is not always possible. Fortunately, recent research suggests that the contingencies need not be continued indefinitely to maintain behavior and need not be extended to each situation in which transfer is desired. Various techniques are available that are likely to postpone or eliminate the loss of behavioral gains made during a reinforcement program and to ensure that changes transfer to situations in which the contingencies have not been in effect (Kazdin, 1975a; Marholin *et al.*, 1976; O'Leary & Drabman, 1971; Stokes & Baer, 1977; Wildman & Wildman, 1975).

Selecting Behaviors That Are Likely to Be Maintained

One strategy that has been suggested to achieve maintenance of behavior is to select target behaviors that are likely to be maintained by the natural consequences in the environment (Ayllon & Azrin, 1968b; Baer & Wolf, 1970; Burchard, 1967; Stokes & Baer, 1977). Baer and Wolf (1970) posed the notion of a "behavioral trap" to refer to environmental support systems (e.g., peer praise or attention) which are likely to result from altering specific child behaviors. Once a behavior is developed, perhaps through contrived reinforcers, it may enter into a "trap" which exerts inexorable control over its subsequent performance. For example, extraneous consequences may be required to develop social interaction in a severely withdrawn child. Once social behavior is developed, the child may be trapped in a social matrix which supports that interaction. Attention from peers, prompts to engage in interaction, and similar events from the natural environment may come to control behavior so that it is performed independently of the original consequence responsible for its inception. According to Baer and Wolf (1970), the goal of a behavioral program is merely to develop a response that ensures entry into the trap. For example, Allen *et al.* (1964) provided teacher attention to a child for interacting with others or standing near them. In an ABAB design, teacher attention was provided for interaction or for solitary behavior (i.e., a DRO schedule). When attention was almost completely withdrawn at the end of the study, the child continued to interact with peers. The maintenance of behavior was hypothesized to result from the "behavioral trap" associated with social interaction (Baer & Wolf, 1970).

A similar notion to developing entry behaviors to a behavioral trap has been recommended by Ayllon and Azrin (1968b). These investigators suggested as a general guideline that "relevant" target behaviors be selected and referred to this notion as the "Relevance of Behavior Rule." The rule states, "Teach only those behaviors that will continue to be reinforced after training" (1968b, p. 49). Presumably, behaviors such as self-care skills (e.g., grooming oneself or eating appropriately), and social behaviors (e.g., conversing with others) would be relevant behaviors because performance as well as the lack of performance of these behaviors is likely to result in particular consequences even after a training program is terminated. Conversely, other behaviors such as those performed only in a treatment program (e.g., attending various activities such as music or play therapy) might develop responses that are not sustained by the supporting environment such as the community, once training is terminated.

There is some evidence that behaviors which are likely to generate their own reinforcing consequences are maintained. For example, Berkowitz, Sherry, and Davis (1971) trained 14 institutionalized profoundly retarded boys to spoon-feed themselves rather than be fed by someone else. The boys were trained through a series of steps involving prompts and fading of prompts until self-feeding was mastered. After training was terminated, the boys were returned to their normal cottage living setting where they ate family style at long tables with relatively little supervision. Yet, after 41 months, 10 of the boys had retained their self-feeding skills. The remaining four boys had maintained their skills for 23–35 months. It is likely that the self-feeding behaviors were maintained by the naturally reinforcing consequences (food) that consistently followed their performance after training.

On the one hand, the notion of selecting target behaviors that are likely to have reinforcing consequences once developed is instructive. Certainly, developing behaviors that are *not* likely to be followed by reinforcing consequences following training appears to be unwise. On the other hand, the selection of behaviors that may lead to behavioral traps or that are relevant to subsequent functioning has not proved particularly useful in most behavior modification programs. Recently, the assumption that behaviors which appear to be important in everyday functioning will be maintained by the environment and transfer to everyday situations has been brought into question. Behaviors which appear to be the kind that the environment would normally support such as appropriate social interaction skills in adults (Kazdin & Polster, 1973) and children (Buell *et al.*, 1968) or even appropriate eating skills (O'Brien, Bugle, & Azrin, 1972) are not invariably maintained once a program is terminated.

It is difficult to select behaviors which one can be assured will be maintained after the program is withdrawn. A given behavior such as paying attention to the lesson is maintained in some classrooms (e.g., Hewett *et al.*, 1969; Kazdin, 1973g) but not in others (e.g., Bushell *et al.*, 1968; Craig & Holland, 1970; Kazdin & Klock, 1973). Similarly, smiling in individuals who appear depressed may be maintained at a high level after programmed reinforcement is withdrawn (Hopkins, 1968) or it may decline (Reisinger, 1972). Thus, there is little consistency, at least at present, on the *behaviors* which will be maintained automatically by the environment.

It is unlikely that very many behaviors would be automatically maintained by reinforcement from the social environment. The everyday social environment usually does not provide consequences as systematically as does the programmed environment in which a token program is conducted. Typically, desirable behaviors go unreinforced in the natural environment. Rather, punishment is delivered for not performing desirable behaviors. For example, one rarely receives reinforcing consequences (at least immediately) for coming to work or class on time. These behaviors are expected and their nonoccurrence may result in social censure and additional undesirable consequences. A client who leaves a setting in which token reinforcement was delivered is not likely to respond in a similar fashion in the community where most forms of reinforcement are very intermittent and unsystematic. Also, deviant or disruptive behaviors are likely to receive attention or notice in everyday settings such as at home and at school and be reinforced rather than extinguished in the natural environment (Allen *et al.,* 1964; Wahler, 1969). The lack of consistent reinforcement for routine appropriate behaviors in the natural environment may account for the failure of behavior to be maintained or to transfer to everyday situations once a program is withdrawn.

Specific procedures usually have to be employed to achieve response maintenance and transfer of training. Several procedures can be employed to serve this purpose after a behavioral program has been withdrawn. The evidence for many of the procedures does not allow unequivocal statements to be made about their efficacy. Nevertheless, preliminary evidence suggests that they may be effective in ensuring response maintenance and transfer of training. Although some procedures to be discussed are more suited to either response maintenance or transfer of training, in most instances these goals and the procedures used to obtain them are difficult to separate.

Substituting One Program for Another

One way to develop maintenance and transfer of training is to substitute one program for another in a given setting or to extend the program to settings in which transfer of training is desired. For example, Walker and Buckley (1972) evaluated the effect of different maintenance strategies with elementary school students. After disruptive students were treated in a special classroom using token reinforcement to increase appropriate classroom deportment, they were returned to their regular classrooms. Different procedures were implemented in the regular classrooms to determine whether behaviors developed in the special classroom would be maintained and transfer to the regular class. Children were returned to regular classrooms using one of four procedures. One procedure was "peer reprogramming" in which the child from the special class could earn points that were used to purchase reinforcers for the group (e.g., field trip). A second procedure was "equating stimulus conditions" in which the conditions of the special classroom (e.g., a token reinforcement program, similar academic materials) also were used in the regular classroom. The third procedure was training the regular classroom teacher in behavior modification. The final procedure was a control condition in which students were returned to the regular classroom that had no special program associated with it. The results indicated that during the two-month follow-up period in the regular class only peer reprogramming and equating stimulus conditions were associated with higher levels of appropriate behavior than the control condition.

In a number of programs in the home, behaviors are maintained by substituting one behavioral program (e.g., token economy) for another which is easily implemented. For example, O'Leary et al. (1967) altered the behavior of an aggressive, hyperactive child in the home by providing candy, social, and token reinforcement for cooperative behavior. After cooperative behavior had been established, the mother was trained to continue the program on her own. Although some deviant behaviors remained, the cooperative behaviors were maintained.

Programs in institutional settings sometimes train relatives of the clients so that behavioral techniques can be continually applied after the clients' release. For example, Kallman et al. (1975) treated a patient who complained of being unable to walk. The patient, diagnosed as having a conversion reaction, was hospitalized and confined to a wheelchair. Social reinforcement was used to develop standing out of his wheelchair and walking during brief sessions conducted each day. Eventually, the patient walked, first with a walker and then by himself. These gains were maintained for a 4 week follow-up period after the patient had returned home. However, after the follow-up assessment the patient again indicated he was unable to walk and was readmitted to the hospital. Walking was shaped at a high level again and the patient was returned home. To ensure that walking was maintained at home, the patient's family was trained to reinforce walking and ignore complaints of being unable to walk. Videotaped interactions of the patient and family revealed that the family previously had been supporting the disability by ignoring attempts to walk. Follow-up after training the relatives indicated that the behavior was maintained up to 12 weeks after treatment.

The importance of training relatives in ensuring response maintenance and transfer was emphasized in a report of the treatment of autistic children (Lovaas, Koegel, Simmons, & Long, 1973). In this report, follow-up assessment of the behavior of autistic children 1 to 4 years after treatment showed that children whose parents were trained to carry out behavior modification procedures maintained their gains outside of the treatment facility and, indeed, slightly improved. In contrast, children who had been institutionalized showed a relatively great loss of those gains previously achieved in treatment.

Overall, the above investigations suggest that continuing the program can ensure maintenance and transfer of training. Some of the authors who have substituted one program for another have discussed their results in terms of "maintenance" and "transfer" (e.g., Walker & Buckley, 1972). Strictly, this type of demonstration does not address the salient issues associated with maintenance of behavior and transfer of training as usually discussed in token economy research. For example, showing that appropriate behavior can be "maintained" and "transfer" to a new classroom when a behavior modification program is implemented in that classroom does not investigate maintenance and transfer at all. Rather, the program only shows that implementing a behavioral program alters performance in the setting in which the program is implemented. The issue is the extent to which behaviors are maintained and can transfer when a complete program cannot be established in another situation. Relocating the program from one setting (e.g., a special classroom, prison, or psychiatric facility) to another (e.g., regular classroom, the community) is not an option available in most token programs. Thus, the research on developing programs in new settings does not address the issue of maintenance or transfer directly after a program is withdrawn.

While establishing programs in new settings does not by itself address the issue of maintenance of behavior change, this technique may be relevant to maintenance when all contingencies finally *are* removed. Walker *et al.* (1975, Exp. 1) addressed this directly by treating "highly deviant" children in a special education classroom based upon a token economy. For some of the children, a behavioral program was implemented in the regular classrooms to which they returned after the special-education token program. This program consisted of consulting with the teacher about the child's performance, providing feedback on child behavior, and reinforcing the teacher's behavior (with a course grade) based upon how well the student maintained his behavior. Children who returned to a classroom where a behavior modification program was in effect "maintained" the gains developed in the special classroom to a greater extent than those who returned to a class with no program in effect. More importantly, data were collected over a 4-month period after all contingencies in the regular classroom had been terminated. The subjects who had experienced the "maintenance" strategy showed a higher rate of appropriate behavior than those who had not received the maintenance strategy in their regular classroom. This study suggests that implementing programs across settings may facilitate ultimate maintenance once the specific contingencies are withdrawn. Of course, in the Walker *et al.* (1975) study maintenance may have resulted from the protracted treatment program rather than from the differential settings in which the programs were conducted.

In several programs, an attempt is made to sustain performance levels achieved during a token program with consequences that are more likely to naturally follow behavior. After behavior is developed, maintenance is achieved by eliminating tokens and bringing behavior under control of "naturally occurring" reinforcers such as approval, praise, or attention from others. Essentially, consequences normally available in the environment are used to replace tokens which have initially changed behavior. For example, depression was decreased in a female patient by providing tokens for smiling and by taking tokens away for crying (Reisinger, 1972). At the end of treatment, tokens were eliminated entirely. Social reinforcement alone was delivered for smiling whereas crying was merely ignored. This effectively maintained smiling for 4 weeks at which time the patient was released from the hospital.

If "naturally occurring" consequences can be programmed to maintain behavior, the question arises whether they could have been used to change behavior in the first place without introducing extraneous events. Although events such as praise, activities, and privileges which are readily available in the normal environment usually effectively alter behavior, a more powerful extrinsic reinforcer sometimes is needed. For example, in a report by Wahler (1968), parents praised their children for playing cooperatively in an attempt to reduce obstreporous behavior. Because praise was not effective initially, the parents provided the children with points that could be exchanged for toys. Praise was paired with and eventually replaced the tokens. Praise had become a reinforcer after being paired with points. Thus, extrinsic reinforcers sometimes are necessary to develop the reinforcing properties of naturally occurring events.

Fading the Contingencies

Behavior can be maintained by gradually removing or fading the token reinforcement contingencies before the program is completely withdrawn. The gradual removal of consequences is less likely to be discriminable to the clients than the abrupt withdrawal of the consequences. Eventually, the consequences can be eliminated entirely without a return of behavior to its preprogram rate. The idea of gradually withdrawing the contingencies is not new in token-economy research. Many token programs are structured into levels through which the clients progressively move. As discussed earlier, progression through levels of the program is associated with fewer specific contingencies that control behavior. For example, if an individual consistently engages in target behaviors for a protracted period, he may receive the reinforcing consequences in the setting without specific contingencies. Leveled token systems have been used frequently in institutional settings where the ultimate goal is to return the client to the community as in many programs for psychiatric patients, hospitalized drug addicts, or delinquent youths (e.g., Atthowe & Krasner, 1968; Kelley & Henderson, 1971; Lloyd & Abel, 1970; Melin & Gotestam, 1973; Phillips et al., 1971).

Although gradual withdrawal of token-reinforcement contingencies has been used rather extensively, few studies have evaluated the effect of this technique on long-term maintenance of behavior. Classroom token economies have examined different techniques to fade contingencies and the effect of fading on short-term maintenance. For example, Drabman et al. (1973) used token reinforcement with an "adjustment" class to decrease disruptive behaviors such as being out of one's seat, creating noise, not working, and other behaviors. Tokens (ratings by the teacher) were provided for appropriate deportment and for completion of academic tasks. The contingencies were gradually faded to transfer control over behavior from the teacher to the students.

After token reinforcement had developed appropriate behavior, the children determined their own points by rating their behavior. They received points if their ratings in a given period matched those (within one point) made by the teacher. If the ratings of a given child matched the teacher's ratings, he received the points he had given himself. If the ratings were exactly the same as those made by the teacher, he received a bonus point. In subsequent phases, the children continued to rate their behavior but the teacher only checked some of the children. Those children who were checked could receive the bonus for exact concurrence with the teacher's ratings but all children received the number of points they rated themselves as earning. Initially, several children were randomly checked (i.e., one-half of the class that had been divided into teams). Eventually, a coin toss was used to determine whether anyone would be checked. In the last phase, no one was checked. Praise was provided by the teacher for the child's accurate self-rating.

The results indicated that token reinforcement was associated with a marked reduction in disruptive behavior. Disruptive behavior remained at low levels during the fading phases in which the children only were periodically checked by the teacher. During the final phase in which no checks were made, behavior did not return to baseline levels. Because this phase was relatively brief (only 12 days), the effect of fading the program on long-term maintenance cannot be inferred.

In a related study, Turkewitz *et al.* (1975) extended the fading procedure. As in the previous study, fading of the contingencies was accomplished by gradually decreasing the percentage of students whose ratings were checked by the teacher. After all checks were faded, a separate procedure was used to fade the exchange of self-rewarded points for back-up reinforcers. Across various phases, a progressively lower percentage (50%, 33.3%, 12.5%) of randomly selected students was allowed to exchange their points for back-up events on a given day. Eventually, none of the students were allowed to exchange points for consequences. When all back-up reinforcers were eliminated, disruptive behavior remained at a substantially lower level than the original baseline. Regrettably, the final period without back-up reinforcers lasted only 5 days so the maintenance of behavior without specific contingencies could not be assessed. Interestingly, the effects of the program carried over to a brief in-class period (15 minutes) in which the program was not conducted. However, there was no transfer of behavior to regular public school classrooms as shown by similar rates of disruptive behavior of subjects who received tokens relative to those who did not.

In general, the research on fading external token reinforcement and the delivery of back-up reinforcers is promising. Although the research has yet to evaluate long-term maintenance, the consistency with which behaviors ordinarily return to baseline levels in the majority of investigations makes this research notable. Fading the contingencies is likely to prove useful because the token program is made to increasingly resemble the natural environment in which specific token reinforcement contingencies are not in effect. Gradually withdrawing the contingencies prepares a client for the conditions under which he must normally function. In everyday living, the contingencies often are unsystematic and the consequences for performance are delayed, if present at all. The gradual withdrawal of contingencies provides a transition between a highly programmed environment and one which is less well programmed with regard to given target behaviors.

Expanding Stimulus Control

One way of conceptualizing the failure of behavior to be maintained or to transfer to new settings is based upon stimulus control. Clients readily form a discrimination between conditions in which token reinforcement is and is not delivered. Behavior becomes associated with a narrow range of cues. As soon as the program is withdrawn or the setting changes, clients discriminate that the desirable behavior is no longer associated with certain consequences. Thus, responses are not maintained and do not transfer across situations.

Individuals make extremely fine discriminations in their performance and may respond differently in the presence of different adults and peers even in the same setting (Johnston & Johnston, 1972; Redd, 1969; Redd & Birnbrauer, 1969). Indeed, in some programs behavior comes under very narrow stimulus control. For example, Rincover and Koegel (1975) showed that imitative behavior developed in autistic children sometimes comes under control of ancillary stimuli associated with training such as incidental gestures on the part of the trainer or special prompts. Only when these stimuli are introduced into a new situation does behavior transfer outside of training.

Several studies have systematically attempted to expand the range of stimuli that control behavior during training as a means to achieve transfer of training. The stimuli introduced into training include specific components of the environment to which transfer is desired and various behavior-change agents. For example, Jackson and Wallace (1974) used tokens to increase vocal loudness of a female mentally retarded adolescent. Although the procedures were conducted in a laboratory situation, daily generalization checks were made in the lab during a nontraining period and in a classroom situation. In an ABAB design, token reinforcement was shown to systematically increase loudness although there was little or no change in the transfer situations. Thus, in the final phase, a procedure was used to develop transfer. To promote transfer to the classroom, environmental stimuli associated with training and classroom situations were made increasingly similar. This included having some of the girl's classmates present in the training session, reducing the privacy of the training session by opening up the booth where the sessions were conducted, scheduling classroom events in the booth for all of the students, and having the teacher temporarily lead the group of students in the class while the girl was in her training session in full view of the class. In addition, the girl's response requirements for tokens were increased so that increased loudness across more responses was maintained. The extent to which generalization training was successful can be seen in the classroom data in Fig. 22. The figure shows that vocal loudness had not transferred to this setting until the last phase when an attempt was made to expand stimulus control.

Emshoff, Redd, and Davidson (1976) used praise and points (exchangeable for money) to develop positive interpersonal comments with four delinquent adolescents. To develop generalization of behavior, two clients were trained under varied stimulus conditions across activities (e.g., during games or discussions), trainers, locations in the facility, and time of the day. Clients who received training under varying stimulus conditions rather than under constant conditions showed greater positive comments under specific tests for generalization when tokens were delivered noncontingently, when a new trainer was introduced, and when the activity, setting, and trainer all varied. Also, clients who were trained under varying stimulus

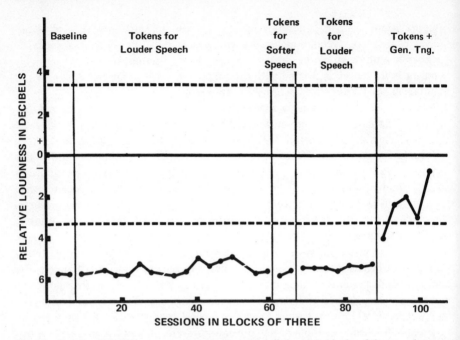

Fig. 22. Results of the "classroom generalization checks" shown in blocks of three sessions per data point. The data are shown in relation to the average loudness of the subject's peers in this setting (zero point on the ordinate). The range for these subjects is shown by the dashed lines. No reinforcement contingencies for loud speech were used until the Tokens Plus Generalization Training Phase. (From Jackson & Wallace, 1974; Society for the Experimental Analysis of Behavior.)

conditions showed greater maintenance of behavior 3 weeks after training was terminated than those trained under constant conditions.

Stokes *et al.* (1974) expanded stimulus control during training to achieve transfer of behavior across several staff in an institutional setting. Praise and food were delivered for greeting responses (handwaving) of four institutionalized severely and profoundly retarded children. Although one individual served as the trainer in experimental sessions, data were gathered across up to 14 staff members daily in the institution to determine whether the greeting response transferred to other individuals. The staff members systematically approached the target subject, remained in his presence briefly, and recorded whether a greeting response occurred. The results showed that behavior usually was under stimulus control of single individual who conducted training and did not generalize across staff members. However, after a second individual was also included in training, thus increasing the individuals in whose presence the response was performed, three of four subjects increased markedly in their greeting responses across other staff members. The generalized effects remained up to a 6-month period during which training was continued for one of the subjects.

Koegel and Rincover (1974, Exp. 1) also achieved expanded stimulus control by varying both the environmental conditions and individuals present during training. Various behaviors such as attending to a task, imitating, speaking, and recognizing words were altered in autistic children. Individuals were trained in a one-to-one situation with the goal of returning them to a classroom setting. Although the target behaviors were readily acquired, the behaviors did not extend beyond the one-to-one situation. When a child was placed with others, even with only one other child, the trained behaviors decreased markedly. To develop transfer of training to the stimulus conditions of the classroom situation, the classroom conditions were gradually introduced into training (Koegel & Rincover, 1974, Exp. 2). New behaviors were developed in a one-to-one situation while data on classroom behavior were gathered concurrently. To promote transfer, conditions which approximated the classroom were introduced such as training two individuals together, introducing the teacher and teacher aides into training and having them administer prompts and reinforcers, and forming larger groups of children while training continued. In addition, candy reinforcement was delivered on an increasingly intermittent schedule. Expanding the stimuli conditions of training and thinning the delivery of candy resulted in transfer of behavior to the classroom situation.

In some programs, an attempt has been made to establish broad stimulus control over behavior by having the client's peers administer or be involved in the contingencies (Kazdin, 1971b; Stokes & Baer, 1976). For example, Johnston and Johnston (1972, Exp. 3a, 3b) developed correct articulation in two girls in a classroom setting. The girls were trained to monitor each other's pronunciation. Correcting the other child for inappropriate speech and providing feedback for correct speech were reinforced with tokens. This procedure led to a reduction of incorrect speech sounds. Interestingly, when tokens were withdrawn for monitoring the other person's speech, the children continued to monitor behavior and correct speech responses tended to be maintained.

The stimulus control exerted by the presence of the peer who had monitored behavior was revealed by placing the children in the presence of others. When the two children were with each other, their rates of correct articulation were markedly higher than when they were placed with children who had never monitored speech. Even when monitoring was virtually eliminated, being in the presence of the peer who had previously monitored speech was associated with a reduction of articulation errors. These results suggest that stimulus control over the target behavior was maintained by the presence of peers who had previously administered the contingencies. A logical extension of these results would be to intentionally establish control by one's peers so that behavior is maintained in those situations in which the peers are present. Other investigations have shown that peer-administered contingencies sometimes lead to maintenance and transfer of behavior to situations outside of treatment (e.g., Bailey et al., 1971).

The above studies and others have demonstrated that transfer of training and in some cases response maintenance can be established by introducing various components of the situation in which transfer is desired or by increasing the number of individuals who administer the contingency (Allen, 1973; Corte, Wolf, & Locke, 1971; Garcia, 1974; Goocher & Ebner, 1968; Horner, 1971; Lovaas & Simmons,

1969; Reiss & Redd, 1970). It appears that the stimulus conditions of the transfer setting need to be introduced into the training situation while training conditions are still in effect to establish a consistent pattern of responding across situations and individuals. The stimuli which exert control over behavior become increasingly broad. Once behavior is consistently performed in a few situations or in the presence of a few individuals, the likelihood that the behavior will be performed in a variety of new situations or in the presence of a variety of individuals is increased.

Scheduling Intermittent Reinforcement

No longer presenting reinforcing consequences after a token economy usually results in an abrupt decline in performance. Unless some maintenance strategy is used, there is little resistance to extinction. As is well established in animal laboratory research, altering the schedules of reinforcement can enhance resistance to extinction. Although the manner in which events can be scheduled in the laboratory and applied settings differ, the notion of intermittent reinforcement applies well in both situations. Research in token economies and other applied operant programs has shown that intermittent reinforcement can postpone extinction, and in some cases, apparently maintain the response.

Intermittent reinforcement has been employed to maintain behavior in one of two ways. One use is to continue to use intermittent reinforcement to maintain behavior at a high level rather than to eliminate all reinforcer delivery. Intermittent reinforcement is extremely effective in maintaining behavior while reinforcement is still in effect. For example, Phillips *et al.* (1971) gave or withdrew points on the basis of the degree to which pre-delinquents cleaned their rooms. When the points were given, daily room-cleaning was maintained at a high level. After the behavior was well established the reinforcing and punishing consequences became increasingly intermittent. The rooms were checked daily but reinforcement or punishment occurred only once in a while. When the consequences were delivered, the number of points given (or lost) on the basis of that day's performance was multiplied by the number of previous days that consequences had not been given. Although the consequences were delivered on only 8% of the days checked, behavior was maintained at a high level. The consequences were not completely withdrawn so the effect of intermittent reinforcement on extinction was not evaluated.

The second way in which intermittent reinforcement has been used to maintain behavior is to gradually thin the reinforcement schedule until all reinforcement is eliminated. For example, Kazdin and Polster (1973) provided token reinforcement to two withdrawn adult male retardates for conversing with their peers in a sheltered workshop setting. During three daily work breaks, clients received tokens for peer interaction. An interaction was defined as a verbal exchange in which a client and peer made at least one statement to each other. Each client received a token for each peer with whom he had interacted during a work break. Figure 23 shows that during the first phase in which tokens were delivered, the daily mean number of interactions increased for both clients. Yet, during a reversal phase in

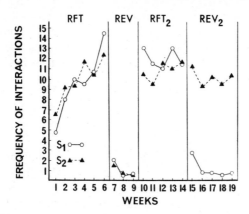

Fig. 23. Mean frequencies of interactions per day during Reinforcement, Reversal, Reinforcement$_2$, and Reversal$_2$ Phases for subject 1 (S$_1$) and subject 2 (S$_2$) (from Kazdin & Polster, 1973; Academic Press).

which tokens were withdrawn, interactions decreased. When reinforcement was reinstated, one client received tokens for each peer interaction as he had before (i.e., continuous reinforcement). The other client was told that he would receive tokens only once in a while (i.e., intermittent reinforcement). At first, this latter client received tokens at two of the three work breaks and after 2 weeks at only one of the daily breaks according to a schedule that varied each day. Eventually, all token reinforcement was withdrawn. As Fig. 23 shows, the client who had received tokens on a continuous schedule showed an immediate decline in social interaction. However, the client who had received tokens intermittently maintained a high rate of interaction for the final 5 weeks of the project. Thus, intermittent reinforcement appears to have helped maintain behavior.

The effect of intermittent reinforcement was evaluated for a longer maintenance period in a study by Kale *et al.* (1968). In this investigation, three schizophrenics were trained to greet staff on the ward. Whenever patients greeted a staff member (e.g., "Hello, Mr. ———" or "Hi!"), they were given cigarettes and praised. After greetings were performed at a high rate, reinforcement was thinned so that not all responses were reinforced. For example, for one patient a variable ratio schedule was used so that on the average every second response was reinforced. The ratio was gradually increased so that many responses (on the average 20) were required for a cigarette. After reinforcement had been delivered intermittently for a few weeks, reinforcement was withdrawn completely. Yet, behavior was maintained at its reinforced level. A follow-up conducted approximately 3 months after treatment showed that greeting responses were maintained at the high level which was developed during treatment.

The above research suggests that intermittent reinforcement can be used to maintain behavior while the consequences are delivered or after they are withdrawn completely. The extent to which intermittent token reinforcement can effect long-term maintenance is unclear. Altering schedules of reinforcement would be expected only to forestall extinction effects temporarily rather than to eliminate these effects altogether. A few studies suggest that withdrawing consequences after intermittent reinforcement does not result in a loss of the behavior. It may be that the extinction periods studied have not been sufficiently long to show the decline of behaviors toward baseline levels.

Delaying Reinforcer Delivery

Behavior might also be maintained by gradually increasing the delay between the target behavior and delivery of the reinforcer. When behavior is initially developed, immediate reinforcement usually is essential to ensure a high rate of responding. As behavior stabilizes and is well established, the delay between behavior and the reinforcing consequences can be increased without a loss in performance.

In token economies, it is possible to delay reinforcement in one of two ways. First, the delay between behavior and the delivery of tokens can be increased. Initially, the client receives tokens as soon as the target response is performed. Eventually, the delivery of tokens is delayed. Second, the delay between token delivery and exchange of tokens for back-up reinforcers can be increased. Initially, the client can exchange tokens for back-up reinforcers almost immediately after receiving them. Eventually, the delay between receiving tokens and their exchange is increased.

The effectiveness of delaying reinforcement after a response in maintaining behavior has not been widely studied. In a classroom program (Cotler *et al.*, 1972), tokens were delivered to the class for studying and completing academic tasks. The tokens could be cashed in for toys or candy after only studying for a 30-minute period. After study behavior stabilized at a high level, the delay between token earning and exchange of tokens for other reinforcers was increased to every other day and then to every 4 days. In spite of the delay, study performance was maintained at a high level. Of course, it is unclear what effect increasing the delay beyond 4 days or eliminating reinforcement altogether would have.

Greenwood *et al.* (1974) provided group consequences contingent upon appropriate classroom deportment for the group in three classrooms. After behavior had achieved relatively high levels (80% appropriate behavior), a delay was introduced in earning the group consequence. The classes had to perform high levels of appropriate behaviors for an increasing number of consecutive sessions before earning the group reinforcer. A maximum delay of 10 sessions was attained before earning the reinforcer. Interestingly, follow-up data 3 weeks after the program was terminated revealed maintenance of appropriate behavior on the part of the students. Because teachers tended to continue to administer contingent praise at a high level during the follow-up period, it is not possible to attribute maintenance on the part of the students to any particular component of treatment administered while the intervention had been in effect.

In a classroom token program by Turkewitz *et al.* (1975), the delay between token delivery and exchange of tokens for back-up reinforcers was manipulated as part of a more elaborate procedure to fade the program, as discussed earlier. During the fading phases, only some of the children on a given day were selected to exchange their tokens for back-up reinforcers. On days in which a person's tokens could not be exchanged, the tokens were not associated with back-up events. Fading was continued so a decreasing percentage of students per day could exchange their tokens. Essentially, the procedure introduced increasing amounts of time between token earnings and exchange of tokens for back-up events. At the end of the fading phase, tokens were earned but not exchanged for back-up events. Disruptive behaviors continued at a relatively low rate.

Because most reinforcers available in the social environment are delayed, it is important to wean a client from immediate token reinforcement. Behavior should be well established prior to invoking long delays. When delayed reinforcement is employed performance should be observed to ensure that there is no loss of behavior gains. Eventually, reinforcement may be withdrawn entirely or delivered only after a long delay.

Self-Reinforcement

If reinforcement programs are administered by an external agent (e.g., teacher or psychiatric aide) in a restricted setting (e.g., classroom or ward), one might expect the individual to discriminate in his or her performance across these conditions. Thus, behavior may not be maintained across situations in which the agent does not administer tokens. However, if the client is trained to control his own behavior, behavioral gains may not be lost once a program is withdrawn. The client can continually monitor his own behavior and provide consequences to himself across situations. If the individual conducts the program with himself, one might expect maintenance over time and transfer of behavior across situations. In many programs, an attempt has been made to gradually transfer control of the contingencies to the clients. In such programs, clients can evaluate their own behavior, set the criteria for token reinforcement, and determine the number of tokens they should receive (e.g., Felixbrod & O'Leary, 1973, 1974; Fixsen, Phillips, & Wolf, 1973; Glynn, 1970; Lovitt & Curtiss, 1969).

Training individuals to reward themselves for their performance has been used in a number of programs. In general, the purpose of these programs has been to demonstrate that individuals can perform as well under conditions in which some-one else administers reinforcing consequences as when they self-administer the consequences. As noted earlier, clients frequently become increasingly lenient in rewarding themselves. For example, Santogrossi *et al.* (1973) found that when the teacher determined the number of tokens to be awarded for appropriate classroom behavior, disruptive behavior markedly decreased. However, when the students were given the opportunity to determine the amount of tokens they should receive, they rewarded themselves noncontingently and disruptive behavior increased. Other studies have found that permitting individuals to self-reward results in increasingly lenient standards for self-reward although this is not invariably the case. (See Chapter 3 for a review of these studies.)

From the standpoint of developing response maintenance and transfer, the issue is whether individuals can be trained to reward themselves and to maintain performance standards. If this can be accomplished, the self-administered program might be maintained and extended to new settings. In fact, few programs have examined the utility of self-reward strategies on long-term maintenance. A few investigations have provided suggestive evidence in this area.

For example, as part of procedures to fade the contingencies, Drabman *et al.* (1973) allowed students to self-evaluate their own performance. Although students initially received tokens only if their estimate of tokens earned for a given period closely matched the estimate of the teacher, they eventually had complete control over administration of the consequences. Effectively, the teacher-managed token program was eliminated. Low levels of disruptive classroom behavior achieved with the orginal teacher-administration token program were maintained in the 12-day period in which the children had complete control over the contingencies. The program suggests that students might be able to have responsibility for the contingencies to maintain their behavior. In this sense, the self-reward procedures might be conceived of as substituting one program (self-administered tokens) for another (staff-administered tokens) to sustain performance.

Certainly, a question of interest is the relative efficacy of externally administered and self-administered tokens after the program is withdrawn completely. A few studies have compared teacher- versus student-administered contingencies on resistance to extinction. For example, Bolstad and Johnson (1972) found that after teacher-administered or self-administered points were withdrawn with elementary school students, disruptive behavior immediately increased. Those children who had observed their own behavior and determined the number of points they received in a previous phase tended to show slightly greater resistance to extinction than those who had received externally administered consequences. However, the differences were extremely small and appeared to be of little applied significance. Also, the maintenance period lasted only 7 days. Other studies have shown that there is little or no difference in response maintenance when an individual has previously received tokens from someone else or self-administered them after the contingencies are completely withdrawn (Felixbrod & O'Leary, 1974; Johnson & Martin, 1972).

The effects of self-reinforcement on long-term maintenance and transfer remain to be carefully evaluated. Developing self-reward would seem to be invaluable to ensure that the program is constantly in effect across time and situations. At present, it appears that individuals might not be able to self-administer consequences adequately without some sources of external reinforcement. That is, some consequences may be required from external agents to sustain accurate self-reward and to ensure that the reinforcers are not merely taken noncontingently.

Self-Instruction Training

Self-instruction training consists of a technique in which individuals influence their own behavior by instructing themselves to perform particular responses. Specifically, individuals are trained to make suggestions and directives to themselves. Self-instruction training has not received attention in the area of response maintenance and transfer of training in behavior modification programs. Nevertheless, the research that has been conducted suggests that the technique holds promise in each of these areas.

Meichenbaum and Goodman (1971) demonstrated the effects of self-instruction in training "impulsive" and "hyperactive" children to work deliberately and methodically on various tasks. Training began by having the experimenter model careful performance on tasks such as coloring figures, copying lines, and solving problems. As the experimenter performed the tasks, he spoke aloud to himself. These verbalizations included questions about the nature of the task and its requirements, answers to the questions, planning what to do next, statements guiding performance, and self-praise. The experimenter modeled "thinking aloud." The impulsive children were trained to do the tasks while instructing themselves aloud as the experimenter had done. Eventually, they were trained to perform the task while only whispering the instructions and then saying them privately to themselves without lip movements or sounds. Not only was the self-instruction training program effective in improving the work of these children relative to control children who practiced the tasks without self-instruction training, but the gains made in training were maintained up to 1 month. Similarly, self-instruction training in other studies has been shown to maintain levels of performance over follow-up periods and to transfer to tasks that are not included in training (cf. Meichenbaum, 1969, 1973; Meichenbaum & Cameron, 1973).

Although a number of studies have shown that behaviors developed through self-instruction are maintained and transfer to new stimulus conditions, there are exceptions. For example, Robin, Armel, and O'Leary (1975) used self-instruction training to develop writing of children with writing deficiencies. While self-instruction developed the target writing responses, the skills did not transfer to nontarget tasks which served as a measure of generalization. Interestingly, the authors reported that individuals sometimes self-instructed correctly, although their nonverbal responses were not correct. Thus, the verbalizations that are developed through self-instruction do not invariably control nonverbal behavior.

Self-instruction has received little attention in general and has not been explicitly studied as a maintenance or transfer strategy in the context of token economies. The bulk of the research has been conducted in laboratory settings with molecular motor and verbal tasks rather than with more molar target behaviors frequently examined in token economies. Aside from the ultimate efficacy of self-instruction for purposes of maintenance and transfer, attention will need to focus on the feasibility of this method for several populations. Some authors have alluded to the difficulty in teaching self-instruction skills to children and noted the extended training that is required to develop self-instruction (Higa, 1973; Robin *et al.*, 1975).

PROGRAMMING MAINTENANCE AND TRANSFER: GENERAL COMMENTS

The discussion of separate procedures that are useful in developing response maintenance and transfer of training implies that the procedures are distinct and mutually exclusive. In fact, the techniques might be subsumed under a few general principles. For example, the techniques might be considered as different ways to gradually fade the token program, so that fewer contingencies control behavior or as a way of expanding stimulus control by making the training situation increasingly resemble the situation in which performance is ultimately desired.

There is another reason to avoid emphasizing the independence of particular techniques. In maintenance studies, different procedures are sometimes utilized simultaneously rather than separately so it is difficult to attribute maintenance or transfer to one particular technique. Because research on the generality of behavior change has begun relatively recently, investigators have frequently used a "package" of several procedures to maximize the likelihood of response maintenance.

As an illustration, Jones and Kazdin (1975) developed response maintenance in a special education classroom in which four retarded children in a class of several students received token reinforcement for paying attention to their lesson. The effects of the token program were demonstrated on inappropriate motor behaviors (rocking in a chair, facing away from the lesson, engaging in repetitive body movements, sitting slumped in the chair) in a multiple-baseline design across time (morning and afternoon periods of the day). After it was clear that the program was responsible for change, a few of the procedures discussed above were implemented to maintain behavior.

First, "natural" reinforcers (peer praise) were substituted for the tokens. If each child performed well throughout the day, he would come up individually at the end of the day and receive applause from his peers. Second, a group contingency was also implemented so that desirable performance by all four children on a given day earned an activity for the class as a whole at the end of the day. This procedure was used to increase the likelihood that peer reinforcement would be given for appropriate behavior after tokens were withdrawn.

Third, back-up reinforcement was delayed before being eliminated entirely. The delay was made between the delivery of tokens and their exchange for back-up reinforcers at the end of each day. Before eliminating tokens, exchange was shifted to every other day only. On days in which tokens could not be exchanged, they were simply collected by the teacher. Essentially, with this procedure the contingencies were gradually removed since the consequences (back-up reinforcers) for behavior were faded.

After the maintenance procedures had been in effect for less than 2 weeks, all contingencies (tokens, peer-group applause, group activity, delayed back-up reinforcers) were eliminated entirely. No specific program remained in effect. However, follow-up data were collected to determine whether inappropriate behavior was maintained at the low rate achieved during the token program. Behaviors were maintained at their low rate (and slightly lower than during the token program) up to 9 school weeks (and 12 calendar weeks) after the program had been completely terminated. These results suggest the efficacy of the maintenance procedures although the effect of any specific component on response maintenance cannot be determined.

Another example illustrates the use of several procedures to promote response maintenance and transfer. Ayllon and Kelly (1974) restored speech in an 11-year-old retarded girl who had not spoken in class for over 8 months. Training consisted of providing candy and social reinforcement for components of speech (e.g., opening the mouth, blowing air out of the lips, making a sound, and so on). Eventually, verbal responses to specific questions were reinforced. Training was conducted in sessions in a counselor's office outside of the classroom. To ensure that the behaviors would be maintained and transfer to the classroom, several different techniques were used.

First, over the course of training the schedule of reinforcement was altered. Initially, continuous reinforcement was provided but this was replaced by increasingly intermittent reinforcement. Second, primary reinforcement (candy) was used initially to develop verbal responses. Praise was paired with the delivery of candy and eventually was substituted for candy to approximate more naturally occurring reinforcers. Third, stimuli associated with the classroom were introduced into the counselor's office so that the training situation would increasingly resemble the classroom. Thus, other children, a blackboard, and desks were added to the training situation. The trainer stood in front of the children and subject to simulate the teacher's relation to the student. Fourth, to maximize the occurrence of verbal responses in the presence of her peers, a group contingency was introduced. The contingency consisted of providing candy if each of the three children in the session verbally responded to a specific question asked by the trainer. Fifth, training was continued in the classroom situation in which the trainer rather than the teacher initially administered candy on an intermittent schedule for verbally responding to instructions. Eventually, the teacher continued training in class.

Overall, training consisted of eight sessions outside of the classroom followed by seven sessions in the classroom. The child's verbal responses to questions increased markedly over the course of treatment. A follow-up assessment 1 year after training had been terminated indicated that the verbal responses were maintained at level achieved during training and transferred across three new teachers and settings within school.

Other studies that focused on response maintenance or transfer of training have combined various techniques discussed earlier including expanding stimulus control and intermittent reinforcement (Jackson & Wallace, 1974; Koegel & Rincover, 1974), fading the contingencies, self-reinforcement, and increasing the delay between token earning and exchange of back-up events (Turkewitz et al., 1975), substituting one program for another and expanding stimulus control (Walker et al., 1975), fading the contingencies and expanding stimulus control (Henderson &

Scoles, 1970; Kelley & Henderson, 1971), and others. As research increasingly examines the reliability of techniques in developing maintenance and transfer, the focus is likely to become more analytic so that specific components of generalization strategies are carefully analyzed.

CONCLUSION

Major considerations in evaluating the effectiveness of the token economy, and indeed, any treatment strategy are whether behaviors are maintained once treatment is terminated and whether they transfer to extratreatment settings. Occasionally, behavior changes are maintained and do transfer across settings in token-reinforcement programs, although these are exceptions. The literature shows rather clearly that as a general rule maintenance and transfer must be programmed directly.

Only recently has research systematically begun to investigate techniques to promote maintenance and transfer. Different techniques have been effective in accomplishing these ends. Sometimes, selecting behaviors that will generate their own reinforcing consequences after training is terminated is sufficient to effect sustained performance across settings. This, however, appears to be the exception. Specific procedures need to be implemented to program maintenance and transfer. First, substituting a program in the naturalistic setting for the one used in training can maintain performance. Implementing contingencies across settings ensures continuation of behavioral gains and augments response maintenance after the contingencies are withdrawn completely. Second, the reinforcement contingencies can be faded during training so that performance is maintained under decreasing influence of the contingencies. Third, stimulus control over behavior can be expanded in training so that behaviors transfer broadly across situations, therapists, and settings. Fourth, parameters of reinforcement can be varied such as scheduling increasingly intermittent reinforcement or increasing the delay between behavior and the reinforcing consequences. Fifth, self-reinforcement techniques may be used to promote maintenance and transfer, at least for relatively short periods. Finally, self-instruction training also has been suggested as a technique to promote generalized behavior change. Studies attempting to develop response maintenance and transfer usually combine several procedures to maximize the likelihood of achieving these ends. The existing literature suggests that there is a technology of establishing the generality of behavior change. Treatment effects achieved with reinforcement techniques are no longer routinely transient or restricted in their generality.

CRITICAL EVALUATION
OF THE TOKEN ECONOMY

8

The accomplishments of the token economy have been impressive. Although early programs in the field effected notable change, the techniques for changing behavior, for ensuring responsiveness of patients who do not initially respond to the contingencies, and for maintaining behavior and developing transfer have evolved considerably. Overall, the technical advances have been marked.

Along with a discussion of the gains made in developing an effective treatment modality, it is useful to critically evaluate the characteristics of the token-economy research. A critical evaluation of the token economy permits assessment of the limitations of current research and outlines future areas of investigation. This chapter evaluates several areas of token-economy research including the behaviors, settings, and populations that have been studied, the variables which contribute to the efficacy of the token economy, the possible deleterious effects of token reinforcement, and the relative efficacy of the token economy versus alternative techniques. The final section evaluates the effects of token economies and the extent to which token economies have achieved clinically and socially important changes in behavior. Overall, the issues raised with respect to each of these areas generate the greatest controversy over the value of the token economy and are the most fertile areas of empirical research.

FOCUS OF TOKEN PROGRAMS

Target Behaviors

One source of criticism of token-economy research, and reinforcement programs in general, is the target behaviors focused upon. Relatively little criticism has been made from within behavior-modification research along these lines. As an exception, Winett and Winkler (1972) criticized the selection of target responses for behavioral programs in classroom settings. Specifically, these authors criticized the preoccupation of investigators with classroom deportment, particularly in classes where behaviors may not be severe or even necessarily incompatible with learning. The criticism is valid even though the article was based upon a somewhat narrow sampling of the literature and omitted many investigations focusing on other behaviors such as academic performance (O'Leary, 1972).

The general criticism of the focus of behavior-modification programs can be applied widely across all treatment populations. Considerable attention has focused upon target responses that may be more convenient rather than of clear therapeutic significance. If a naïve observer were asked to infer the problems of select treatment populations based upon the focus on token programs, there might be considerable distortion of the population's defining characteristics. For example, a naïve observer might infer that psychiatric patients consist of individuals whose main problems are the failure to bathe, dress, and groom themselves, get up on time, and attend activities punctually. The plethora of investigations that have focused on self-care would not make this inference unjust. Similar inferences might be made about other populations. For example, pre-delinquents might be construed as individuals who have difficulty in cleaning their rooms, in watching the news, or attending activities on time. Prisoners might be construed as individuals who are in need of education and self-care skills. And, children and adolescents are individuals who cannot remain in their seats in school.

While this reflection on the target behaviors selected in token economies is intentionally hyperbolic, there is some basis for criticizing the focus of many token programs. The focus frequently seems to be on behaviors that appear convenient from the standpoint of the research or from the standpoint of staff's interest in management rather than necessarily the focus required for client change. Many of the behaviors are selected because they can be easily observed and readily fit into an observational coding system for which reliability can be easily achieved. Classroom token economies perhaps are more guilty of focusing on behaviors of unclear significance than any other area. A large number of investigators have studied deviant, disruptive, or inattentive behavior (e.g., Kazdin, 1973g; O'Leary et al., 1969; Walker & Buckley, 1972). In fact, there is considerable question about the need to focus on inappropriate classroom behavior given that improvement in academic performance often is sufficient to decrease inappropriate behavior, as discussed in Chapter 4. Also, academic gains sometimes can be effected even without great reductions of "inappropriate" behaviors. Few of the classrooms referred to as disruptive seem to evince insurmountable inappropriate behaviors that need to be the sole or major intervention focus.

Although much of the research can be criticized for the target behaviors focused upon, it would be misleading to imply that significant behaviors of therapeutic or educational import have been completely ignored. As discussed in the review of token programs, symptomatic behaviors of psychiatric patients, use of drugs by addicts, and complex academic skills have all been altered with token reinforcement. Moreover, follow-up data do support the efficacy of token programs on measures that are of obvious import. For example, token reinforcement programs for delinquent youths have been shown to be associated with reduced recidivism rates (Alexander & Parsons, 1973; Fixsen et al., 1976; Liberman & Ferris, 1974). Yet, there are many cases where follow-up data reveal that treatment has little or no effect on important measures. For example, prisoners who leave a prison ward based upon a token economy have upon follow-up been found to show no fewer contacts with the courts than individuals who were not in a token-economy program (Jenkins *et al.,* 1974).

For many programs, treatment focuses on behaviors that are not always clearly related to the problems usually ascribed to the population. Greater attention might be given to those behaviors that are likely to ensure adequate functioning in extratreatment settings. Perhaps the reason for the focus of most programs on adaptive or routine behaviors appropriate to the treatment facility is more a criticism of the settings in which programs are conducted than their therapeutic focus.

Settings

A criticism against the target behaviors focused upon in many token economies is related to the settings in which programs usually are conducted. Token economies often are implemented in facilities which themselves may interfere with therapeutic change. It may be difficult to treat behaviors in settings far removed from those in which the problematic behaviors have occurred. Thus, programs in psychiatric hospitals, prisons, and other settings attempt to alter behavior in situations in which the problem behavior does not ordinarily occur. For example, if a psychiatric patient is not adjusting adequately to community life, removal and treatment in a noncommunity setting is not likely to resolve these problems unless training eventually is developed again in the community (Kazdin, 1976b). Similarly, developing behaviors of a convicted inmate in a prison may have inherent limitations because training does not develop behavior under the stimulus conditions in which the individual fails to adequately function.

The criticism against programs in various "treatment" settings cannot be directed at token economies alone. As new treatments develop, they tend to be integrated into existing treatment facilities even when the treatment facilities may violate some of the principles upon which the treatments are based. Token programs in many facilities focus on behaviors that may not be of direct therapeutic interest but need to be developed to ensure adjustment of the client to the setting. For example, it is generally accepted that psychiatric patients develop a variety of debilitating behaviors as a function of adjusting to institutional life (e.g., Goffman, 1961; Paul, 1969). Token programs often focus on precisely those behaviors that seem to be developed or exacerbated by institutional living such as "apathy" and lack of participation in activities. In many psychiatric hospitals, token economies seem to be viewed as a way to manage patients rather than as a way to program rehabilitation. While there is little question that token reinforcement contingencies can add to the efficiency of the institution, the therapeutic end toward which the program is directed is not always clear.

The criticism of token-economy research is not that it has been conducted in institutional environments. Indeed, alternative settings have not been widely available. The criticism is that investigators have implicitly endorsed current institutional structure by designing programs for management purposes and have failed to stress the inadequacy of institutional structure based upon the principles of behavior modification. Behavioral programs have not only implicitly endorsed traditional treatment settings but sometimes *rely* on their inadequacies to effect change. For example, the deprivation states of many individuals in institutional settings make the usual back-up reinforcers quite effective in altering behavior. Certainly a legitimate question is whether the program would be as effective if individuals had access to a wide variety of events that might otherwise be available in a community setting. For example, walks on the grounds, overnight passes, recreational activities, and similar events are used as reinforcers for psychiatric patients. The efficacy of a token program would seem to depend upon the institution *not* ordinarily providing many of these reinforcers. Or another way, token programs may partially depend upon the deprivation that many treatment settings normally impose on their clients (Hersen, 1976). The reinforcers then made available in a token program are great incentives because of the deprivation.

The possibility that behavioral programs depend heavily upon the inadequacies of existing treatment practices can be well illustrated in educational programs. For example, in some programs leaving school early is used as a back-up reinforcer and staying after school as an aversive event (Harris & Sherman, 1973). If escape from school is a reinforcer, this suggests that school is an aversive event. Relying upon the aversiveness of the system to effect therapeutic change suggests that the undesirable features of existing environments may be important to sustain the token programs.

One might expect the greatest impact of token economies to be achieved in the natural environment or in treatment facilities integrated with community life. Indeed, such an approach has been taken in token-economy programs for pre-delinquent and behavioral-problem youths (e.g., Fo & O'Donnell, 1974; Patterson *et al.*, 1972; Phillips, 1968; Tharp & Wetzel, 1969) and psychiatric patients (Henderson & Scoles, 1970). There are several reasons for conducting reinforcement programs in the natural environment. First, individuals in everyday situations have access to problematic behavior and to the consequences that can be used to alter behavior and can be effective behavior-change agents (Guerney, 1969; O'Dell, 1974). Second, bringing clients into an artificial situation does not guarantee alteration of the conditions that maintain deviant behavior. Thus, when a client returns to the situation in which the target behaviors developed, he or she is likely to respond to these contingencies again. Also, as mentioned earlier, an institutional environment may exacerbate behavioral deterioration because of existing contingencies. Third, bringing clients into institutional settings may obscure those behaviors that led to institutionalization. The behaviors that need to be altered may not appear as the individual moves from one environment (e.g., community) to another (e.g., prison and hospital) (cf. Wahler, 1975).

Some of the problems that are considered to plague token economies in institutional settings may partially result from focusing on behavior within a particular treatment setting rather than in the natural environment. The problem of achieving maintenance and transfer of behavior might result from focusing on behaviors *in* settings removed from the situation to which the client returns. If behavior is to be performed in a given situation (e.g., at home, in the community), it should be developed in that situation or settings which closely simulate the situation.

Of course, as reviewed earlier, token-reinforcement contingencies are not restricted to institutional settings. Outpatient applications with adults and children have increased in recent years. So the settings in which token programs are implemented *have* encompassed the natural environment. Also, many individuals for whom token economies are designed may not necessarily return to any particular setting. For example, token reinforcement contingencies for the severely and profoundly retarded or for geriatric residents in a nursing home may have as their sole purpose increased performance in the setting. In these cases, of course, the criticism of the setting in which the program is conducted is less cogent. Yet, for some institutionalized patients and residents, return to the community is a viable aim and the implementation of contingencies in an institutional environment may not be the most desirable focus.

Populations

Probably little criticism of token programs could be levied against the populations focused upon. A wide range of clients has been exposed to token reinforcement. Indeed, often the most intractable clients in a given setting such as a prison, psychiatric hospital, or school are especially selected for the program. Despite the use of token economies with diverse populations, some questions can be raised about the appropriateness of the focus.

Token programs usually identify as clients those individuals who are being treated, rehabilitated, or educated. The clients have something done to them as part of the program. There are alternative sources of focus that might be more appropriate than the client. For example, the behavior of individuals who influence the clients such as parents, teachers, prison guards, and aides may be a more appropriate focus than the clients. Behavioral programs need to be designed for behavior-change agents because their behavior is so related to client performance.

It is difficult to alter behavior of select participants of a system and sustain these alterations if there are not methods built into the system to ensure that the means of change are sustained. System here refers to the complex hierarchy of individuals and positions that make up a treatment, rehabilitation, or education program. For example, in schools, focusing on the students' behavior may be less ideal than focusing on the teachers' behavior. If the teachers receive reinforcing consequences for behaving in a particular fashion, behavior changes in students are likely to occur (Cossairt *et al.,* 1973; McNamara, 1971). If the behaviors of the teachers are not focused upon directly, it is unlikely that the changes in the students will be maintained. Yet, the teacher probably is not the appropriate focus of a behavioral intervention unless some changes are made in those individuals who reward teachers so that teacher behaviors are maintained.

As the focus continues up the hierarchy, it may well be that a system of accountability needs to be built into treatment systems so that specific achievements (behaviors) by any level (e.g., students, teachers, principals, superintendents) are associated with contingent consequences from a higher level. If there is any value to this approach, it suggests that changes in systems are needed to ensure that desired behavior changes at all levels are achieved and maintained. The behavioral approach has recognized the importance of changing systems in which programs occur but has made relatively few inroads in this area (Harshbarger & Maley, 1974).

FOCUS OF TOKEN PROGRAMS: SUMMARY

Despite the diversity of token programs in terms of target behaviors focused upon, settings in which programs have been implemented, and populations treated, some questions were raised about the orientation and focus. Many of the target behaviors that have received considerable attention (e.g., self-care skills in patients, attentiveness in nonproblem classes) are of unclear relevance to the apparent goals or the ideal goals of the system (e.g., patient adjustment in the community and academic advancement).

The implementation of token economies in many existing treatment facilities may be counterproductive. Many of the problems for which token economies are implemented may result from the setting in which the clients are placed. The problems might be eliminated by altering the setting in which treatment is conducted rather than trying to strengthen the contingencies to achieve behaviors within the system.

The focus on clients rather than the systems in which clients function may be somewhat misplaced. For example, in the long run, the most effective technological advance may not be introducing carefully programmed contingencies in the classroom but by altering the functioning of educational systems so that accountability and contingency management pervade all levels. If there is any merit in this position, it calls for a new or expanded focus of behavioral interventions.

While the focus of the token economy may be criticized in light of some idealized form of treatment, the overall evaluation of this treatment modality must be viewed more generally in the context of current research. In this light, it is clear that no form of treatment seems to have been applied so generally as the token economy. There is no other technique that can claim such a widespread applicability across behaviors, settings, and populations, at least among treatment, rehabilitation, and educational techniques that have been empirically validated. Consequently, there is no clear competing technique that has demonstrated empirically the breadth of changes that can be effected with token reinforcement. (This will be especially clear in the next chapter where the extension of reinforcement techniques to societal problems is discussed.) Generally, the above criticisms of the token economy suggest areas where research needs to be conducted. Yet, discussion of areas that remain to be researched is not tantamount to criticizing the existing accomplishments.

ANALYSIS OF THE TOKEN ECONOMY AND ITS EFFECTS

Variables Contributing to the Token Economy

The token economy has been discussed as a unified entity. In fact, it includes several component variables some of which are essential to the definition of a token economy (e.g., use of conditioned reinforcers) and others which are not (e.g., instructions about the contingencies). Because there are several variables which may contribute to the token economy, it would be of interest to determine their individual and combined effects on behavior.

The variables that may influence the effect of a token economy might include instructions about the contingencies, feedback or praise provided for performance, social interaction with staff, various subject and demographic variables, and others. In many programs, investigators have referred to the effectiveness of the token economy with a strong implication that the administration of tokens contingent upon behavior accounted for change. In these programs, several other procedures have been used which may have contributed to or even completely accounted for the efficacy of the program.

Before briefly discussing the role of some variables in the token economy, it is important to note some distinguishing features of applied research. Given a goal of many clinical interventions, programs often employ a treatment or intervention "package" which consists of a group of procedures. Although the components of the procedures might be separable from an experimental standpoint, such a separation usually is not made when a clinical intervention is applied. As various researchers have suggested, analysis of the specific components of treatment may not be as essential for applied as for basic research (Wolf, Sidman, & Miller, 1973). Applied research is more concerned with determining whether the intervention is responsible for change rather than analyzing components of the intervention. While this goal is acknowledged, one can raise questions about the efficacy of a given intervention that has not been analyzed for its component features (Kazdin, 1973f). The present section discusses variables which may contribute to the effectiveness of a token program. The variables are important to review because in some programs the token economy may not be any more effective than one or two of the components which usually are viewed as incidental. More likely, various components of a token program contribute to the overall impact. For purposes of analysis, it is essential to examine that contribution.

Instructions

Instructions, as employed in token-economy research, refer to a complex set of manipulations that rely on verbal or written cues. Instructions are used to describe the desired behavior and the consequences associated with performance, to prompt behavior, to convey threats or promises stating conditional statements, and others. The role of instructions in contributing to behavior change has received attention.

The potential importance of instructions in behavior change was suggested by Ayllon and Azrin (1964, Exp. 1). These investigators found that providing a tangible reinforcer (choice of extra food, drink, or cigarettes) to psychiatric patients for picking up utensils for eating their meal did not affect the number of patients who responded. However, when verbal instructions stating the contingency were provided, behavior immediately increased. These results suggested that instructions were essential to develop responsiveness to the contingency. Similarly, Herman and Tramontana (1971) found that providing tokens to Head Start children for appropriate classroom behavior did not markedly alter behavior until the children were told the responses required to earn tokens. Generally, instructions about the contingency and task requirements augment the effects of reinforcement although there are exceptions (Kazdin, 1973g; Resnick et al., 1974).

In many token programs, the effects attributed to the contingencies might be due to specific instructions how to perform. For example, Ayllon and Azrin (1965, Exp. 2) instructed psychiatric patients that they could continue to work on their jobs even though they would receive tokens for not working (i.e., a "vacation with pay"). Work performance abruptly declined relative to the previous reinforcement phase, an effect that was attributed to the delivery of noncontingent consequences. However, the effect of instructions or the combination of instructions and reinforcement may have accounted for the pattern if not the magnitude of change. Indeed, instructions about the contingencies facilitate making discriminations across reinforced and nonreinforced responses (Resnick et al., 1974).

The effect of instructions in changing behavior might depend upon whether the clients have the target response in their repertoire. When clients do not have the response, subsequent performance attributed to reinforcement effects could be more parsimoniously accounted for by instructions how to perform the response or by acquisition of information about the response. For example, in one token program, psychiatric patients were exposed to training for grooming skills prior to receiving contingent tokens (Suchotliff et al., 1970). Training consisted of instructing the individuals how to execute the behaviors that constituted grooming. Subsequently, a token program was implemented to reinforce performance of these skills. Interestingly, grooming behaviors did not increase during token reinforcement relative to what they were during instructional training. The increments in performance over baseline were due to training and instruction rather than to token reinforcement.

In a large number of studies, mostly conducted in classroom settings, instructions have been evaluated in relation to token reinforcement. In these studies, instructions have been provided for behaviors that appear to be in the repertoire of the students (e.g., studying, attending to the lesson, remaining in one's seat, and other rules of deportment). There has been consistency in most of these studies showing that when children are merely instructed in the rules of the classroom or are asked to "behave," there is little or no change in actual performance (e.g.,

Greenwood *et al.*, 1974; O'Leary *et al.*, 1969; Packard, 1970). Similar findings have been obtained with psychiatric patients (Aitchison & Green, 1974; Liberman, 1972).

In some cases, instructions to perform appropriately and noncontingent delivery of tokens have altered behavior. For example, Kazdin (1973g) provided rules of appropriate classroom behavior to elementary school students and a statement that appropriate behavior would be rewarded with tokens. The teacher administered tokens on a prearranged schedule based upon a child's hair color, sex, and place of first initial in the alphabet. Although tokens were delivered independently of performance, significant increments in appropriate behavior were obtained over 3-week intervention period. This finding supports laboratory research that has shown that instructions can overcome the actual contingency to which subjects are exposed (Baron, Kaufman, & Stauber, 1969; Kaufman, Baron, & Kopp, 1966).

In general, instructions appear to play an important role in token economies. Stating the desired behaviors or the consequences associated with these behaviors often contribute to some of the effects attributed to token reinforcement alone. In most programs, the extent to which reinforcement contributes to behavior change over and above instructions has not been evaluated. As stated earlier, this is not necessarily a criticism of programs because the goal has been to design an intervention package to effect clinical change. However, from the standpoint of evaluating the effect of the token economy in general, relatively little credit has been given to instructions as an ingredient related to behavior change.

Social Reinforcement and Interpersonal Interaction

Social reinforcement in the form of praise, approval, attentiveness, and other interactive variables may contribute to the efficacy of token economies. Indeed, in most programs, the delivery of tokens is explicitly associated with verbal praise by the staff, in part to augment the reinforcing properties of praise (cf. Cohen & Filipczak, 1971; Stahl, Thomson, Leitenberg, & Hasazi, 1974). Social consequences may be increased by implementing group contingencies that bring peer social consequences to bear on the clients (Axelrod, 1973; Feingold & Migler, 1972). Thus, social consequences would appear to play an important role in some token economies.

The importance of social interaction in the administration of a token economy was suggested anecdotally in a report of treatment facilities for pre-delinquents (Phillips *et al.*, 1973). Specifically, an early extension of the Achievement Place program established a token system in new home-style facility. Although the new facility was similar to Achievement Place, the program did not appear to function as smoothly or as effectively as the original program. The initiators of the program attributed the apparent differential efficacy of the programs to the types of interactions made by the staff. Approval, instructions, positive attempts to prompt behavior, and feedback had been a part of Achievement Place program but were not transferred to the new facility. The authors attributed major differences in the effects of the two programs in part to the more reinforcing and less aversive interpersonal relationships (Phillips *et al.*, 1973). While this anecdotal report does not establish the importance of interactive factors in determining the efficacy of a token economy, it is suggestive and warrants further research.

Aside from anecdotal reports, evidence on social consequences controlled by the staff suggests variables that might mediate the effects of a token program. For example, the frequency of staff-client interaction in general or the proximity of a staff member to a client may increase during a period in which tokens are delivered. As noted earlier, in a classroom token program, teacher contact with the students was shown to increase during periods in which tokens were delivered contingently (Mandelker *et al.*, 1970). Given this finding, the question arises about the effect of token delivery independently of extra contact with the staff with which it might be associated.

Not only might the quantity of staff-client interaction be altered during token reinforcement, but the quality of that interaction might change as well. Some studies have suggested that staff evaluate patients more favorably after participating in a token program, which might be associated with different kinds of staff-client interaction (McReynolds & Coleman, 1972; Milby *et al.*, 1975). More pertinent, participation in a token economy is associated with a reduction in the staff's use of discipline techniques such as reprimands and disapproval, a reduction in attention to inappropriate behavior, and an increase in attention and approval to appropriate behavior (Breyer & Allen, 1975; Chadwick & Day, 1971; Parrino *et al.*, 1971; Trudel *et al.*, 1974).

Of course, if administering a token economy reliably alters the delivery of approval and reprimands, it might well be the social consequences rather than the tokens that account for behavior change. Indeed, the use of tokens may be important precisely because it provides a way to structure and prompt staff behavior such as approval and feedback rather than as a unique incentive which is responsible for behavior change (Liberman *et al.*, 1975). The literature has shown that social consequences in the form of verbal or nonverbal approval and attention are sufficiently effective to alter a wide range of behaviors (cf. Kazdin, 1975a; Ulrich, Stachnik, & Mabry, 1966, 1970, 1974). However, a number of studies have compared contingent social and token reinforcement. Typically, the delivery of tokens contingent upon behavior is more effective than the delivery of social consequences (Kazdin & Polster, 1973; O'Leary *et al.*, 1969; Zifferblatt, 1972).

To say that token reinforcement is more effective than social reinforcement does not imply that they are usually independent. In most programs, tokens are provided in addition to rather than as a substitute for praise. Given the effects of praise independent of tokens, social consequences may constitute a considerable part of the efficacy of token reinforcement. To examine this question, Walker, Hops, and Fiegenbaum (1976, Exp. 1) studied the effects of adding specific components of a token economy in a sequential fashion in an experimental classroom of elementary school students. After baseline, social reinforcement was provided for appropriate academic and social behavior. In the next phase, token reinforcement was added to the social-reinforcement contingencies. The introduction of social reinforcement led to an average gain in appropriate behavior of 12% over baseline rates. The addition of token reinforcement increased appropriate behavior by another 19%. Thus, social reinforcement made a definite contribution to the overall program. (Incidentally, the addition of a response cost contingency to the social and token reinforcement program led to additional increments in appropriate behavior.)

Feedback

Token reinforcement includes another variable that may be distinguished from the incentive value of tokens. The delivery of tokens provides feedback for performance or knowledge of results of how someone is doing. This effect can be separated from the delivery of tangible conditioned reinforcers that are redeemable for back-up events. It is meaningful to inquire into the role of feedback in token economies because many applied behavioral programs have explicitly employed feedback to alter behavior (e.g., Drabman & Lahey, 1974; Leitenberg et al. (1968).

In a few token-reinforcement studies, the effect of feedback has been evaluated independently of delivering tokens that can be exchanged for back-up events. For example, Zimmerman et al. (1969) showed that merely telling retardates how many tokens they would have earned if a token program were in effect substantially increased production in a sheltered workshop setting. In other programs, tokens have increased performance even though the tokens cannot be exchanged for back-up events (Deitz & Repp, 1974; Hall et al., 1972, Exp. 2; Haring & Hauck, 1969; Jens & Shores, 1969; Sulzer et al., 1971). In most of these studies, feedback has been followed by a phase in which tokens (backed by various events) are delivered for performance. In these cases, token reinforcement usually is superior to feedback alone. This would be expected given that tokens provide feedback plus an additional incentive.

It is important to mention the effect of feedback because evidence suggests that it does contribute to change. Although token reinforcement probably cannot be delivered without providing performance feedback, feedback can be delivered independently of token reinforcement. Studies that do not evaluate the effect of feedback often assume that the contingent delivery of tokens and their back-up value are responsible for change. While this may be accurate, the magnitude of change that tokens achieve over feedback alone cannot be determined. From a more pragmatic standpoint, it is worth noting that feedback has been quite variable in its effectiveness in applied settings (Kazdin, 1975a). Thus, feedback alone might be insufficient to alter behavior in many cases where token reinforcement procedures have been implemented.

Modeling Influences

The effects of observing another individual's performance and receipt of reinforcing consequences has been discussed in a previous chapter as a possible means of enhancing responsiveness to the reinforcement contingencies. In that context, vicarious reinforcement and punishment were discussed as adjunctive procedures which might be explicitly programmed into a situation. However, even if vicarious influences are not intentionally introduced, they are likely to be present in token programs (Bandura, 1969).

The influence of a model might be considered unlikely in applied settings because the behaviors modeled are not as clearly delineated, made as salient, or attended to as readily as they are in well-controlled laboratory settings. Yet, many of the conditions of a token economy would seem especially conducive to modeling effects. First, the program almost always takes place in a group context where models who perform appropriately receive reinforcing consequences. Thus, participants in a token economy have the opportunity for extensive modeling experiences by viewing others earn tokens. Second, various parameters important in modeling such as the performance of multiple models and by models who are similar to the observers (i.e., several peers performing the modeled response) are present in token economies. Thus, the conditions which increase the salience of modeling cues are likely to be present. Finally, the delivery and receipt of reinforcing consequences often is conspicuous in reinforcement programs so that the vicarious effects are likely to be maximized (cf. Kazdin et al., 1975). For example, in many programs tokens are posted, delivered, or displayed in full view of others (Glickman, Plutchik, & Landau, 1973; Koch & Breyer, 1974; O'Brien et al., 1971).

Direct evidence from token-economy research suggests the operation of vicarious effects. Hauserman, Zweback, and Plotkin (1972) found that when adolescent psychiatric patients received tokens for initiating verbalizations in group therapy, others who did not receive tokens also increased in the target response. Tracey et al. (1974) found that psychiatric patients who received tokens for discussing their activities in a group increased their actual participation in activities. Interestingly, some individuals who never received tokens for verbalizations about activities increased in their participation in activities. These results suggest that modeling influences operated in the situation. Also, McInnis et al. (1974) found that some patients apparently increased their utilization of back-up reinforcers after observing others utilize these events.

In addition to the token-economy research, programs using verbal or nonverbal reinforcement or feedback have demonstrated reliable changes in the behavior of individuals who have merely observed their peers receive contingent consequences (Broden et al., 1970; Brown & Pearce, 1970; Christy, 1975; Drabman & Lahey, 1974; Kazdin, 1973b, 1973c, 1977c; Kazdin et al., 1975). In general, vicarious effects seem to contribute to the effects of reinforcement programs. The extent to which behavior changes in token economies depend upon these effects is not clear because relatively few studies have attempted to separate direct from vicarious consequences on behavior.

Subject and Demographic Variables

A relatively unexplored set of factors that may mediate the efficacy of a token economy is subject (individual difference) and demographic variables. Other variables previously discussed such as instructions, social interaction, feedback, and modeling were posed as interventions which may account for the efficacy of a given program or interact with token reinforcement in achieving behavior change. Subject and demographic variables refer to characteristics of the population that may interact with the procedures. These characteristics refer to variables that may affect the *external validity* of the token program, i.e., the generality of its effects across subjects, rather than affect the *internal validity* of the program, i.e., whether the contingencies account for change (cf. Campbell & Stanley, 1963).

There is a paucity of investigations of subject variables that contribute to the effect of a token economy, therefore few affirmative statements can be advanced. The diversity of applied research suggests that sex, age, intelligence, patient diagnosis, years of institutionalization, and similar factors are not clearly related to the efficacy of reinforcement procedures in general, or to token reinforcement in particular. Some studies that have reported the ineffectiveness of token programs with select clients have found consistent differences to distinguish those individuals who respond from those who do not.

Golub (1969) found that psychiatric patients with more education and more months of hospitalization were more likely to gain from the ward token program. Other authors have reported that the degree of patient withdrawal and isolation (Atthowe & Krasner, 1968; Ayllon & Azrin, 1968b) and length of hospitalization (Panek, 1969; Ulmer, 1971) are negatively correlated with improvement. Contradictory evidence was provided by Allen and Magaro (1971) who found that age of psychiatric patients and length of hospitalization were not related to improvement or responsiveness to the contingencies. Similarly, while some investigators working with chronic psychiatric patients have found no differences in responsiveness of patients according to psychiatric diagnosis (Ayllon & Azrin, 1965; Lloyd & Abel, 1970), other studies have (Panek, 1969).

One investigation found that patient physical attractiveness was related to performance and outcome measures in a token economy (Choban, Cavior, & Bennett, 1974). Psychiatric patients who were judged as more physically attractive tended to earn more tokens and showed a higher discharge rate from the hospital than less attractive patients. As the authors noted, physical attractiveness of a patient may result in differential staff treatment and account for group differences.

Relatively few studies have examined the influence of subject variables in token programs. In general, the prevailing assumption of operant programs is that reinforcement procedures are applicable across virtually all subject populations. There has been little question about the suitability of programs of clients differing according to subject variables (but see Davison, 1969). Moreover, even if a given subject variable interacts with the reinforcement contingencies in altering behavior, this effect may be a function of specific features of the program rather than a lawful relationship across all token programs. In general, the utility of examining subject variables needs to be explored. Given the existing evidence, it is unlikely that token reinforcement procedures would be effective with some clients but not at all with others simply as a function of subject variables. However, there are questions about effective treatment interventions which might be answered by examining subject variables. The types of contingencies, back-up reinforcers, ideal parameters for reinforcer delivery, and similar program options might be influenced by characteristics of the treatment population.

Variables Contributing to the Token Economy: Summary

This section discussed some of the major components of a token economy. There are other components as well that were not mentioned. These refer to many of the options that were discussed in Chapter 3 as special procedures such as response cost, different types of contingencies, variations in the back-up events, and so on.

Research on the components of token economies needs to examine the contribution of a particular component to behavior change and the interactive or synergistic effects of the components. Some research suggests that specific components produce additive effects. For example, Walker *et al.* (1976, Exp. 1) found that sequentially introducing social reinforcement, token reinforcement, and response cost led to gradual improvements in appropriate classroom behavior. When all components were in effect an extremely high rate of appropriate behavior was achieved. In a second experiment, the simultaneous implementation of each of these procedures led to the same high rate of appropriate behavior, although the change, of course, was more immediate. In any case, the amount of change associated with each component of the token economy warrants additional investigation. From a practical standpoint, it is important to isolate those components that account for change that require the least intrusion in the natural environment. Thus, behavior-change programs might be implemented without requiring the complex package that characterizes many token economies. Yet, it is important to know the salient components of a token program and their relative contributions to behavior change so that they can be implemented when needed (McLaughlin & Malaby, 1975c; Walker *et al.*, 1976).

POSSIBLE DELETERIOUS EFFECTS OF EXTRINSIC REINFORCEMENT

While the effects of token programs are widely accepted as generally positive, some investigators have been concerned about undesirable effects of providing extrinsic reinforcers for behavior. Various concerns have been voiced including the belief that giving tokens to someone for behavior will make him perform responses only when tokens are provided or teach him to be "greedy," or to use "bribery" to control the behavior of others. These objections do not seem to portray the usual effects of token reinforcement and have little support (Eitzen, 1975; Kazdin, 1975a; O'Leary *et al.*, 1972). However, there is one area of research that has been raised as directly relevant to token-economy research which does demonstrate potential deleterious effects of administering extrinsic consequences for behavior.

Although token economies might alter behavior while reinforcement is in effect, some investigators have argued that withdrawal of the reinforcers actually may have a deleterious effect on performance. That is, not only will behavior fail to be maintained, but withdrawal of tokens may actually decrease performance below the level that was achieved during baseline conditions (see Deci, 1975).

The basic tenet of this position is that individuals are "intrinsically motivated" to perform certain activities. Intrinsic motivation is defined as engaging in an activity for "its own sake" rather than for extrinsic consequences. For example, children may play because of the enjoyment of the activity itself without any other apparent extraneous rewards. Recent research has examined whether providing extrinsic consequences for activities which are performed for their own sake can undermine motivation.

The prediction that performance can be undermined by extrinsic consequences comes from self-perception theory in social psychology (Bem, 1967) and relies upon an individual's interpretation of the locus of control over his or her own behavior (deCharms, 1968). Specifically, an "overjustification" hypothesis has been posed to predict that providing strong extrinsic consequences for a behavior may alter an individual's perception of the reasons he performs the behavior (Lepper, Greene, & Nisbett, 1973). According to the overjustification hypothesis, if an individual expects extrinsic consequences for a task, this will cause him to reattribute the cause of his behavior to extrinsic rather than to intrinsic factors. The intrinsic motivation for performing the task is decreased as the extrinsic consequence becomes the primary incentive for performance. The extrinsic reward, as it were, replaces the intrinsically motivating factors of the task. When the extrinsic consequences are no longer provided, the individual engages in the task less often because intrinsic reinforcers that previously maintained performance are no longer present. Thus, performance of the task deteriorates well below the levels found prior to the administration of extrinsic consequences.

The implications of this view are great because they suggest that tokens, an extrinsic consequence, may actually harm individuals because their delivery will reduce in performance of some target behavior when the contingencies are ultimately withdrawn. Thus, tokens may lead to short-term improvements but long-term decrements in behavior (Levine & Fasnacht, 1974).

Investigations in social psychological experiments have demonstrated that providing extrinsic reinforcers such as money for a task may temporarily increase performance of the task but later decrease performance when the reinforcers are no longer provided relative to a baseline rate. Much of the initial systematic work in this area was completed by Deci (1971, 1972a, 1972b). In the basic paradigm of many of these studies, college students received money for completing various puzzles (using the Soma puzzle that can be made into different shapes including a cube). After individuals received money for completing the puzzles within specified time periods, they had the opportunity to work on the puzzles during a free-time period in which no consequences were provided. Interestingly, subjects who had previously received payment, worked on the puzzles less than control subjects who had not been paid (Deci, 1971, 1972b). These results were interpreted to suggest that extrinsic rewards led to a subsequent reduction in interest in the task and interfered with performance after the rewards were withdrawn.

Several other studies have demonstrated that when extrinsic consequences are provided subsequent performance may be reduced after the consequences are withdrawn (see Deci, 1975; Notz, 1975; Staw, 1975, for reviews). The effects of undermining subsequent performance sometimes depends upon whether individuals are told in advance that extrinsic consequences are provided, although this is not always the case (Kruglanski, Alon, & Lewis, 1972; Lepper et al., 1973). Also, the effect has been shown whether the consequences are administered contingently or noncontingently, although there is current debate about this point (Deci, 1972b; Greene & Lepper, 1974; Kruglanski, Friedman, & Zeevi, 1971).

The research on undermining performance with extrinsic consequences recently has received increased attention. One reason for this is that the research which shows a reduction in performance following withdrawal of extrinsic reinforcement stands in sharp contrast with the overwhelming majority of token-economy programs, with rare isolated exceptions (Meichenbaum et al., 1968). As noted earlier, termination of token reinforcement usually results in a return of behavior to near baseline levels. However, the level rarely drops below baseline and frequently is maintained at a level higher than the original baseline.

Research has attempted to explain why laboratory studies show that extrinsic reinforcement can impede performance. Some evidence suggests that characteristics of the laboratory studies showing a deleterious effect of extrinsic rewards differ in significant ways from token economies which do not usually show such an effect. For example, Reiss and Sushinsky (1975a, Exp. 1) found that promising children an extrinsic reward (opportunity to play with a doll) for performance did decrease performance after withdrawal of the reward relative to subjects who had never received the reward. The effect was obtained when there was a single trial (performance opportunity) in which the reinforcer was provided independently of a specific target response. In their second experiment, extrinsic reinforcers (tokens exchangeable for a toy) were administered across *several trials* in which specific behaviors were *differentially reinforced*. In this situation, withdrawal of extrinsic consequences was not associated with a decrease in performance. Indeed, the positive effects of token reinforcement carried over to the situation in which tokens were no longer presented, thus showing an effect opposite of the overjustification prediction.

Similarly, Feingold and Mahoney (1975) attempted to test the overjustification hypothesis in a situation which more closely resembled token economies than the usual previous laboratory investigations. Children received extrinsic consequences (tokens exchangeable for prizes) for several sessions contingent upon response rate on a laboratory task. Performance immediately after reinforcement was withdrawn and up to 6 weeks of follow-up was consistently *higher* than the original baseline. Hence, token reinforcement enhanced subsequent performance.

The results from the above experiments show that when reinforcement is administered repeatedly across several trials or several sessions and is contingent upon specific responses, withdrawal of consequences may enhance rather than suppress performance relative to an initial baseline. Of course, repeated reinforcement trials and the contingent delivery of reinforcers for differential responding have not been included in many of the laboratory investigations showing response suppression following withdrawal of extrinsic consequences. However, both extended periods of reinforcement and contingent delivery of tokens characterize token economies. Because of some of the important differences between the way in which reinforcers are delivered and the duration over which they are delivered between laboratory work on "intrinsic motivation" and token economies, the generality of the overjustification effect has been questioned (Bornstein & Hamilton, 1975; Reiss & Sushinsky, 1975b).

The controversy regarding the precise situations in which performance of an activity is decreased subsequent to the delivery of extrinsic consequences cannot be fully treated here. The controversy has been actively argued in several sources with no clear resolution (Bornstein & Hamilton, 1975; Deci, 1975; Lepper & Greene, 1974, 1976; Reiss & Sushinsky, 1975a, 1975b, 1976). Investigators who have argued the overjustification hypothesis have recognized that a large number of conditions might increase rather than decrease "intrinsic motivation" (Lepper & Greene, 1974, 1976).

For present purposes, it is important to mention briefly features which make some of the laboratory research on overjustification of unclear relevance to token economies. Differences between the laboratory and token economies such as the duration of the reinforcement period and the contingent delivery of consequences have already been mentioned. There are other distinguishing features. First, most of the studies of "intrinsic motivation" have examined the effect of extrinsic consequences in situations which bear little resemblance to situations in which token economies are conducted. Contrived laboratory tasks have been used (e.g., completing puzzles, connecting dots, listening to tapes) that differ from the usual behavior reinforced in token economies (e.g., correct academic performamce. attentiveness, rational talk, social interaction, self-care). The differences are potentially important because many of the target behaviors in token economies have some influence or operate on the natural environment after they are performed. For example, mastery of an academic assignment after token reinforcement is likely to bear some rewards of its own. Thus, behavior might be maintained after extrinsic consequences have been withdrawn.

Second, in laboratory studies, the effect of extrinsic consequences is evaluated on a response that already is performed. This has been a necessary requirement because the studies have sought to demonstrate that extrinsic consequences can undermine performance of some behavior that already is maintained without extrinsic consequences. In token economies, of course, tokens are used to develop behaviors that are not performed frequently or perhaps are never performed by the clients. In the terminology of the overjustification literature, the clients do not have the "intrinsic motivation" to perform the target response. For example, psychiatric patients may not attend therapy, students may not complete academic assignments, and retardates may not engage in toileting or other self-care responses. The notion that token reinforcement can in some way undermine performance is not relevant because performance may not exist at all without incentives based upon extrinsic consequences.

A third and related point is that in order to develop performance of some behavior for its own sake (i.e., independently of specifically programmed consequences), the behavior must be performed. Extrinsic consequences can be used to help initiate behavior so that natural consequences which may follow from the behavior can control continued performance (Baer & Wolf, 1970). In most token programs, the goal is to develop behavior so that the tokens can be withdrawn and behavior can be maintained by other events. The delivery of tokens is viewed as a necessary albeit transitory step toward this end.

COMPARISON OF THE TOKEN ECONOMY WITH OTHER TREATMENT TECHNIQUES

Although token economies have been shown to effect marked behavior changes, there is relatively little research comparing token economies with other treatment strategies. Perhaps, one reason for the paucity of treatment studies is that between-group designs are routinely required for comparing different treatments. Token economy research has developed within the intrasubject-design tradition and has avoided many aspects of group research and all that is entailed by such comparisons (Kazdin, 1973f; Sidman, 1960). Nevertheless, an important question is the extent to which token economies are superior to alternative treatment strategies. For purposes of organization, the studies might be grouped according to the specific comparisons made.

The Token Economy versus Standard or Routine Treatment

A large number of studies have compared token reinforcement with the usual nonspecific interventions that are used. Comparison studies have been conducted most frequently in psychiatric hospitals where the token economy is compared to custodial ward treatment in which no specific procedures are invoked to alter particular target behaviors.

The earliest comparison study was reported by Schaefer and Martin (1966) in which chronic schizophrenic patients were randomly assigned to experimental or control groups. Experimental subjects received tokens for self-care, social interaction, and work performance, whereas control subjects received routine ward treatment and tokens noncontingently. Over a 3-month period, the experimental subjects decreased in "apathy" ratings, defined on the basis of the amount and type of activities performed, whereas control subjects remained at baseline levels.

Gripp and Magaro (1971) compared token reinforcement with routine hospital treatment with psychiatric patients on a variety of dimensions including several behavioral rating scales. Token reinforcement was delivered for job performance on and off the ward as well as for behaviors individually selected for each patient. Six months after the program had been in effect, patients on the token-economy ward made several improvements in scales reflecting cognitive and affective concomitants of psychoses. Fewer gains were evidenced by control subjects.

Shean and Zeidberg (1971) compared token reinforcement and traditional ward treatment in developing self-care and social behaviors, participation in activities, and reduction of bizarre behaviors. Token-reinforcement subjects showed greater cooperativeness on the ward, communication skills, social interaction, participation in activities, time out of the hospital, and reduction in use of medication than did control subjects. In a similar comparison, Maley et al. (1973) found that patients in a token-economy ward were better oriented, more able to perform a discrimination task, follow complex commands, and handle money in making business purchases than custodial control patients. Moreover, ratings of behavior indicated that token-economy patients needed less hospitalization and were more likeable than control patients.

Birky, Chambliss, and Wasden (1971) compared the efficacy of a token economy versus traditional ward care on discharge rates from the hospital. Although there were no differences in the number of patients discharged (i.e., remaining out of the hospital for at least 6 months) from the token economy and two control wards, the patients discharged on a token-economy had a greater length of previous hospitalization. Thus, the token economy had a greater effect on discharging more chronic patients.

Heap et al. (1970) combined behavior-milieu therapy consisting of token reinforcement, ward self-government, and several other adjunctive procedures with traditional ward care in developing self-care skills. The reinforcement group showed greater performance of self-care skills and rate of discharge over a 35-month period than did control subjects. (The conclusions of this study have been criticized on the basis of several methodological flaws [Carlson et al., 1972; Hersen & Eisler, 1971; Kazdin, 1973f].)

Overall, the research comparing token-economy treatment with custodial ward care supports the superiority of the reinforcement procedures. It comes as little surprise that a token program, or perhaps, any specific treatment can increase appropriate performance relative to custodial care. Indeed, custodial care might not be regarded as another treatment but closer to a no-treatment control condition. All sorts of practices including even a change in wards or a move to improved facilities can improve performance of patients in a psychiatric ward (DeVries, 1968; Gripp & Magaro, 1971; Higgs, 1970). That is, relatively minor changes in the ward environment can effect durable behavior changes relative to custodial ward care.

More convincing evidence of the efficacy of token reinforcement might be drawn from settings other than the hospital where standard treatments constitute specific techniques. For example, in the classroom, token reinforcement is usually compared with traditional educational methods that involve use of specific curricula, delivery of letter grades, teacher feedback, reprimands, approval, and so on. Some of these procedures such as approval are used in a token economy but are more carefully programmed than is routinely the case in the classroom.

In token programs in academic settings, baseline conditions constitute specific traditional treatment methods. The superiority of token economies in studies in the classroom must be viewed against a baseline which already is applying a traditional educational technology. Aside from drawing inferences from within-group differences across baseline and treatment phases, between-group comparisons have been made showing the superiority of token reinforcement. For example, Rollins *et al.* (1974) compared 16 inner-city school classes that received a token program with 14 control classes that did not. Experimental classes were superior on measures of on-task and disruptive behavior, intelligence, and reading and math achievement relative to control classes. The superiority of the token economy to traditional classroom practices comes as no surprise. Many consequences in the usual classroom situation are not systematically programmed. Even if they were, frequently used consequences such as letter grades are not as effective as tokens backed by other reinforcers (McLaughlin & Malaby, 1975b).

Various investigations have compared token reinforcement and standard treatment conditions in quasi-experiments (Campbell & Stanley, 1963). In these comparisons, subjects who complete the token program are compared with nonrandomly assigned subjects who are similar but have experienced some other treatment regimen. For example, and as noted earlier, Fixsen *et al.* (1976) reported that pre-delinquents who participated in a token program at a home-style facility showed fewer police and court contacts and higher grades and school attendance at a two-year follow-up than individuals who had served in an institution or were placed on parole. Similar comparisons with delinquent soldiers have supported the efficacy of the token economy at follow-up assessment (Stayer & Jones, 1969).

The Token Economy versus Other Specific Treatments

Several studies have compared token reinforcement with specific treatments rather than standardized practices such as custodial ward care. In these studies, interventions such as insight therapy, milieu treatment, and other active treatment strategies are tested against or combined with token reinforcement.

Some of the studies have been conducted in hospital settings with psychiatric patients and have looked at diverse treatments. Marks, Sonoda, and Schalock (1968) compared reinforcement and relationship (individual) therapy with chronic schizophrenics. Each of two groups received both treatments but in a counterbalanced order. Reinforcement consisted of the delivery of tokens by the staff at their discretion for improvement in individualized problem areas (e.g., personal appearance, expressing feelings). On several rating scales and tests of mental efficiency, language and social skills, self-concept, and others, there were no differences between treatments. The authors noted the difficulty in keeping the procedures distinct (e.g., avoiding some reinforcement in relationship therapy). The lack of well-specified criteria for administering reinforcement and the failure to evaluate the programs on specific target behaviors make the results somewhat difficult to interpret.

Hartlage (1970) compared traditional "insight"-oriented therapy with reinforcement (social, activities, and consumables) for adaptive responses. After 7 weeks of treatment, the reinforcement group showed greater improvement on a measure of hospital adjustment and on therapist ratings than the traditional-treatment group. However, groups did not differ on self-concept rated after treatment.

Olson and Greenberg (1972) exposed psychiatric patients to one of three treatments for 4 months: milieu therapy, interaction (milieu plus weekly group therapy), and token reinforcement. The reinforcement group earned tokens for making group decisions regarding ward administration and for attending activities. The token-reinforcement group attended more activities in the hospital and spent more days on town passes and more days out of the program. However, not all measures favored the token system (e.g., ratings of social adjustment).

Greenberg et al. (1975) compared a token economy with a token economy combined with a milieu approach. While both programs included tokens for self-care, social, and work behaviors, the latter group included a group-incentive contingency which was designed to develop decision-making responsibilities and to use the peer group as a source of influence. Individuals were assigned to groups of patients that were given responsibility to design treatment programs for their members and to make recommendations that would eventually result in discharge. Although patients in both programs received tokens for individual performance, group-incentive patients also received tokens for the treatment proposals devised by their groups. During the 12-month experimental period, patients in the group-incentive program had spent more days out of the hospital than patients without the group condition. This study suggests the advantages of including group incentives and milieu features in a token program.

Not all comparative studies have been completed in hospital settings. For example, Fo and O'Donnell (1974) used a "buddy" system in which individuals recruited from the community individually treated youths who were referred for behavior and academic problems. The youths received one of four different treatments on an individual basis with their "buddies," namely, relationship therapy in which a warm positive relationship was established, social approval in which approval contingent upon appropriate behavior was administered, social and material reinforcement in which contingent approval and monetary reinforcement were administered, and a control condition in which individuals received no treatment. Both social and social-material reinforcement conditions decreased school truancy rates significantly more than did relationship and control conditions. The reinforcement group also showed reductions in assorted problem behaviors such as fighting, not doing chores, staying out late, and not completing homework assignments.

Alexander and Parsons (1973) compared the effects of a behavioral, client-centered, and psychodynamic treatment to alter the interaction patterns of families of juveniles referred by the court. The behavioral treatment was designed to develop positive family communication patterns and contingency contracts to alter behaviors of the delinquents. Appropriate communication patterns were modeled and socially reinforced by the therapist. (Occasionally, contingency contracts utilized token reinforcement with the youths.) In the client-centered and psychodynamic groups, attitudes and feelings about family relationships and insight into the problem behaviors were focused upon, respectively. On process measures, the behavioral treatment led to more positive family interaction patterns as reflected on several measures of verbal behavior. Most importantly, for a 6- to 18-month interval after treatment, delinquents who had participated in the behavioral program showed lower recidivism (i.e., re-referral to the courts) than the other treatment groups and a no-treatment control group.

Jesness (1975) compared the effects of a token economy with transactional analysis with over 900 adjudicated delinquents randomly assigned to the different treatments (conducted at different facilities). In the token economy, routine ward, academic, and specific problematic behaviors were related to earning points. In transactional analysis, the client met with a counselor to identify treatment objectives and also met in group-therapy sessions. Across an extremely large number of measures, the results indicated generally that behavior ratings tended to favor the token program whereas self-report and attitudinal measures tended to favor transactional analysis. Interestingly, follow-up data up to 2 years after treatment indicated no differences in reconvictions after parole.

The Token Economy versus Pharmacotherapy

The token economy has been compared with pharmacotherapy in several investigations. These comparisons are of particular interest because of the widely held view that drugs and reinforcement techniques are ways of managing problem behavior. Aside from the comparison of drugs and token reinforcement, many of the investigations have explored the efficacy of combined drug and reinforcement procedures.

Ayllon, Layman, and Kandel (1975) studied the relative efficacy of methylphenidate (Ritalin) and token reinforcement in altering the behavior of three children diagnosed as chronically hyperactive and enrolled in a learning disability class. Hyperactive behaviors (e.g., gross motor behaviors, disruptive noise, disturbing others) were assessed across math and reading periods. Performance was assessed while medication was provided and after medication was withdrawn. A token system, implemented after medication was withdrawn, was introduced in a multiple-baseline design across math and reading periods. In this program, points were earned for correct academic performance on assignments.

For each of the three subjects, discontinuing administration of Ritalin led to an increase in hyperactive behaviors. However, implementing the token program increased academic performance and reduced hyperactive behaviors. The reduction in hyperactive behaviors matched the level achieved with Ritalin alone. Thus, the drug and token economy were equally effective in controlling hyperactive behavior. Yet, when the drug was used alone, academic performance was low and improved slightly when the drug was withdrawn. Under the token program, academic performance accelerated dramatically for each subject. Thus, the token economy was superior to drug treatment in accelerating academic performance. Overall, these results suggest that reinforcement was superior to drug treatment because of the differential effects of these two treatments on academic performance.

Christensen and Sprague (1973) altered hyperactive behavior of socially maladjusted and conduct-problem children. Twelve children, divided into two groups, either received Ritalin or a placebo. In a class setting, a reduction of in-seat movements (recorded automatically by stabilimetric cushions) was rewarded with tokens. Token reinforcement was administered concurrently with drug and placebo conditions. The results indicated that the token program reduced in-seat movements for both drug and placebo groups. Yet, the effects were more immediate, somewhat greater, and slightly more resistant to extinction for the group that received the drug. Overall, token reinforcement was superior to drug treatment alone (measured during a drug-only baseline phase) but not as effective as a combination of drug and reinforcement treatments.

A subsequent study by Christensen (1975) compared the effects of a token program administered concurrently with Ritalin and placebo conditions with hyperactive institutionalized retarded children in a classroom situation. The reinforcement program consisted of providing tokens for appropriate classroom behaviors such as paying attention, working, completing assignments, and doing work correctly. Across several phases of an ABAB design, the token program was paired twice with drug and placebo conditions. The token economy decreased inappropriate behaviors as reflected in teacher ratings, stabilimetric seat measures of activity, and observations of several inappropriate behaviors. The effects of the program were not consistently related to whether it was administered concurrently with Ritalin or the placebo.

McConahey (1972) investigated the effects of a token economy and drug treatment with behavioral-problem adult women diagnosed as profoundly to moderately retarded. Residents on the ward participated in a program during the morning where they received tokens (redeemable at a "store" in the facility) for performing self-care behaviors, working on the ward, completing constructive activities (e.g., reading, doing puzzles), following instructions, and other behaviors. No contingencies were operative across afternoon periods each day. While the token program was in effect, chlorpromazine (Thorazine) and a placebo were administered in an ABAB design. Daily observations were made across a large number of behaviors including social interaction (e.g., talking with residents or staff, assisting others), aggressive behaviors (e.g., hitting, kicking, and biting others), disruptive behaviors (e.g., pestering others), and self-care responses (e.g., dressing, soiling).

The token program was markedly effective in improving a variety of appropriate behaviors and in reducing inappropriate behaviors on daily behavioral observations and posttreatment behavioral checklists. The token program was not differentially effective across drug and placebo phases. Behavior changes generally were restricted to the morning periods when the contingencies were in effect. In those phases in which the drug was administered (and in effect all day), performance of inappropriate behavior was not controlled during the afternoons. Thus, when token-reinforcement procedures were utilized alone (i.e., with only placebo treatment) in the morning, behavior changes were marked whereas when the drug was utilized alone in the afternoon, no consistent changes were found.

A subsequent extension of the above project replicated the finding that token reinforcement led to markedly greater gains than did drug treatment across a wide range of specific overt behaviors (McConahey, Thompson, & Zimmerman, 1977). This latter report also demonstrated that higher drug doses were slightly more effective than lower doses in controlling aggressive responses although the effectiveness of the high doses did not approach the dramatic improvements achieved with the token-reinforcement program. Other investigations have demonstrated that reinforcement programs effectively alter the behavior of institutionalized retardates and psychiatric patients and that the addition of psychoactive drugs such as chlorpromazine does not enhance the effects of the program (McConahey & Thompson, 1971; Paul, Tobias, & Holly, 1972).

The results of the drug studies have shown that token reinforcement is a viable alternative to drug therapy in treating hyperactive, aggressive, and a wide variety of general ward behaviors. Token reinforcement has been shown to be more effective than drug therapy in the relatively few studies in which comparisons have been made. Given the paucity of studies across diagnostic groups and drugs, unequivocal conclusions cannot be made. Although the studies currently available usually use "double-blind" procedures, other methodological features sometimes limit the conclusions that can be drawn. For example, studies occasionally have confounded the interventions with sequence effects so that one condition (e.g., drugs or token reinforcement) consistently follows or precedes the other. Also, the duration of the nondrug phases sometimes may not be sufficiently long in select comparisons to ensure that there are no residual drug effects operative when the reinforcement program is implemented. Yet, of the comparative studies, the drug versus token-reinforcement comparisons have yielded relatively consistent results.

Overall, the comparative research suggests that the token economy is more effective than standard treatments such as custodial care, traditional education practices, and institutional correctional treatment. The evidence comparing the token economy to any other particular procedure is relatively sparse. Select studies show the token economy to be superior to variations of insight-oriented treatment or pharmacotherapy. Yet, the paucity of comparative studies evaluating these techniques makes any conclusions tentative.

Comparative studies of the effectiveness of token economies and treatment programs in general require extensive methodological efforts. In many cases, the studies cannot meet the demands of experimentation that are required to render unambiguous conclusions. For example, comparative studies involving token reinforcement often have confounded treatment with other variables such as a change in setting (Birky *et al.*, 1971; Heap *et al.*, 1970), selection of special staff to administer the token economy (Gripp & Magaro, 1971), initial differences in the client's disorders (Shean & Zeidberg, 1971), selection of subjects on the basis of consent regarding particular aspects of treatment (Christensen & Sprague, 1973), and others. In most comparisons, investigators have conducted quasi-experiments because control over all of the variables essential for experimental control was not possible. While research under such circumstances is admirable and to be actively endorsed, the conclusions that can be drawn of necessity are tentative. It should be noted that the lack of clear experimental evidence in favor of the token economy relative to other specific interventions applies to other treatment techniques as well. Comparative studies are rare independently of the techniques studied.

EVALUATION OF THE EFFECTS OF TREATMENT

Token-economy research has been sensitive to the importance of program evaluation. In most of the programs, intrasubject changes in behavior are evaluated as a function of alteration of the reinforcement contingencies. These often are supplemented with between-group comparisons. Notwithstanding the general tendency to evaluate the effect of treatment, the data obtained as part of this evaluation might be criticized. Specifically, data obtained reflect the extent to which behavior change has been achieved. Relatively little evidence has been obtained showing that the changes achieved are clinically important.

Applied behavior analysis, of which the token-economy research is a part, has stressed the clinical significance of treatment as a criterion of evaluation. This criterion includes at least two characteristics, namely, that the behavior focused upon and the magnitude of change achieved for that behavior are of clinical or social significance (Baer *et al.*, 1968; Hersen & Barlow, 1976; Kazdin, 1977b; Risley, 1970). In many programs, the target behaviors selected such as academic performance, psychotic symptoms, stuttering, and others obviously are clinically important. However, target behaviors sometimes focused upon such as performance of duties to maintain the psychiatric or prison ward may bear an unclear relation to the client's ultimate rehabilitation. In cases in which the behaviors are of clear clinical relevance, a major issue is the extent to which the changes made are clinically important.

Unfortunately, applied work has not specified the criteria that can be invoked to determine whether a change is clearly of applied or clinical significance. The inappropriateness of statistical significance has been widely discussed but positive objective alternatives for evaluating clinical change have not been readily offered (cf. Kazdin, 1976c). In some cases, it is obvious that even a dramatic change effected with an intervention is not clinically significant. For example, if the self-destructive behavior of an autistic child were reduced from 80% to 40% of the times observed, the finding would be of questionable clinical significance. In fact, self-destructive behavior of any rate is maladaptive and an intervention would have to eliminate the behavior to effect a change of unambiguous clinical import.

Attempting to achieve behavior changes of clinical significance presupposes a criterion toward which treatment can strive and against which the program's effects can be evaluated. Select investigations have attempted to provide some clinical or social validation of program effects by showing that the extent of behavior change places the client within the range of acceptable levels of behavior evinced by his peers or leads others to evaluate behavior more favorably.

Patterson (1974) demonstrated that behavioral interventions with deviant boys in the home and at school altered behavior. The clinical significance of the behaviors focused upon are not at issue because the conduct problems altered served as a basis of client referral to the program. Interestingly, data were collected which indicated that the magnitude of change was of clinical importance. The intervention and follow-up data revealed that the performance was brought into the range of nondeviant control subjects (i.e., peers). At least for the target behaviors selected, the clients came within the acceptable range although their pretreatment behavior was outside of this range.

Walker *et al.* (1971) used token reinforcement to increase the attentive behavior of elementary school students who were behavioral problems in the classroom. The program improved attentive behavior. Normative data revealed that the program increased attentiveness to a level which approximated but was still below the level of their nonproblem classmates.

Stahl *et al.* (1974) developed eye contact, verbalizations, and other behaviors in psychiatric patients. Interestingly, the frequency of eye contact was increased substantially for one subject although the rate was approximately 24% below the "normal level" for intelligent well-adjusted subjects observed in a similar situation. Similarly, for another subject, verbal responses were increased but still remained approximately 35% below the level for normally functioning individuals. These data show the extent to which the program effected change and more importantly the extent to which further changes might be necessary.

Aside from showing that behavior change associated with the program brings the client to the level of some normative group, social validation also has relied upon the opinions of others to determine the importance of the change. Maloney and Hopkins (1973) used token reinforcement to alter various discrete characteristics of compositional writing in elementary school children. Responses such as the use of different adjectives, action verbs, and sentence beginnings were reinforced. To determine the importance of changes in these responses, individuals unfamiliar with the study rated the creativity of the compositions associated with baseline and intervention phases. Superior ratings of creativity were associated with compositions written under experimental phases in which specific writing responses were reinforced. The ratings add weight to any claims that altering discrete responses are of applied significance.

Quilitch (1975) increased the participation of mentally retarded residents in recreational activities. To evaluate whether the target behaviors selected were clinically important, a large number of staff (mental health technicians, nurses, social workers, secretaries) and some parents of the residents rated the activities. All indicated that the activities would be a worthwhile goal. After the changes had been effected in the behavior, ward staff rated the results of the program and indicated that marked improvements had occurred.

In a community extension of behavior modification, Briscoe, Hoffman, and Bailey (1975) trained lower-socioeconomic-status adults who participated as board members on a rural community project to engage in effective problem-solving behaviors at the board meetings. Training was conducted to help individuals identify sources of problems under discussion, state and evaluate alternative solutions, and decide on the action to be taken. The training program consisted of social reinforcement, shaping, prompting, and fading in individual training sessions. To determine whether training had impact on the meetings rather than merely making changes in discrete verbal responses, two professors and two community leaders not associated with the project but familiar with group problem solving and policy setting, rated sections of baseline and intervention videotapes of the board meetings. Generally, the intervention tapes were rated as showing greater problem-solving ability than were the baseline tapes.

A number of studies conducted with pre-delinquent boys and girls at Achievement Place have assessed the social validity of treatment. Minkin *et al.* (1976) trained pre-delinquent girls in conversational skills by increasing the frequency that they asked questions and provided positive feedback to another person during conversation. The authors attempted to socially validate the importance of the behavior change by having a panel of judges rate conversational ability. Judges rated posttreatment conversations as evincing greater conversational ability than pretreatment conversations. In addition to the judges' ratings, normative data showed that posttreatment ratings of conversational ability fell within the range of university and junior high school females in a similar conversational situation.

Another study focusing on conversational behavior increased volunteering of information and appropriate posture of pre-delinquent girls during conversation (Maloney, Harper, Braukmann, Fixsen, Phillips, & Wolf, 1976). Subsequently, judges (e.g., a social worker, a teacher, a student, a probation officer, and a counselor) rated selections of baseline and treatment conversations. The girls were rated as more polite and cooperative in their training conversation relative to baseline.

Phillips *et al.* (1973) evaluated the effect of different token-reinforcement procedures on room cleaning with pre-delinquent boys at Achievement Place. A criterion for providing or withdrawing consequences was that 75% of the bathroom tasks were completed (e.g., clean sink and counter top, towels hung appropriately). To determine whether this criterion represented a satisfactory level of overall cleanliness independently of the specific codes, individuals who did not participate in the project rated the bathroom under conditions in which a different number of tasks were completed. When 75% of the tasks were completed, there was consensus among five raters that an adequate level of cleanliness had been achieved. When only 50% of the tasks were completed, the next lower percentage rated below 75%, there was consensus that the bathroom was less than adequately clean.

IMPORTANT CONSIDERATIONS IN VALIDATING
THE EFFECTS OF TREATMENT

Establishing the clinical value of the change achieved in reinforcement programs has been accomplished either by determining whether the behavior of the client falls within the range of acceptable behavior among the client's peers who are adequately functioning in the environment or by assessing the opinions of others regarding the importance of the change achieved. Each method of evaluation raises important concerns.

Using one's peers as a criterion for evaluating the importance of clinical change raises an argument frequently voiced in the study of psychopathology, namely, that conformity with one's peers is not necessarily an adequate criterion for adjustment and adaptive behavior. While there is merit in not advocating conformity, behavioral techniques usually *are* implemented for individuals whose behaviors deviate from normative levels. The implicit goal of the intervention is to develop appropriate behavior as defined by some social norm. That is, behavioral interventions in many cases have been trying to achieve social integration of the client by developing normative levels of performance. For example, developing appropriate eating, toileting, grooming, social, work, and other behaviors focus on areas that deviate from normative levels. Behavioral interventions have attempted to achieve "conformity" with existing norms for many target behaviors but have failed to evaluate the extent to which this is achieved. This is a major criticism of token economies and other behavioral programs. The extent to which changes in performance are important has not been carefully evaluated.

Social validation requires examination of the normative group (Walker & Hops, 1976). Some attempts have been made to develop normative data that can be used as a criterion in various situations. For example, Johnson, Wahl, Martin, and Johansson (1973) studied how deviant the "normal" child is in the home. Across a diverse set of behaviors including demanding attention, violating a command, not complying, having tantrums, whining, making threats, and other behaviors, the level of deviant behaviors was shown to be negligible (i.e., 96% of the average child's behavior was nondeviant). Examination of the extent to which nonproblem groups engage in the behaviors focused upon in behavioral interventions represents a worthwhile beginning in establishing criteria for social validation. Of course, this approach is not without problems. For example, deviant behavior among various treatment populations is not sole criterion for referral to mental health professionals or treatment facilities (Scheff, 1966; Ullmann & Krasner, 1975).

The criterion used in select studies and advocated here uses a normative group as a basis for deciding whether treatment has effected sufficient change. There are a number of situations in which this criterion may be inadequate. For example, many behavioral programs are applied to individuals who may never return to normal peer levels such as the severely and profoundly retarded. Here the focus of behavioral intervention on clinically important behavior is sufficient and the criterion for continuation of a program is that competing procedures are not as effective. Thus, in cases in which programs are conducted on a quasi-permanent basis with populations who are not likely to be returned to normal functioning, the comparison of performance against a normative group is not appropriate.

As another problem, it might be somewhat arbitrary to specify those individuals who constitute one's peers and who should be the normative group. For example, what is the normative group for prisoners who have completed a token economy? Presumably, one group would be same-aged and -sexed peers who are functioning without contact with the law. However, this normative group might be unrealistic in light of diverse differences in background, socioeconomic standing, and other variables.

Finally, programs sometimes develop behavior well beyond the level of a normative group. For example, it might be unfortunate to use a criterion of academic achievement performance of individuals functioning in a typical classroom setting. Normative levels of performance in education could be accelerated. To use normative levels as the sole criterion does not challenge the possible lack of programming for nonproblem individuals.

Overall, many token programs have not critically examined the extent to which changes effected with treatment are of clinical importance. Although comparing clients to a normative group might not be relevant in many programs and selecting an appropriate group presents problems for other programs, many token programs do focus on behaviors (e.g., conduct problems, academic performance, social withdrawal, bizarre behaviors) among individuals who are returned to or already exist in the natural environment. In these cases, the extent to which changes of clinical significance have been achieved can be evaluated.

The clinical importance of the change sometimes is assessed by soliciting the opinions of individuals to whom the change might be relevant or who are in a position to evaluate the behavior. A major issue with this method of validating the extent of change is in assessing opinions. Certainly, an essential criterion to determine whether the changes effected are important are the opinions of those who are in frequent contact with the client. Yet, opinions about whether behavior changes are adequate are unsatisfactory as the sole means of evaluation because they often are more readily altered than the actual target behaviors upon which they are supposedly based (Kazdin, 1973g; Schnelle, 1974). However, after change in the target behavior has been effected, evaluation of the adequacy of that change by individuals in the environment adds important information. Relatively few investigations have determined whether the changes made are perceived as important by others.

CONCLUSION

The token economy might be criticized along several lines. First, the specific behaviors focused upon frequently do not appear to reflect the most salient problems of the target population. For example, in institutional settings, behaviors adaptive to a particular treatment environment rather than to community life serve as the major focus. Second, many programs seem to be implemented in settings which are counterproductive to the aim of the contingencies. In many applications, token economies are implemented in facilities that may contribute to the problems that the programs are designed to ameliorate. Third, the population focused upon in token economies warrants additional consideration. While an extremely wide range of populations has been studied, the focus might be more profitably placed upon the behavior-change agents rather than the clients themselves. Focus on the behavior of hospital staff, teachers, parents, peers, and the contingencies that support staff behavior may lead to more durable and broader changes in the clients.

The criticism against the token economy should be viewed in the context of other treatments. Relative to existing treatment alternatives, the token economy reflects an extremely broad focus across behaviors, settings, and populations. Also, the specific behavior changes achieved cannot be faulted on the basis of their superficiality. Select studies have shown that specific behavior changes within a given treatment setting are associated with improvements in generalized measures of attitude and achievement and socially relevant measures such as posttreatment community adjustment (e.g., recidivism of delinquents).

The study of variables that contribute to the effects of the token economy is an area where research could be more analytic. The token economy is a complex intervention that has multiple components including instructions about the target behaviors and contingencies, social reinforcement and interpersonal contact, feedback, and modeling influences. The contribution of these components to behavior change has been infrequently assessed. A related area of research that might be related to treatment outcome is the study of subject variables. Specific components or features of the token economy might be differentially effective along dimensions related to individual differences.

One area of criticism of the token economy that has received recent attention is the effect of providing extrinsic consequences on behavior. Some authors have proposed that extrinsic consequences can *undermine* intrinsic motivation. Undermining intrinsic motivation is assumed to result from the subject's reattributing the reasons for performing a response from the intrinsic benefits to the extrinsic consequences. Removing the extrinsic consequence is thought to reduce behavior below the original baseline level because there is no longer a reason (intrinsic or extrinsic) to perform the response.

Laboratory research has demonstrated the deleterious effects of extrinsic consequences on behaviors. The relevance of this research for token economies is unclear. The overwhelming evidence in token-economy research shows that performance following the withdrawal of tokens does not decrease below original baseline levels. Usually, performance following withdrawal of token reinforcement is superior to baseline levels of performance.

There are substantial differences of the effect of extrinsic consequences in laboratory research and token economies in terms of the behaviors focused upon, the operant rates of these behaviors, the relationship between behavior and the extrinsic consequences, the delivery of unprogrammed (e.g., social) reinforcers, the duration that the contingencies are in effect, the range of back-up events, and other aspects. As it stands, there is no clear basis for stating that extrinsic consequences undermine behavior in applied work and vast evidence against this proposition.

A major issue for the evaluation of the token economy is its effect relative to other treatment, rehabilitation, and educational techniques. Generally, the token economy has been shown to be more effective than standard treatments such as custodial care in psychiatric hospitals and traditional educational practices. Comparative studies with specific treatments suggest that the token economy fares better than insight-oriented and relationship therapy, milieu therapy, pharmacotherapy, and a few other techniques. Although the paucity of studies comparing specific techniques and the quasi-experimental nature of some of these studies argue against drawing unequivocal conclusions, the pattern of results across several studies tends to favor the token economy.

Evaluation of treatment effects is another area that warrants scrutiny in token-economy research. While most programs provide supporting data to evaluate the effect of the program, the clinical importance of the changes remains to be determined. The criticism, of course, applies to all clinical treatments and is certainly not more applicable to token reinforcement. Yet, now that behavior change has been so readily achieved in token economies, the importance of these changes needs to be considered more analytically. Two means of evaluating the importance of the change have begun to be used, namely, whether changes in the target behaviors place the client within the range of acceptable behavior evinced by his peers or lead others to evaluate the behavior more favorably in the client's everyday environment. Invoking these criteria is not without problems such as identifying an appropriate peer group and relying upon the opinions of others. Yet, searching for means to evaluate the clinical significance of change is an important area of research that warrants additional attention.

EXTENSIONS OF REINFORCEMENT
TECHNIQUES TO SOCIETY

9

As reviewed earlier, token reinforcement has been applied widely across treatment, rehabilitation, and educational settings and with populations whose behaviors often are regarded as deviant or deficient (e.g., psychiatric patients, delinquents, special-education students). Treatment and educational applications of reinforcement constitute the vast majority of token programs. However, there have been extensions of reinforcement techniques to various problems of social living that transcend the usual applications. This chapter reviews the extensions of reinforcement techniques, particularly token reinforcement, to socially and environmentally relevant concerns including pollution control and energy conservation, job performance and unemployment, community self-government, racial integration, and military training. Aside from these extensions to specific areas, the design of communities based upon reinforcement also is discussed.

POLLUTION CONTROL AND ENERGY CONSERVATION

Considerable attention has focused upon environmental conditions that affect the quality of human life. Various factors have made this attention salient including advances in technology, increases in the population, and a continued drain on natural resources. The concerns pertaining to human life have been reflected in such areas as pollution control and the utilization of limited environmental resources. Recently, operant techniques have begun to be applied to aspects of these environmental concerns. Although the results are preliminary, they suggest potential solutions for large-scale problems. Five areas that have received attention are the control of littering, recycling of waste materials, energy conservation, the use of mass transit, and excessive noise.

Littering

Littering in public places is a matter of great concern not only for the unsightly pollution it creates but also for the monetary expense to remove it. Litter control is expensive and it costs millions of dollars to clean major highways alone (Keep America Beautiful, 1968). Moreover, the expense appears to be increasing (Clark, Burgess, & Hendee, 1972). The usual procedure to reduce littering has been to make public appeals through national campaigns. Although campaigns sensitize individuals to the problem of littering and may change attitudes, there appears to be no relationship between antilitter attitudes and littering behavior. Public appeals, general instructions to dispose of litter, and national campaigns are commonly employed but often do not alter littering (Baltes, 1973; Burgess, Clark, & Hendee, 1971; Ingram & Geller, 1975; Kohlenberg & Phillips, 1973). Some studies have shown that instructional prompts, antilitter messages, and directions for disposal written directly on the material that is likely to be littered (e.g., handbills) can decrease littering (Geller, 1973, 1975).

Recently, positive reinforcement techniques have been used to control littering in a variety of naturalistic settings. In one project, littering was controlled in a national campground during a summer weekend (Clark *et al.*, 1972). The campground (over 100 acres) was divided into separate areas so that litter could be counted on foot. One weekend served as baseline to determine the amount of accumulated litter. Litter (bags, beverage cans, bottles) was planted in the campground to provide a consistent level for the weekends of baseline and the program. During the program, seven families were asked whether the children would help with litter accumulation. The children were told they could earn various items (e.g., Smokey Bear shoulder patch, comic book, a ranger badge, gum, and so on). Children were given a large plastic bag and told that they had one day to earn the badge. The children were not told where to look for litter or how much litter had to be collected to earn the reward. The amount of litter was markedly reduced during the incentive program. The children collected between 150 and 200 pounds of litter. The rewards given to them cost approximately $3. To accomplish the clean-up using camp personnel would have cost approximately $50 to $60. These results show that a relatively inexpensive incentive system can contribute to the reduction of litter.

Chapman and Risley (1974) decreased litter in an urban housing project area with over 390 low-income families living in a 15-square-block area. Litter was collected in such areas as residential yards, public yard areas, streets, and sidewalks. Three different experimental conditions were evaluated over the 5-month period of investigation including making a verbal appeal (asking children to help clean up the litter and giving them a bag for litter deposit), providing a small monetary incentive for collecting a certain amount of litter (10¢ for filling up a litter bag), and providing the incentive for making yards clean rather than for merely turning in a volume of trash.

The effect of the conditions was evaluated on 25 randomly selected yards in the housing project. As shown in Fig. 24, providing payment for filling bags of litter or for clean yards decreased the number of pieces of litter relative to verbal appeal and no payment conditions. Payment for clean yards tended to result in less litter than payment for volume. The authors noted that payment for cleanliness reinforced picking up the unsightly litter that may have been left behind in lieu of the bulkier litter when payment for volume was in effect. In any case, the results clearly demonstrate the utility of reinforcement in a community setting in maintaining low levels of litter.

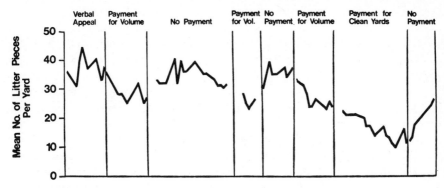

Fig. 24. The overall effects of antilitter procedures on the litter present in 25 yards sampled daily (from Chapman & Risley, 1974; Society for the Experimental Analysis of Behavior).

Other investigations have shown the effects of incentive systems in reducing littering in a movie theater (Burgess *et al.,* 1971), forest ground (Powers, Osborne, & Anderson, 1973), school (Baltes, 1973), and zoo (Kohlenberg & Phillips, 1973), athletic stadium (Baltes & Hayward, 1974), and a youth correctional facility (Hayes, Johnson, & Cone, 1975).

In most of the above programs, someone is required to carefully monitor behavior and to administer the reinforcing consequences. The supervision required to administer the program raises the question of feasibility of conducting antilitter programs on a large scale. Powers *et al.* (1973) devised a program that suggests a way to resolve the problem of surveillance and supervision. These investigators placed litter cans and bags on the grounds of a camp site with no one to monitor behavior and administer consequences. Individuals in the setting who participated were able to label their bags filled with litter with their name, address, and similar information. Also, individuals indicated the reward they wanted for collecting litter which included either 25¢ or a chance to win $20 in a lottery. No one was required to administer the consequences because individuals received a letter with their money. Thus, it appears feasible to devise a system that does not require someone to administer immediate consequences.

Recycling Wastes

A major concern related to littering is with the accumulation of waste products in the environment. The accumulation is not only undesirable for aesthetic purposes but also for the development of masses of waste products. Much of waste material might be recycled, particularly if consumers could purchase those products which were specifically suited to recycling. If waste products (e.g., metal cans, paper, returnable bottles) can be recycled in some form, they become a resource rather than an accumulated waste. The use of biodegradable products such as detergents and food containers represent innovations to ameliorate the problem of accumulated waste. Reinforcement techniques have altered consumer purchase of materials that can be recycled such as returnable rather than throwaway or nonreturnable containers (Geller, Farris, & Post, 1973; Geller, Wylie, & Farris, 1971) and have increased collection of waste for recycling (Geller, Chaffee, & Ingram, 1975; Ingram & Geller, 1975).

Geller *et al.* (1973) conducted a project in a small "quick stop" grocery store. The proportion of customers who bought returnable containers (defined as individuals who purchased at least 50% of their soft drinks in returnable bottles) was recorded across conditions where prompts or no prompts (baseline) for such purchases were provided. During the prompt condition, customers were given a handbill urging them to purchase returnable bottles to save money (because there was a refund) and taxes (because less money would be spent collecting littered bottles), and to fight pollution (because nonreturnable bottles are permanent pollutants). In a modification of this prompt procedure, a note was placed on the handbill stating that the purchase of the customer would be recorded on a chart and reported in a local newspaper and to the American Psychological Association. The "pollution chart," placed at the front of the store, indicated whether the customer had bought 50% of his drinks in returnable bottles. The results indicated that the prompting procedure increased utilization of returnable bottles. The variations of the prompting condition were not differentially effective.

Geller *et al.* (1975) increased the recycling of paper on a university campus. Several dormitories had collected waste paper from the students to be sold at a paper mill at $15 per ton and eventually recycled. Over a 6-week period, each of six dormitories received three different phases. During baseline, no special contingencies were invoked and students who brought paper to the collection room were merely thanked. One experimental intervention involved making a contest between a men's and women's dormitory. "Recycling contest" rules were posted on the collection-room door of each dorm and specified that the dorm with more paper would receive a $15 bonus. Another experimental intervention consisted of making a raffle in which each student who brought paper to the collection room received a ticket. Tickets were drawn each week and earned prizes including monetary certificates redeemable at clothing or grocery stores or for a piece of furniture, sleeping bag, and other items ranging in monetary value.

Both contest and raffle contingencies increased the amount of paper brought to the collection rooms of the dormitories. The raffle contingency was superior to the contest contingency in the total amount of paper received and in the number of individuals who brought paper. Subsequent research has replicated the efficacy of a raffle contingency in controlling paper collection (Ingram, Chaffee, Rorrer, & Wellington, 1975; Ingram & Geller, 1975; Witmer & Geller, 1976). Overall, the research suggests that contests, raffles, and lotteries might be utilized on a larger scale to increase the amount of paper and, indeed, other materials that are recycled.

Extrinsic incentives are not always necessary to increase recycling. Reid, Luyben, Rawers, and Bailey (1976) merely informed residents of three apartment complexes of the location of recycling containers to be used to collect newspapers. Near the containers, which were placed in the laundry rooms of the apartment complexes, were instructions explaining the use of containers. Also, residents were briefly interviewed and prompted about the use of the containers prior to beginning the program. Increases in the amount of newspapers collected were marked relative to baseline levels over a several-week treatment period. Luyben and Bailey (1975) replicated the effects of providing prompts and convenient collection containers in recycling newspaper in mobile home parks.

Energy Conservation

Recently, consumption of energy in the home has received increased attention. Electricity, natural gas, and oil, all limited resources, serve as the basis of heat and as sources of energy for diverse appliances. Reinforcement techniques have been applied to control energy consumption in the home in several studies. In a project with one family, Wodarski (1975) reduced electrical energy consumption by providing points to the husband and wife for decreasing the number of minutes that specific major appliances were used (e.g., stereo, television, oven, heater). Energy-saving behaviors such as turning off the dishwasher before the "dry" cycle began, using the charcoal grill instead of the oven, opening the oven door after use to circulate the heat in the room, and others also were reinforced with points. The points could be exchanged for food, money in the bank, entertainment, camping and hiking opportunities, and other events. In an ABAB design, contingent points were shown to reduce the number of minutes that electrical appliances were used. However, a one-week follow-up after the program was withdrawn indicated that energy consumption returned to baseline levels.

Similarly, Palmer, Lloyd, and Lloyd (1977) reduced the consumption of electricity of four families using variations of feedback or prompting techniques. Feedback consisted of providing daily information to each family of their meter readings indicating electrical consumption and a comparison of current electrical consumption with baseline levels. A second variation of feedback translated daily electrical consumption into monetary cost and provided the projected monthly bill based on that rate of consumption. Prompting consisted of telling individuals each day to conserve energy or providing a personal letter from the government encouraging conservation. The feedback and prompting conditions were equally effective in reducing consumption of electrical energy relative to baseline. A one-year follow-up indicated that electrical consumption tended to return to baseline levels for two of the three families still available.

Kohlenberg, Phillips, and Proctor (1976) reduced the consumption of electricity during peak periods of energy use with three families. Instructions to the families about the problems on consumption during peak periods and a request to avoid "peaking" did not reduce consumption during these periods. Feedback consisting of a light placed in the house which came on during excessive energy consumption moderately reduced consumption during peak periods. Relatively large reductions in consumption during peak periods were made when monetary incentives for a reduction in consumption was added to the feedback. Similarly,

Hayes and Cone (1976) decreased use of electricity in four units of a housing complex for married university students. Feedback was effective but not as effective as monetary incentives for reducing energy consumption.

Winett and Nietzel (1975) compared the effects of information on how to reduce energy consumption with information, feedback, and monetary incentives for reduction in a study including 31 homes. Households in the feedback-incentive condition received information about weekly energy consumption and fixed payments depending upon the percentage reduction in energy (gas or electricity) relative to baseline. Also, special bonuses were given for the greatest reductions made in a weekly period. After a 4-week experimental period, the results indicated that the feedback-incentive condition led to greater reduction in electrical consumption than the information-only condition. Groups did not differ in gas consumption which seemed to be closely related to weekly temperature. Group differences in electrical energy consumption were maintained at a 2-week follow-up but not at an 8-week follow-up assessment.

On a much larger scale than the above studies, Seaver and Patterson (1976) studied 180 homes randomly drawn from rural communities in a 4-month project designed to reduce fuel-oil consumption. Homes were assigned randomly to one of three conditions: feedback, feedback plus commendation, and no treatment. In the feedback condition, households received information at the time of fuel delivery about the amount of fuel consumed relative to the same period for the previous year and the dollars saved (or lost) in light of this reduced (or increased) consumption. In the feedback plus commendation condition, households received the above feedback information and, if there was a reduction of fuel consumption, a decal saying, "We are saving oil." The decal could not be exchanged for other events but served as a source of social recognition for fuel savings. In the no-treatment group, no intervention was implemented. Feedback plus commendation homes consumed significantly less fuel in the delivery period after the intervention than did the other two groups. The feedback and control groups were not different from each other.

Overall, reinforcement techniques have been effective in reducing energy consumption across diverse settings. The results suggest that token and social reinforcement are much more effective than feedback or instructional prompts designed to decrease energy consumption. Additional research is required to extend interventions on a larger scale and to ensure that the responses are maintained.

Mass Transportation

A major source of energy consumption stems from the extensive reliance upon automobiles. In many situations other transportation is available. For example, many individuals drive their automobiles to work when mass transportation is available. Recently, the beneficial aspects of mass transportation have been stressed because of the savings in fuel and the reduction of pollutants in the air. Use of mass transit in place of automobiles also is likely to reduce the expense of law enforcement because of the number of individuals that are required for traffic control can be reduced.

Token reinforcement has been extended to increase use of mass transit on a large university campus (Everett, Hayward, & Meyers, 1974). Buses on the university campus traveled a number of routes. Two buses were used, one of which was designated as an experimental bus and the other as a control bus. After baseline observations of the number of riders per day, a token system was implemented on the experimental bus. When the program was implemented, a newspaper advertisement announced that the bus marked with a red star would provide tokens to each passenger. Tokens consisted of wallet-sized cards that listed the back-up reinforcers available. An exchange sheet that listed the cost of the back-up events was presented with the tokens. The tokens (each worth from 5¢ to 10¢) could be exchanged for a free bus ride, ice cream, beer, pizza, coffee, cigarettes, movies, flowers, records, and other items made available at local stores in the community. The effect of the token system on bus ridership is illustrated in Fig. 25. The results indicate that the ridership on the experimental (token system) bus increased when the contingency was in effect and returned to baseline levels when the contingency was withdrawn. Ridership did not change systematically on the control bus which never received the token system.

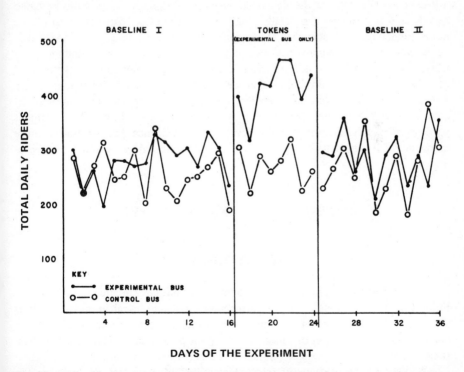

Fig. 25. Daily ridership for the Experimental Bus and the Control Bus across conditions (from Everett, Hayward, & Meyers, 1974; Society for the Experimental Analysis of Behavior).

The demonstration shows that the use of mass transit can be increased with a token system. The initial investigation is preliminary because the influence of the reinforcement system was particularly pronounced with those individuals who walked rather than those who used cars. Yet, the main concern with encouraging the use of mass transit is to increase ridership of those who ordinarily use cars. The results are extremely encouraging and suggest the utility of a wider application.

Excessive Noise

Noise from diverse sources is sometimes regarded as a source of pollution in its own right. Although noise pollution usually refers to noise caused by transportation, industry, and machinery, people also create distracting and disruptive noise.

Select investigations have focused on altering noise in situations where noise can be distracting such as elementary and junior high school classrooms (Schmidt & Ulrich, 1969; Wilson & Hopkins, 1973). Recently, excessive noise was altered in the dormitories at a university. Meyers, Artz, and Craighead (1976) decreased noise in four floors of university residence halls that housed female undergraduates. The frequency that discrete noises above a decibel criterion (either 82 or 84 dB across different halls) served as the dependent measure. Although the procedures varied slightly across residences, generally treatment consisted of instructions to reduce noise, modeling, feedback, and reinforcement. Modeling consisted of having residents view students communicating quietly. Feedback included sounding a bell when noise surpassed the decibel criterion and providing charts informing residents of noise transgressions and points that were earned for low rates of transgressions. The points (exchangeable for money or credit toward a course grade) were delivered for decreasing the frequency of transgressions over time. The treatment package markedly decreased noise level across residence halls although noise immediately returned to baseline or near baseline rates when treatment was withdrawn.

JOB PERFORMANCE AND UNEMPLOYMENT

On-the-Job Performance

Job performance is an area where operant techniques are already in wide use. Employees receive money for job performance and additional incentives such as promotions and recognition for superior performance and receive aversive consequences such as demotion or loss of job for poor performance. Systematic applications of operant techniques have enhanced job performance in industry.

Well-known applications of reinforcement in industry were conducted at the Emery Air Freight Corporation (*Business Week*, 1971). In diverse areas of company operation, feedback, praise, and recognition were used to reinforce job performance. For example, in one application, the cost of delivering air freight was reduced by shipping several smaller packages into a single shipment rather than separately. Employees who worked on the loading dock received feedback using a checklist and praise for shipping in large containers and thus saving costs. Use of the containers increased to 95% and remained at a high level for almost 2 years. Savings to the company as a result of the increased use of large containers was estimated to approximate $650,000 per year.

Additional applications of operant principles have extended to select areas of job performance. For example, in one investigation, token reinforcement was used to increase punctuality of workers in an industrial setting in Mexico (Hermann, de Montes, Dominguez, Montes, & Hopkins, 1973). Of 12 individuals who were frequently tardy for work (one or more minutes late), six were assigned to a treatment group and the other six to a control group. The treatment group received a slip of paper daily for punctuality. The slips could be exchanged for a maximum of 80¢ at the end of each week. The control group did not receive slips or money for punctuality. The results, presented in Fig. 26, revealed that tardiness decreased

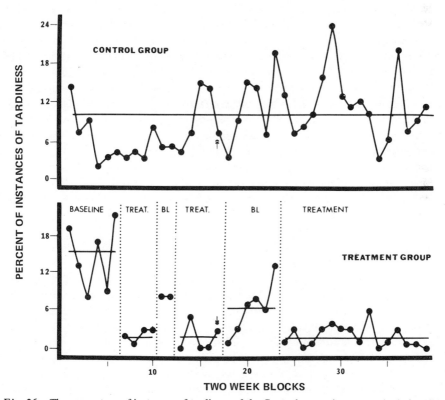

Fig. 26. The percentage of instances of tardiness of the Control group (upper portion) that did not receive treatment and of the Experimental group that received Baseline and Treatment Conditions (lower portion). (The arrows indicate that the points were based upon 1 week instead of 2 weeks of data collection. Horizontal lines represent means for each condition.) (From Hermann, de Montes, Dominguez, Montes, & Hopkins, 1973; Society for the Experimental Analysis of Behavior.)

when tokens were provided for punctuality, as demonstrated in the treatment group in an ABAB design (lower portion of the figure). The control group did not decrease in tardiness (upper portion of the figure).

Pedalino and Gamboa (1974) reduced absenteeism in a manufacturing-distribution center consisting of several plants. One plant was assigned to the behavioral intervention while four other plants served as controls (no-treatment). The intervention consisted of a lottery in which an employee who came to work and was on time received an opportunity to choose a card from a deck of playing cards. At the end of each week, the individual would have five cards which were used as a poker hand. Several individuals who had the highest poker hands received $20 each week. The lottery reduced absenteeism. The effect was maintained even though the lottery was eventually conducted every other week rather than every week. The effect of the lottery was convincingly demonstrated in an ABA design and in between-group comparisons with nontreated groups. A 22-week follow-up indicated that termination of the lottery was associated with a return of absenteeism to baseline rates.

Performance on the job has been altered in select studies. For example, Pierce and Risley (1974) applied token reinforcement (wages) to improve job performance of adolescent Neighborhood Youth Corps workers. The individuals worked as aides at a community recreation center where they supervised play activities. The jobs entailed several specific tasks that were carefully defined such as enforcing rules of the center, recording the number of individuals engaging in activities, preparing the game rooms, operating a snack bar, and several others. In an ABAB design, different interventions were implemented to increase the number of tasks completed. Merely providing explicit job descriptions and explicit instructions to perform the tasks or providing full wages independently of the completion of various tasks led to relatively low work performance. In contrast, when pay was delivered for the number of tasks completed, performance increased markedly. In one phase, a threat was made to fine any individual whose performance did not improve. While the threat led to some improvement, the effects tended to be transient.

Another aspect of job performance that has been studied is the shortage of cash in the register of a small business (Marholin & Gray, 1976). In this program, the cash register receipts were monitored in a family-style restaurant. Cash shortages were defined as a lower amount of cash in the register at the end of the day relative to the amount of cash received, as automatically recorded on an internal record of the register. After a baseline period of observation, a response cost contingency was introduced. Any cash shortage equal or greater than 1% of the day's total cash receipts was subtracted from the cashiers' salaries for that day. The total cash shortage was simply divided by the number of individuals who worked at the register that day and the fine was paid. In an ABAB design, the response cost procedure was shown to markedly reduce cash shortages.

The investigations of reinforcement in industry suggest that inadequate performance on the job might easily be altered by providing contingent consequences. The alternative for inadequate performance usually is to terminate employment. This is undesirable for several reasons. With some populations, such as the one studied by

Pierce and Risley (1974), unemployment rate is already very high. Firing individuals solves the problem of inadequate performance for the employer but not for the employee. Developing desirable performance enhances the likelihood that the employee will retain a particular job, develop skills that otherwise might not be learned, and may decrease employee turnover.

Obtaining Employment

Developing adequate work performance presupposes that the individual has a job or can obtain one once adequately trained. However, procuring a job is difficult in its own right. Jobs are not obtained by merely applying for well-advertised opportunities. Rather, a major portion of jobs are obtained through informal contacts regarding job opportunities which are not made public (Jones & Azrin, 1973, Exp. 1). Employees often provide advance information about job openings to friends, acquaintenances, and relatives before the information is generally available. Thus, a major task of job procurement is to determine existing job opportunities so that qualified individuals have access to the position.

Jones and Azrin (1973, Exp. 2) evaluated the effect of token (monetary) reinforcement in obtaining information about jobs that actually resulted in employment. During baseline, an advertisement was placed in the newspaper under the auspices of a state employment agency noting several occupational categories for which individuals registered at the agency were qualified (e.g., stenographer, bookkeeper, sales clerk, mechanic, janitor) and asked readers to phone in any positions which were known to be available. After one week, the ad was changed to indicate that anyone who reported a job opening that led to employment would receive $100. After one week of the ad with the monetary incentive, the original ad was reinstated. Over the 2 weeks of the baseline ad, only two calls were received and only one of these led to employment. During the one incentive week, 14 calls were received which resulted in eight instances of employment. Thus, the incentive program was markedly successful in procuring jobs.

Although the incentive condition may seem extravagant, the relative costs were less for the incentive than the no-incentive condition. In the incentive condition, the average cost per successful placement was $130 (which included the cost of the ad and the reward to persons who called in the jobs). In the no-incentive condition, the cost per successful placement was $470 (which included only the cost of the ad). These results strongly support the utility of monetary incentives for revealing job opportunities.

Azrin, Flores, and Kaplan (1975) expanded the job-procurement program by developing a Job-Finding Club. Unemployed individuals who participated in the club attended group meetings which reviewed methods of obtaining employment, shared job leads, and offered opportunities for job interview role-playing. Also, club members were paired so that each member would have a buddy to provide support and encouragement, were trained to systematically search for jobs from a variety of sources and to communicate with former employers, relatives, and friends about possible job opportunities, received instructions and training in dress and grooming and in developing a résumé, were encouraged to expand their job interests, and in general worked to enhance their desirability as an employee. In a careful experimental evaluation, job seekers who participated in the club found work within a shorter period and received higher wages than individuals who did not participate in the club. Indeed, after 3 months of participation in the club, 92% of the clients had obtained employment compared with 60% of individuals who had not participated in the club.

Overall, the above studies suggest the utility of reinforcement techniques with the spectrum of employment-related issues. Obtaining job leads, procuring jobs, ensuring punctuality, attendance, and high levels of job performance have all been reliably altered with reinforcement techniques. Because employment systems are naturally enmeshed with monetary reinforcement, this area would seem to be an obvious candidate for further extensions of operant techniques.

COMMUNITY SELF-GOVERNMENT

A major issue in government is to increase the involvement of individuals in decisions that affect their welfare. Individuals can have some impact in decisions made about their own communities. Preliminary attempts have been made to increase involvement of lower socioeconomic individuals in community issues. For example, one program increased the attendance of welfare recipients to self-help meetings (Miller & Miller, 1970). The participants, individuals who were recipients of Aid to Families with Dependent Children, met monthly to discuss and to resolve problems including receipt of welfare checks, medical allotments, and various community problems covering such areas as urban renewal, school board policy, police problems, and city government. In the initial phase of the project, the welfare counselor mailed notices to self-help group members indicating when the next meeting was scheduled. During the reinforcement phase, group members were informed that they could select various reinforcers if they attended the meeting. For the first meeting, individuals could select two free Christmas toys for each child of their family if they attended. In subsequent meetings, attendees could select from a wide range of events including toys, stoves, refrigerators, furniture, clothing, rugs, assistance in negotiating grievances, locating a suitable house, resolving court or police problems, finding jobs, and others. In an ABAB design, reinforcement for attendance increased the number of individuals who came to the meetings. Moreover, while the contingency was in effect, new members were attracted to the self-help group.

Briscoe et al. (1975) developed problem-solving skills in lower socioeconomic adolescents and adults who participated in a self-help project. The participants attended weekly board meetings which focused upon identifying and solving community problems such as arranging to have repairs made at the community center, organizing social and educational events, discovering and distributing social welfare resources to the community, and providing medical care. Training developed problem-solving skills at the meetings which included identifying and isolating problems under discussion, stating and evaluating alternative solutions, selecting solutions, and making decisions. Verbal responses which reflected each of these problem-solving skills were developed such as "The problem seems to be . . ." "One solution is . . ." or "What action are we going to take?" Initially, an attempt was made to increase these behaviors by instructing the members as a group in the methods of solving problems and the statements that would achieve this end. Because this was ineffective, training was conducted on an individual basis outside of the meetings with select members. In training, prompts and social reinforcement as well as modeling and role-playing opportunities were used to develop problem-solving statements.

The effect of training was evaluated in a multiple-baseline design across three different problem-solving skills (stating a problem, selecting solutions, and deciding on the action to be taken). The results, presented in Fig. 27, indicated that the specific problem-solving statements were altered only when training was introduced. Interestingly, merely instructing the group to help formulate the problem ("pilot" phase in the figure) had little impact and was not carried out across all behaviors. During follow-up, problem-solving behaviors were maintained even though training had ceased. For one meeting during the follow-up period, several

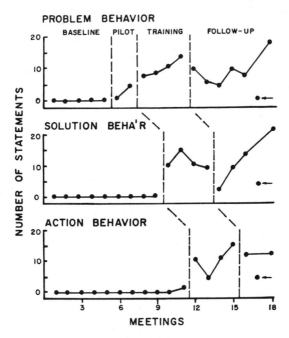

Fig. 27. Number of problem-solving statements made during the board meetings by all nine subjects. (The arrow at meeting 17 indicates that several outside individuals and the first two authors attended and participated in this meeting.) (From Briscoe, Hoffman, & Bailey, 1975; Society for the Experimental Analysis of Behavior.)

outsiders attended. During this meeting the individuals who had been trained markedly decreased their problem-solving statements. This suggests that training might need to be extended to situations involving larger and perhaps constantly changing groups.

The above investigations suggest that reinforcement techniques can increase self-help skills in the community. Increasing participation in activities and developing problem-solving skills provide entry responses to effective self-government. The effect of these and similar interventions on community change needs to be explored further.

RACIAL INTEGRATION

Racial integration represents a complex social issue that has received little direct intervention from behavior-modification techniques. The precise targets for resolving segregation are multiple and encompass diverse behaviors across a range of settings and target groups. Procedures to facilitate racial integration usually have been directed at changing attitudes. However, changes in attitudes toward racial groups does not necessarily result in corresponding behavior change (Mann, 1959).

Some research has focused upon changing how individuals describe and evaluate others against whom they show racial bias. For example, Best, Smith, Graves, and Williams (1975, Exp. 1) used reinforcement techniques to alter the tendency of children to evaluate light-skinned (Euro-American) persons positively and dark-skinned (Afro-American) persons negatively. This racial bias has been demonstrated in a large number of studies with both Euro- and Afro-American children (see Williams & Morland, 1976, for review). Preschool children were trained with a teaching machine in which selection of light- or dark-skinned figures were differentially reinforced in the context of stories in which positive and negative descriptions were presented. Responses associating Afro-American characters with positive adjectives and Euro-American characters with negative adjectives were reinforced either with advance of the machine to the next story or by tokens (pennies exchangeable for candy). Racial bias was altered by the reinforcement procedures. The changes in bias were maintained up to one week and, to a lesser extent, up to one year after training. These results, and similar findings from other studies, show that operant techniques can alter children's tendencies to associate negative adjectives with Afro-American children (Edwards & Williams, 1970; McMurtry & Williams, 1972). Although these studies clearly demonstrate the change of attitudinal and laboratory-based responses involving racial bias, they do not reflect whether interracial social behaviors would vary. Additional research is necessary to more clearly show that behaviors in nonlaboratory settings can be altered.

An initial attempt to alter overt interracial social behavior was reported by Hauserman, Walen, and Behling (1973) who developed a program to socially integrate five black children in a predominantly white first-grade classroom. The students in class isolated themselves along racial lines for all activities of a social nature. To increase interracial interaction, all students in the class received tokens and praise for sitting with "new friends" during lunch. Sitting with new friends included interracial combinations of students although this was not specified to the students. Tokens were redeemable for a snack after lunch. Interracial interactions increased during the token- and social-reinforcement phase. Interestingly, interaction transferred to a free-play period immediately after lunch even though reinforcers were never provided for interracial socialization during this period. In the final phase of the ABA design, baseline conditions were reinstated and interracial interactions decreased. Although the effects were not maintained, this study suggests the amenability of interracial interaction to change as a function of consequences. Positively reinforcing racial interaction might provide another alternative to public appeals and forced integration.

MILITARY TRAINING

An interesting application of operant principles has been in training individuals to function in the military. Of course, the military normally incorporates operant principles as in providing promotion for desirable performance. However, structure of military training, particularly basic training, has been criticized because of its heavy reliance upon aversive control (Datel & Legters, 1970). Investigators have questioned the necessity of placing recruits under severe psychological stress in the form of arbitrary punishments and unnecessary penalties for minor offenses. Moreover, there are features of the military that may resemble the total institution by making individuals submit and surrender to authority, suffer seemingly unreasonable demands, and lose their own identity (Goffman, 1961). In an attempt to ameliorate the aversive properties of military training, training has been restructured with greater emphasis on positive reinforcement.

In one facility, a token economy was used during basic training (Datel & Legters, 1970). In the token economy, referred to as the merit system, merits were earned for behaviors such as performance at daily inspections, training formations, proficiency in physical combat, written test performance covering areas of basic training, and work. The merits consisted of punches on a card that each recruit carried with him. The card specified the behaviors that needed to be performed, the back-up reinforcers, and the prices of these reinforcers. The back-up events consisted of privileges that were highly valued by the recruits such as attending a movie or taking an overnight pass. Different privileges were available at different points in training as recruits accumulated an increasing number of merits. At the end of eight weeks and the completion of training, the total number of merits earned determined whether the individual was considered for promotion. Although several categories of behavior resulted in merits, usually provided by the drill sergeant, bonus merits could also be administered for exceptional performance of one's duties, as designated by the commanding officer. Punishment was also built into the system. Misbehavior (e.g., returning late from a privilege) could result in temporary suspension of the privilege of cashing in the merits. However, no demerits were invoked. Although no evidence was provided regarding the overall success of the program, the report represents an interesting extension of the token economy to military training.

The military has used token reinforcement frequently in the rehabilitation of military personnel. For example, psychiatric facilities which are part of the Veterans Administration have frequently relied upon the token economy (Chase, 1970). In addition, one of the more well-known token programs was at Walter Reed Hospital where behavioral-problem or "delinquent" soldiers were treated (Boren & Colman, 1970; Colman & Baker, 1969; Ellsworth & Colman, 1970; Stayer & Jones, 1969). As noted in the review of token programs, this latter program reported favorable follow-up data in returning problem soldiers to completion of their tour of duty.

BROADER SOCIAL EXTENSIONS: THE DESIGN OF COMMUNITIES AND SOCIETY

The extrapolations of reinforcement techniques to society, reviewed to this point, have focused upon specific social problems. Although each of the problems is multifaceted (e.g., pollution control, racial integration, and others), impact can be achieved by focusing on specific and relatively circumscribed behaviors. Extrapolation of reinforcement principles has extended to even more complex social situations where several behaviors need to be dealt with simultaneously and where the complexity of the applications becomes multiplied. These extrapolations refer to the design of communities or society at large on the basis of reinforcement principles.

The extension of reinforcement principles to community living is consistent with the token-economy programs in restricted treatment and educational settings. In these latter applications, diverse features of community life are restructured around the earning of reinforcers and purchase of back-up events. In these applications, usually there is an implicit hierarchy of members in the community defined minimally as those who are in the program (clients) and those who administer or design the program (staff), although there may be some overlap of functions. In the extension of reinforcement principles to community life, all individuals are clients and the appeal to an outside group not part of the system is not readily available. The extension of reinforcement to the design of cultures is a logical extension of the material covered throughout previous chapters. The present discussion reviews the design and implementation of behavioral techniques in community living. Extensions on three different planes are discussed including the utopian community described in *Walden Two* (Skinner, 1948), the Twin Oaks community developed on the basis of *Walden Two* (Kinkade, 1973c), and an Experimental Living Project in a university setting (Miller, 1976).

Walden Two

The first extension of behavioral principles to society at large was described by Skinner (1948) in his utopian novel, *Walden Two*. Although the society described was fictional, it has stimulated attempts to design communities based upon operant principles. The main tenets of *Walden Two* have been the source of extensive discussion so it is necessary only to highlight basic features of the community here.

For present purposes, a significant feature of the community Skinner described was that it was structured as a token economy (Krasner, 1968). Members of the community received credits for their work in the form of labor credits. The credits were not tangible tokens such as money or tickets but simply a record of one's performance of jobs and accumulation of credits. The jobs required to maintain the community yielded different credit values depending upon the extent to which the job was of value to the community and the appeal of the job to potential workers. Unpleasant jobs which were essential to community functioning had a high credit value. For example, cleaning sewers had a very high credit value and workers could accumulate sufficient credits by working only 2 hours per day. In contrast, working in the flower gardens had a very low credit value and did not by itself provide adequate earnings to achieve sufficient annual credits. Once in a while in the system, the preference value of the job was altered if the job seemed to be avoided.

Each individual in the community had to earn 1200 labor credits per year. The earnings entitled the member to all the privileges in the community including a large variety of services and events such as living quarters, food, medical services, cultural events, instruction for themselves and their children, child-care services, and others. Unlike the usual token economy, at *Walden Two* there were no set prices associated with various privileges and activities. By earning the required number of labor credits, a member remained in the community which automatically entitled him to all the privileges.

The leaders of *Walden Two* consisted of six Planners who received labor credits for planning and governing the community. The planners received only 600 labor credits per year so they had to work on other jobs to achieve their credit total. When the Planners completed their terms (10 years with staggered terms), they selected a replacement board on the basis of the names suggested to them by the Managers, another category of workers. The Managers in *Walden Two* were specialists in charge of managing various services (e.g., food, health, dentistry, nursery school, and several others). These jobs were acquired after gradual progression through the ranks including apprenticeships and increasing responsibility. Another category of workers was the Scientists who conducted experiments primarily on applied questions (e.g., plant and animal breeding, control of infant behavior). As did the Planners, both the Managers and Scientists had to earn some of their labor credits in menial tasks, in part, to guarantee that these individuals were familiar with the problems of those who performed these tasks as their sole means of achieving labor credits.

Outside of the labor credit (token-economy) system, behavioral principles were used to develop a wide range of behaviors. For example, developing moral behavior in children was achieved through successive approximation and exposure to situations where the individual would be rewarded for avoiding temptation. Self-control was developed by exposing the individual to a series of mildly provoking annoyances so that he or she learned to handle frustration. These and other practices reflect the attempt to program specific behaviors on a community-wide basis using operant principles.

Several critiques have appeared evaluating Skinner's recommendations for the design of society as expressed in *Walden Two* and other sources (Skinner, 1961, 1971). These critiques have emphasized the lack of individuality among community members, the extreme subservience of personal interests to the community, the absence of reason by the members because the "right attitudes" have been conditioned, the removal of the individual's ability to make choices, and others (e.g., Chomsky, 1973; Matson, 1973; Wheeler, 1973). The purpose of the present discussion is to provide a conspectus to the token system proposed by Skinner rather than to advocate a particular system.

Many of the criticisms are based upon philosophical grounds challenging the assumptions about human nature; others are based upon objections that reduce to predictions about the course of the proposed society and the reactions of its members. For many of the concerns, empirical evidence about the viability of the proposed system and variations which may be derived from the basic principles would be desirable. Critics occasionally have failed to appreciate the relevance of empirical evidence about the problematic features of the system in addition to the philosophical concerns. Major questions about whether planned societies based upon principles of human behavior could resolve social and personal problems to a greater extent than current social systems, whether individuals would choose to stay in such a system if given an alternative, and whether the potential of individuals and society could be better realized in such a situation remain to be explored. While appropriate experiments may not be done, evidence from the viability of small communities and living units based upon principles of behavior would seem to be highly relevant.

Twin Oaks Community

Twin Oaks is a community based upon *Walden Two*. The community, situated on a farm in Virginia, began with eight individuals in 1967. By 1973 it had grown to about 40 (Kinkade, 1973a, 1973b, 1973c). Membership in the community is attained through application and the community maintains a waiting list. The members average 23.5 years of age and two years of college. Many of the practices were based directly on those outlined by Skinner (1948). Indeed, the overall purpose of the community was to form a community along the general lines of *Walden Two* (Kinkade, 1973c).

A main feature that resembles *Walden Two* is the labor-credit system. Jobs in the community such as cooking, cleaning, farming, bookkeeping, building, and others are selected by the individual members. Although members work a 40-hour week, this may encompass diverse jobs. Job selection is made by signing up for various tasks. If more than the necessary number of individuals wish to perform a particular job, the job is decided on a coin toss or a similar basis. Tasks which are not desired are allocated in a similar fashion.

The credits provided for a given job vary as a function of the desirability of the job. A highly sought job has a lower credit value than a job that has to be assigned. Generally, individuals select the jobs they want with only occasional (e.g., once per week) performance on an undesirable job. The labor-credit system allows individuals to earn surplus credits that are applied toward a vacation from work obligations.

The completion of one's work is essential in Twin Oaks. An individual who does not complete his or her work is asked to make up the credits or to leave the community. The credits allow one access to the privileges of the community. Privileges do not have specific credit values. As in *Walden Two,* completion of one's work entitles one to all of the goods and services. A member earns no money for work but receives a small personal allowance (25¢ to $1.00 per week). Any money an individual owned prior to joining the community is retained by him in a bank account for 3 years unless it is donated to the community. After the 3-year period all assets become communal. The general principle is that no member should enjoy financial privileges that are denied the rest.

The major governing body of Twin Oaks is a three-person board of planners whose job is to appoint and replace managers, to decide difficult questions pertaining to the ideology of the community, and to replace themselves when their 18-month term expires. Most of the decisions, however, are made by the managers who are in charge of various areas of work. The managers are responsible for ensuring that work is completed by those individuals who are assigned specific jobs. The managers make decisions about their specific areas. For example, the garden and food managers together make decisions about the diet and menu. The board of planners can overrule the managers and the community as a whole can overrule the board by a two-thirds vote of the members (Kinkade, 1973c).

As noted earlier, Twin Oaks has been based upon *Walden Two* both in the general philosophy as well as many of the specific procedures. However, as the community evolved several procedures were added which deviated from those described by Skinner. For example, various procedures developed to handle interpersonal problems (Kinkade, 1973b). One procedure entailed providing feedback to other members of the community for behaviors that engendered animosity. In many cases, the individual could be confronted with the objectional behavior. To ameliorate the difficulty that some individuals had in confronting others, a "bitch box," analogous to a suggestion box, was devised. Community members could leave notes of complaint or grievance about a particular individual. A "bitch manager" took the note, toned it down if necessary, and confronted the offender. Another means of resolving interpersonal problems was the use of group criticism sessions or encounter groups. At these general group meetings, individuals received feedback and criticism from others about their behaviors.

Periodically, behavior modification techniques have been used in the community in an attempt to resolve isolated problems (Kinkade, 1973c). In one procedure, called the Behavior Improvement Token System, members who were trying to alter some of their own behaviors requested others to administer cardboard tokens for select behaviors (e.g., smiling, complaining in a group). The tokens had no back-up value but served as feedback. Another procedure, a self-management project, involved altering the behavior of the group toward a single end. For example, to help keep the living facilities clean, individuals provided reminders to each other about not cleaning up after themselves. If someone was caught not cleaning up after himself, he would have to give up an M & M that was purchased from his nominal allowance. Finally, self-observation was periodically used as a behavior-change technique. Individuals recorded and graphed aspects of their own behaviors (e.g., saying nasty things) they wished to change.

The community has experienced a variety of problems and sources of conflict. Interpersonal conflict repeatedly arose in heterosexual relationships. Twin Oaks adhered generally to sexual freedom among its members. Marriage and monogamous relationships were not objected to. Yet, some of the reasons for marriage such as economic security and child rearing were obviated by the structure of the community. Each individual did his own work and was economically self-sufficient. Also, children in the community were raised by everyone. Responsibility for the child (e.g., babysitting, entertainment) was diffuse. Conflicts associated with "love triangles" periodically served as a source of friction and led some individuals to leave the community.

Children have served as a source of conflict at Twin Oaks. The lack of facilities and playmates for young children led to strained child-adult relationships. Also, children seemed to draw adults away from other activities that appeared to be of more immediate use to the community's development (e.g., building living quarters, farming). For a period of 3 years, children were not permitted in the community although plans were made to alter this policy.

As Kinkade (1973c), one of the founders, noted, Twin Oaks has yet to succeed independently as a commune. It depends upon work of commune members in the outside world to provide support for the community. Also, there has been periodic turnover of members so that the stability of the community has not been achieved. Yet, the commune must be viewed against the large number of other attempts to sustain similar types of living conditions whose fate is usually short-term. Twin Oaks community represents a significant attempt to design a community on behavioral principles. While the community attests to the viability of many of the recommendations provided in *Walden Two*, the complexities of human living required developing practices that were not outlined. *Walden Two* described a community that had been in effect for some time with many of the problems of development already solved (e.g., development of elaborate physical facilities, recruitment of individuals for diverse positions). These and other problems have had to be worked out in Twin Oaks and the transition from ordinary living to communal living warranted the development of special procedures in their own right.

Experimental Living Project

Recently, a small community has been developed at the University of Kansas Experimental Living Project (Miller, 1976; Miller & Feallock, 1975; Miller & Lies,

1974). The community consists of 30 male and female college students who live in a large house. The environment is managed on the basis of operant procedures some of which have been subjected to careful experimental scrutiny (Feallock & Miller, 1974).

Three major problems of community living and their resolutions illustrate the use of behavior principles. The problems include worksharing, leadership, and self-government. Worksharing involves dividing the work equally to ensure that basic essentials of living (e.g., preparing food, cleanliness) are provided. Leadership involves ensuring egalitarian rule so that select leaders cannot exert power for their own advantage. Self-government is designed to solve problems democratically.

In the program, a token economy is used to manage worksharing. Basic tasks that need to be completed to maintain the facility were divided into approximately 100 jobs. Each job was described in terms of its expected outcome. Whether the outcome was achieved each day was determined by a community member whose function is to inspect work. To ensure job performance, a token system was developed. Credits are provided to each person for completion of jobs. Each job is assigned a 15-credit per hour value. Full credits are given for correct job completion, partial credit for only partial completion. The credits earned are backed by a reduction in monthly rent (about $105). Earning 100 credits per week results in the maximum rent reduction. Fewer credits decrease the rent reduction (by 10¢ per credit less than 100). The jobs are assigned on a volunteer basis each week. Completion of the job is recorded on a public display sheet so that each individual knows his or her own earning status. At the end of each month, the amount of rent reduction for the upcoming month is determined.

The selection of rent reduction as a back-up reinforcer was based upon the anticipated expense of having the jobs completed in the house by outside help. The monthly rent included the cost of hiring outsiders (e.g., a maid, handyman, cook, janitor). The savings made by not having to hire such positions were passed on directly to the group members. The total size of the rent reduction for the group was calculated to be the entire cost of hiring outside help for the same jobs.

Select features of the system have been evaluated to determine their relation to job performance. Separate investigations have shown that when credits were provided noncontingently or when credits were omitted entirely, rather than provided for the amount of work completed, work output was reduced. Also, noncontingent rent reduction led to less work output than did contingent rent reduction (Feallock & Miller, 1974; Miller, 1976). These findings strongly support the utility of the token system in maintaining work.

Leadership in the experimental community is structured in a way that minimizes the power of the leaders and the possibility of corruption of the system toward the leaders' own ends. Leadership activities involve coordinating the system with such tasks as handling finances, inspecting and coordinating the work of others, providing credits, and several others. At a given time, approximately 10 of the 30 residents are in a coordinating or leadership position. The coordinating jobs are carefully defined in terms of explicit outcomes so that the individual can easily be accountable to others. For example, the job may require that the individual post information about rent at three different dates each month which could easily be monitored by others. This task can be monitored by community members who wait to see the rent information and who, by virtue of their prior performance of the same leadership job, are likely to know how the task should be completed.

Each coordinator receives a salary in credits (50 credits per week) which by itself is one half of the amount necessary to achieve the full rent reduction. Thus, the coordinator needs to perform routine jobs to earn the maximum credits. The coordinator can be fined if the outcomes of a given leadership task were not correctly accomplished. Carefully designed training programs were developed (involving specification of the tasks and practice of the task in simulated situations of varying difficulty) so there is no need to depend upon any one individual for a particular job. A contract is signed by the job holder specifying that the term of office is limited to 5 months, after which another member must assume the job. Thus, the coordinating positions are continually rotated.

The leaders are responsible for ensuring the maintenance of the work system. Their main task is coordinating the existing program rather than deciding how to change the program. Decision-making is handled by the members of the community who meet weekly as a group. Problems of the current system and its efficacy in meeting various needs may be discussed and decisions can be made to alter the system if the members can agree. The initial system was not based upon self-government. Rather, upon joining the group each member was presented with a handbook which described the behaviorally based rules that govern the group (e.g., describes the work system, finances, etc.). Although these rules serve as a basis for government of the group, they can be altered at the group meeting. An attempt is made to have individuals formulate changes, problems, and solutions in behavioral terms. To this end, group members receive a programmed course in operant principles and their application.

Overall, government of the system is in the hands of the group members. Although the original system is based upon behavioral principles and group members are trained in these principles, the members may alter the various practices that govern their behavior. It is difficult to evaluate the long-range effects of the self-government system and to determine whether it could be extended on a larger scale and maintain its apparent efficiency. In any case, the project represents an attempt to develop a functioning community. More than merely applying a few principles, the facility has attempted to research various components of the program to ensure their effect on behavior. Few community projects have tried experimentally validated procedures. The evaluation entails monitoring of job performance to ensure that the requirements of community living are adequately met and having community members rate diverse features of their living conditions to ensure that satisfaction is maintained (Miller & Feallock, 1975). Members rated their community-living situation as superior to other available alternatives (e.g., dormitory living) and expressed high satisfaction with diverse features of the system such as the rules, the food, cleanliness, and others. Interestingly, members rated their influence in determining important aspects of their living conditions as relatively great. This supports the overall goal of self-government expressed by the community designers.

EVALUATION OF SOCIAL EXTENSIONS OF REINFORCEMENT TECHNIQUES

Extension of reinforcement techniques to specific social and environmental problems such as pollution control, energy conservation, job performance, and others represents a promising area of research. Many of the interventions employed to alter these behaviors take advantage of the existing reinforcement contingencies enmeshed in community life, viz., monetary incentives. Token reinforcement in the form of money is a potent reinforcer already available and need not be artificially introduced into the environment.

At present, the programs designed to alter social and environmental problems are preliminary extensions. As yet, there is no evidence that the interventions could be effectively employed on a large scale for protracted periods. The evidence suggests that incentive systems are likely to be widely effective. Also, systems could be developed so that the use of monetary incentives is economically feasible. For example, lotteries may be a feasible way to provide reinforcing consequences to alter socially and environmentally relevant behaviors.

Although the current evidence is promising, there are limitations of the research that are worth noting. Initially, many of the investigations have had a relatively limited focus. In most of the studies, socially relevant behaviors are altered for only a short period (e.g., a few days, weeks, or occasionally months), across relatively few subjects (e.g., a small number of workers, families, housing units), and in restricted settings where experimental control can be maximized (e.g., university dormitories, classroom), and so on. There are exceptions which suggest larger scale applications in industry, the community, and dense urban areas (e.g., Chapman & Risley, 1974; Pedalino & Gamboa, 1974; Seaver & Patterson, 1976).

Current investigations have only begun to extend reinforcement techniques to social problems and, thus, primarily have been intended to serve as demonstrational projects. Various obstacles will need to be overcome for the large-scale application of reinforcement techniques to society. First, specific means of monitoring behavior on a large scale may present its own problems. Solutions to some of these may already be feasible. Presumably, one's energy consumption at home and in one's car are easily monitored. Minor changes in apparatus (fuel meters) would be required to make consumption patterns salient.

Second, procedures would be required to ensure that the behaviors would be maintained for extended periods. Relatively short-term interventions that characterize current applications may be effective because of their novelty. Small monetary incentives for behaviors related to pollution control and energy consumption, for example, may decrease in their effectiveness over time.

Finally, the ethical and legal implications of social extensions of reinforcement techniques need to be adequately resolved. As discussed in the next chapter, applications of reinforcement techniques have spawned concerns over the infringements of individual rights. Social extensions are directed at enhancing the general welfare of society and, thus, seek to control the behavior of large numbers of individuals. The infringements of rights may result from unchecked contingencies intended for the general good but differentially coercive to select subgroups of individuals. Social extensions may increasingly focus on behaviors that are more controversial than the current applications. For example, the general welfare of society may be a function of the birth rate, eating habits, consumption of cigarettes, alcohol, and drugs, use of leisure time, exercise, and so on. Extensions of reinforcement techniques into these areas on a large scale are likely to raise objections on ethical and legal grounds.

The extension of reinforcement techniques to social and community living situations obviously is at a preliminary stage. Current extensions to design living situations make no pretense of offering a means of establishing a new social order on a large scale. The extensions are interesting because they deal with some of the problems that large communities have but on a very small scale. Even if small community living situations could be successfully developed and sustained on the basis of operant principles, there is no guarantee that the techniques could be profitably implemented on a larger scale. Many of the current problems of social living seem to derive from the large size of communities where the behaviors of each member cannot be easily monitored. The extent to which operant principles could feasibly contribute to the design of cultures is unclear and has no strong empirical basis at this time.

CONCLUSION

Reinforcement techniques have been extended to a wide range of socially and environmentally relevant reponses including littering, recycling of waste material, energy conservation, use of mass transit, noise control, diverse aspects of job performance and procurement, community involvement, racial interaction, and performance in the military. Token reinforcement, usually in the form of monetary incentives, has been utilized frequently, in part because money is a highly reinforcing event and already is available in the natural social environment.

Although research in the social extension of reinforcement techniques to society at large is proliferating, the investigations are preliminary in nature. The individual investigations do not suggest a means for the wide-scale applications of reinforcement procedures for specific problems. Indeed, most of the applications have been restricted across a narrow range of subjects and settings and to a brief time period. The extension of relatively simple programs on a large scale is likely to raise multiple problems in such areas as maintaining a data collection system and ensuring that the contingencies are effectively implemented. Despite the possibility of problems in extending operant techniques on a large scale, there are few other techniques that have shown promise in changing the socially significant behaviors mentioned earlier even on a small scale. Thus, by comparison, application of operant techniques would appear to have fewer remaining obstacles for effective implementation relative to other techniques.

The extension or reinforcement techniques to community development raises the idea of social design on a large scale. At present, of course, too little research is available to discuss whether community design is feasible beyond a very limited scale. Yet, the very preliminary attempts at community design, and more recently, empirical evaluation of the effects of specific procedures on performance and attitudes of community members suggests that larger scale applications should not be categorically ruled out.

ETHICAL AND LEGAL ISSUES 10

The application of operant techniques offers promise in achieving various treatment, rehabilitation, and educational goals. Recent applications extend the promise to social and community problems. As the technology of behavior change is extended, increasing concerns are voiced over the misuse of the techniques. Use of the techniques to redesign institutional living of patients or inmates stimulates concern for client rights and mistreatment (Begelman, 1975; Davison & Stuart, 1975; Friedman, 1975; Goldiamond, 1974; Kassirer, 1974; Kittrie, 1971; Lucero, Vail, & Scherber, 1968; Mental Health Law Project, 1973; Stolz, 1976, 1977; Wexler, 1973, 1975a, 1975b). Hypothetical extrapolations of behavioral principles to design society has led to even greater ethical concern because everyone's freedom is potentially jeopardized (London, 1969; Wheeler, 1973).

This chapter discusses ethical and legal issues that surround the application of operant principles both for therapeutic and social ends. The discussion of ethical issues addresses concerns about behavioral control, the purposes for which behavior is controlled, the individuals who exert control, and individual freedom. The discussion of legal issues reviews recent court decisions that have direct implications for implementing reinforcement programs. The issues addressed include the infringements of rights associated with restricting the activity of individuals, limitations of the target behaviors that can be reinforced, the rights of individuals to receive and to refuse treatment, and informed consent. Finally, means to insure the protection of client rights are reviewed.

In some applications of behavior modification, aversion techniques are used in which the client experiences some undesirable event such as prolonged withdrawal from the situation, shock, injection of a drug that produces discomfort, and so on. While these programs are relatively infrequent, they have received attention both in ethical discussions and litigation. Although token economies sometimes invoke aversive conditions (e.g., withdrawal of privileges until the client earns a sufficient number of tokens), they do not usually invoke events which constitute the typical interventions of aversion therapy (e.g., Rachman & Teasdale, 1969). Thus, ethical and legal issues peculiar to the use of aversive events as an end in themselves will not be reviewed. Litigation pertaining to the use of these events and the conditions which might justify use of aversion techniques have been discussed elsewhere (cf. Baer, 1970; Budd & Baer, 1976; Kazdin, 1975a; Opton, 1974; Schwitzgebel, 1972).

255

ETHICAL ISSUES

Behavior Control

Controlling human behavior either in the context of a treatment facility or in society at large has tremendous ethical implications. Behaviors, attitudes, thoughts, and feelings of individuals can be regulated. There is some fear that the control will be used toward despotic ends (London, 1969). The concern that advances in behavioral technology will lead to control over behavior entails several related issues including the purposes for which behavior is controlled, who will decide the ultimate purposes, and the extent to which control over behavior interferes with individual freedom.

The concerns subsumed under behavior control have become salient as the technology of behavior has developed and information about the technology has been widely disseminated. Lay books have presented dramatic accounts of control and manipulation of behavior, often extending the technology well beyond its current level of development or suggesting practices that in some cases may even conflict with current knowledge (e.g., Burgess, 1963). The concerns also have increased because of the number and type of behavioral control techniques that are available and the tendency to view all of these techniques as fundamentally similar. Some of the techniques represent extremely dramatic and irreversible changes in the individual (e.g., sterilization, psychosurgery) and raise vehement protestations. The protestations tend to carry over to areas where the objections might be less well founded.

The tendency to view behavior-control techniques as a unified area has created particular problems for behavior modifiers. For example, medical interventions such as psychosurgery have been grouped with behavior-modification techniques (Subcommittee on Constitutional Rights, 1974). However, behavior modifiers have carefully distinguished behavior modification techniques as those based primarily upon psychological principles of behavior and social or environmental interventions rather than upon biological principles and physiological intervention (Davison & Stuart, 1975; Martin, 1975; Stolz, Wienckowski, & Brown, 1975).

The distinction between behavioral and medical techniques and the precise domain of behavior modification may not be crucial for the discussion of ethical considerations in general. The concerns of control over behavior do not apply uniquely to behavior modification but extend to any procedure which influences behavior (London, 1969; Skinner, 1953b). For example, in sciences other than psychology, technological developments raise many of the same ethical issues of behavioral control. For example, research in biochemistry and pharmacology raise potential ethical questions as evident in applications such as using drugs in chemical warfare, assassination attempts, or even sustaining life "artificially" with drugs for protracted periods. Similarly, advances in electronics and brain research raise the possibility of brain implantation techniques that could influence social interaction patterns and conceivably control society (London, 1969). Other technological advances could be mentioned such as those pertaining to atomic or nuclear energy. In general, technologies share concerns about potential misuses in relation to human life.

Of course, aside from advances in technologies, the existence of social institutions in government and law, business, education, religion, therapy, and the military explicitly attempt to alter behavior and are characterized by specific techniques to achieve their ends (Skinner, 1953a). Behavior control merely refers to influencing the behaviors of others and is present in ordinary mundane interactions with parents, teachers, employers, peers, spouses, siblings, and others. These agents provide or fail to provide consequences for behavior, all of which have some impact on one's own behavior. The prevalence of control in every-day experience means the issue of control is not raised by behavior modification or other advances in technology. However, the concern with technological advances is well founded because technology promises an efficiency in controlling others that may not otherwise be available.

Control for What Purposes and by Whom?

Behavioral technology, as any other technology, is ethically neutral. Technologies characteristically have the potential for use and abuse. The concern of behavioral technology is not with the technology *per se* but rather with the potential misuses. The main concern is the purposes for which behavior modification will be used and whether these are consistent with the overall welfare of society. In one sense, behavior modification has no inherent goal. It consists of a series of principles and techniques and a methodological approach toward therapeutic change. It does not necessarily dictate how life should be led.

As noted in the previous chapter, various authors, particularly Skinner, have discussed the use of behavioral principles to design society. However, this extension presupposes certain goals and values and transcends that which can be derived from the behavioral principles themselves or the scientific research upon which those principles are based. Lay discussions frequently assume specific goals to be an inherent part of the technology. One must separate the scientifically validated principles that constitute behavior modification from applications that entail judgments about the ends toward which they should be directed. Behavior modification specifies how to attain goals (e.g., develop certain behaviors) and not what the goals or purposes should be. A scientist might well be able to predict where preselected means or interim goals will lead and make recommendations to avert deleterious consequences. Yet, initial selection of the goals is out of the scientist's hands.

The scientific study and practice of behavior change is not value-free (Gouldner, 1962; Krasner, 1966; London, 1964; Rogers & Skinner, 1956; Szasz, 1960). Indeed, the values of the individual therapist influence the course of therapy and the values of the client (Holzman, 1961; Rosenthal, 1955). While values enter in the process of behavior change, specialists in science and technology are not trained to dictate the social ends for which their specialties should be used.

In most applications of behavioral principles the issue of purpose or goal is not raised because the psychologist or psychiatrist who uses the techniques is employed in a setting where the goals have been determined in advance. Behavior modification programs in hospitals and institutions, schools, day-care treatment facilities, and prisons have many established goals already endorsed by society such as returning the individual to the community, accelerating academic performance, developing self-help, communication, and social skills, alleviating bizarre behaviors, and so on.

In outpatient therapy, the client comes to treatment with a goal, namely, to develop some adaptive skill or to alleviate a problem which interferes with effective living. The primary role of the behavior modifier is to provide a means to obtain the goal insofar as the goal is consistent with generally accepted social goals. If an individual is free from some behavior which impedes his functioning, the community is either enhanced or is not deleteriously affected. Such functioning is consistent with the democratic value that, within limits, an individual should freely pursue his own objectives.

The effective use of a behavioral technology to achieve social goals requires that these goals be clearly specified. The technology can be used to achieve the goals society selects and its members wish to achieve. The issue of misdirected goals or purposes selected by society is not raised by behavior modification. However, the problem of vague or misdirected purposes is aggravated by the potent technology. The issue of goals has not changed and reduces to questions of the good life and human values.

An issue directly related to determining purposes for which behavioral techniques are used is who shall be in control of society. The fear is raised on the assumption that one individual or a small number of individuals will be in charge of society and misuse behavioral techniques. Who should control society is not an issue raised by behavior modification or advances in technology. Countries differ philosophically in conceptions of how leaders should come to power and execute their rule. A behavioral technology is compatible with diverse philosophies. Indeed, the citizenry can be completely in charge of leadership and, essentially, exert ultimate control. A behavioral technology can help the citizenry achieve goals (e.g., ameliorate social ills, improve education) with no actual change in who "controls" society. Society can determine the goals to be achieved relying on technological advances to obtain them. There always exists the concern in society that a despot will rule. Again, the concern is aggravated by providing the despot with powerful control techniques. Yet, despots already use powerful control techniques to achieve their goals (e.g., execution, imprisonment) (London, 1969).

Individual Freedom

Another aspect of behavior control is the widespread concern that a behavioral technology will necessarily mean an abridgment of individual freedom and choice. The deliberate control of human behavior may reduce an individual's ability to make choices and freely select his own goals. Whether an individual ever is free to behave counter to existing environmental forces has been actively discussed by philosophers, scientists, and theologians (Hospers, 1961; Novak, 1973; Platt, 1973; Rotenstreich, 1973; Skinner, 1953a). However, a source of agreement among those who posit or disclaim the existence of freedom is that it is exceedingly important for individuals to feel they are in fact free independently of whether they are (Kanfer & Phillips, 1970; Krasner & Ullmann, 1973; London, 1964). In any case, the question is whether a behavioral technology will decrease the extent to which an individual can exert control over his environment or to which the individual can feel free to make his own choices.

The fear that a behavioral technology threatens to eliminate or reduce freedom and choice ignores much of the applied work in behavior modification. Applied work usually is conducted with individuals whose behaviors have been identified as problematic or ineffective in some way. The responses may represent deficits or behaviors which are not under the influence of socially accepted stimuli. Such clients ordinarily have a limited number of opportunities to obtain positive reinforcers in their life as a function of their deficit or "abnormal" behavior. Individuals who differ from those who normally function in society are *confined* by their behavioral deficit which delimits those areas of social functioning from which they might choose.

Behavior modification is used to increase an individual's skills so that the number of response alternatives or options are increased. By overcoming debilitating or delimiting behaviors which restrict opportunities, the individual is freer to select from alternatives that were previously unavailable (Ball, 1968). For individuals whose behavior is considered "normal," and even for those who are gifted in some way, behavioral techniques can increase performance or develop competencies beyond those achieved with current practices. As improved levels of performance are achieved, whether or not an individual initially was deficient in some way, response opportunities and choices increase. Thus, behavior modification, as typically applied, often increases rather than stifles individual freedom (Krasner, 1968).

In spite of the potential of behavior modification to increase an individual's freedom and choice, there are cases in which individual rights have been abridged. For example, in cases where individuals are confined for treatment or incarcerated, legal issues have been raised about the abridgment of Constitutional rights. These issues, as discussed later in this chapter, point clearly to the need to define and protect rights of individuals who may be subjected to diverse types of interventions and abuses. Once the rights of individuals are defined, behavioral interventions can work within this framework.

On a larger social scale than treatment within particular settings, there is little question that individual freedom can be abridged by a dictatorial ruler. It is unclear, however, to what extent a behavioral technology would enhance this end. Governments already control strong reinforcers such as food, water, and money and effectively deliver aversive consequences such as imprisonment and immediate execution. There is no counterpart that the individual has to overcome such control techniques. In the structure of almost all governments are explicit means to handle individuals or groups whose purpose is to remove the government's control. Obviously, then, in the case of despotic governments the possibility of behavior control is a problem. Yet this is by no means a new problem or one which is introduced by a behavioral technology. The fear introduced by behavior technology is that an unreasonable leader will have even more at his disposal to control the people. While this may be true, the people still will have more at their disposal to avert such control.

There are sources of protection against large-scale social control that stem directly from the behavioral technology itself. While the behavioral technology may provide would-be controllers with special powers, it also gives those individuals who are controlled greater power over their own behavior (London, 1969). Individuals can use the principles derived from a behavioral technology to achieve their own goals. Hence, self-control is partial deterrent against control by others. Moreover, new methods of control which can be used against the people also can be used for countercontrol against the controllers (Platt, 1973; Skinner, 1971, 1973). The effectiveness of a leader, in part, depends upon the consequences which derive from his or her leadership. If a leader cannot satisfy the needs of a given society, counter movements are likely to grow to remove the leadership.

Awareness of the people may be a partial defense agianst coercive behavioral control and manipulation (Roe, 1959). Individuals who are unaware of those factors which control behavior are easily controlled by others. Thus, people need to know about factors controlling their behavior and the principles upon which such control is based. By itself, awareness does not actively overcome coercion and counter overt types of manipulation (Bandura, 1974). Yet, awareness of controlling factors can allow individuals to submit to or resist their influence more effectively. Knowledge and awareness of controlling factors may be a condition for individual freedom (Ulrich, 1967).

LEGAL ISSUES

The increasing feasibility of behavior control raised by behavior modification has made many perennial ethical issues more salient than has previously been the case. As is often the case with the discussion of ethical issues, there has been little in the way of resolution (Baer, 1970). Yet, several of the questions raised (e.g., individual freedom) have been addressed on a concrete level by the courts. Treatment programs in general have led to increased legal confrontation. Behavioral and nonbehavioral treatments seek to directly control and to modify an individual's behaviors. These goals provide a potential conflict in the values and rights of the individual with those of the institution and society. The conflict stems from individual personal freedom guaranteed by the United States Constitution and situations where the state intervenes and restricts an individual's choice concerning procedures (e.g., treatment, rehabilitation) designed to enhance his own welfare or that of the society. The legal questions have been particularly salient with institutionalized individuals who are involuntarily confined. The general question is whether institutionalized individuals have the right to receive or to refuse treatment and the extent to which treatment can abridge rights normally granted outside of institutional life.

Cases pertaining directly to behavior modification have only recently been brought before the courts. Until relatively recently, the courts have assumed a "hands off" policy in which institutions such as prisons and mental hospitals were free to determine at their discretion the type of treatment that would be used and the conditions under which treatment would be administered (Goodman, 1969; Martin, 1975; Wexler, 1973). Despite the relative paucity of legal decisions and judicial pronouncements, the decisions already have achieved far-reaching implications for behavior modification programs. The present discussion reviews major legal decisions as they pertain directly to the implementation of reinforcement programs. The discussion is oriented toward the conditions in which behavioral programs (particularly token economies) are conducted rather than a review of legal decisions and treatment *per se*.

Contingent Consequences and Environmental Restraint

A basic feature of the token economy is the contingent delivery of reinforcing events to alter behavior. A wide variety of reinforcers usually are incorporated into the system to maximize the value of the tokens, and consequently, the likelihood that clients will perform behaviors that earn tokens. For example, in programs for psychiatric patients, back-up events include such basic items as access to a room, bed, meals, clothes, or more commonly, improvements in each of these areas over very minimal facilities (i.e., a bed instead of a cot, a meal rather than a substitute food substance to maintain nutritive intake). Other events less essential to basic existence are included such as engaging in recreational activities, walking on the hospital grounds, visiting home or a nearby community, access to private space, attending religious services, and other privileges. In most institutional settings, in which behavior modification programs are not implemented, these events are provided on a noncontingent basis. In contrast, patients in token programs must purchase these events with tokens earned for performing particular target behaviors. The constitutionality of withholding events from patients has recently been addressed by the courts.

A landmark decision that outlined the conditions for client rights was *Wyatt v. Stickney* which addressed several features of institutional treatment including the use of contingent events for psychiatric patients and the mentally retarded. The aspects of the decision which pertain to the use of reinforcers specified that the rights of patients included access to a variety of events. Specifically, the decision noted that a patient has a right to many of the items and activities that usually are used as reinforcing events. For example, the patient is entitled to a comfortable bed, a closet or locker for personal belongings, a chair, bedside table, a nutritionally balanced meal, to receive visitors, to wear one's own clothes, and to attend religious services. In addition, a patient is entitled to exercise several times weekly and to be outdoors regularly and frequently, to interact with the opposite sex, and to have a television set in the day room. With other treatment populations such as juvenile delinquents in residential facilities, court cases also have ruled that similar kinds of events and activities need to be provided as part of the clients' basic rights (*Inmates of Boys' Training School v. Affleck; Morales v. Turman*).

The rulings about the basic amenities to which institutionalized populations are entitled have obvious implications for token programs. In token economies, many amenities have been used as *privileges* (i.e., benefits enjoyed by a select group of individuals who earn tokens). The court rulings make these same amenities *rights* (i.e., events to which anyone has just claim). Items usually used for back-up events legally are not as readily available as they once were, at least on a contingent basis. The rulings do not completely rule out use of back-up events to which patients are entitled by right. A patient may voluntarily yield his right by consent and have various events withheld as part of a treatment program. Thus, a back-up event to which a patient normally is entitled might be presented contingently if the patient has provided prior consent. In addition, the rights of the individual may be waived in cases where the state intervenes in behalf of individuals who are considered incompetent (Friedman, 1975). These exceptions, which would allow use of specific events as reinforcers, need to be decided on a case-by-case basis.

The large number of events that legally must be granted as rights may present problems for the token economy. Indeed, some authors have noted difficulty in identifying reinforcers for select patients before restrictions were placed upon exploring diverse back-up events (Ayllon & Azrin, 1965; Mitchell & Stoffelmayr, 1973). As noted in previous chapters, many investigators have added highly attractive back-up events over and above those normally provided in the setting (e.g., Ayllon *et al.,* 1977; Milan *et al.,* 1974). It is likely that the future of the token economy in institutionalized settings will depend upon providing new reinforcing events over and above those basic rights to which the patients are entitled. The recency of the decisions has not led to clear differences in the practices of many token programs. Thus, the precise reinforcing events that will eventually be depended upon for institutionalized populations and whether legally available events in fact will be provided noncontingently remain to be determined.

Court rulings that patients have a right to events in the environment are part of a broader right. This right is that the environment to which an individual is committed should be the "least restrictive alternative" of confinement (Ennis & Friedman, 1973). The individual is entitled to the least restrictive conditions that achieve the purposes of confinement so that the interests of the public (confining someone of potential danger) and the individual (personal liberty) are balanced. The doctrine of the least restrictive alternative was first enumerated in the case of *Lake v. Cameron* which held that a psychiatric patient could be confined only if less drastic but suitable alternatives could not be found. The doctrine has been extended in other cases which noted that confinement of a patient to maximum security or even to full-time hospitalization is justified only when less restrictive alternatives have proven ineffective (*Covington v. Harris; Lessard v. Schmidt*).

The least restrictive alternative doctrine requires an institution to justify its actions with respect to the care given a patient. The restrictions must be shown to be necessary. A number of conditions used in token economies in institutional settings might conflict with the least restrictive doctrine. Maintaining individuals on closed wards, confining them until certain behaviors are performed, limiting social interaction with other patients, or indeed in any way restricting their movement within the hospital might require justification. Also, the extent to which an individual's behavior is controlled by specific contingencies, the number of contingencies invoked, the duration of their implementation, alternatives provided for receiving back-up events outside of the specific contingencies, and similar issues might have to be justified under the least restrictive alternative doctrine.

In many programs, deprivation of access to the hospital grounds or restrictive conditions within the ward serve as the basis for providing back-up events for the tokens. Because various events are frequently withheld, it is important to discuss the issue of deprivation that is used in treatment. Initially, it is difficult to argue against deprivation *per se* in the discussion of treatment of various institutionalized populations. This simply is not one of the available positions. The majority of individuals for whom behavior modification techniques are used are deprived in some significant way by virtue of their failure to perform certain behaviors (Lucero *et al.*, 1968). Institutionalized clients are deprived of "normal" community living, friends, and freedom to choose where and how they would like to live. In educational settings, students with academic difficulties may be deprived of access to employment, additional academic work, and economic opportunities as a direct function of their behavior. Delinquents are socially deprived of the desirable features of social living because of their antisocial acts. Children and adults with debilitating albeit circumscribed problems also are deprived of moving about in life freely because of some behavior which presents obstacles.

The social deprivation that individuals normally experience as a function of problems or deficits has to be weighed against any other deprivation as a function of treatment (Ball, 1968; Cahoon, 1968). The issue of deprivation versus no deprivation would be relatively easy to decide. Yet, weighing the relative disadvantages of different types of deprivation and the duration of each type makes the issue more complex (Baer, 1970). Perhaps, it is important to distinguish the types of deprivation that can be invoked.

Deprivation can embrace both primary and secondary reinforcers. There is some agreement that primary reinforcers and other events essential to existence (e.g., food, water, shelter, physical activity, human contact) should not be restricted. In addition, as noted above, the courts have specified secondary reinforcers and nonessential events that should be provided such as privacy, personal space, and others. The main question is what events can be withheld for the purposes of rehabilitation. "Withholding" events is the appropriate term even in dealing with nonessential events or luxuries introduced into a given treatment facility such as extra recreation, extra furloughs from the facility, extra food, and so on. If these events can be made available as part of a treatment program, the question arises whether they should be given on a noncontingent basis. In a sense, luxury items are already withheld from individuals participating in treatment.

Overall, the question is to what extent individuals can be deprived of events that would serve as reinforcers. The question of deprivation usually raises strong emotional objections in part because the social deficits and limited responses repertoires of individuals who are exposed to treatment are not viewed as deprivation. The issues that need to be considered are the type of deprivation that the client experiences (e.g., nonessential privileges versus social adequacy), the duration of the deprivation (e.g., short-term treatment program versus the remaining portion of one's life outside of the treatment facility), the likelihood that deprivation of one kind will in fact ameliorate deprivation of another kind, and the alternative strategies that can be used for treatment which do not rely on deprivation (Kazdin, 1975a).

Many of the issues pertaining to deprivation such as whether one deprivation or temporary abridgment of freedom justifies the goals of treatment depend heavily upon ethical deliberation. Other questions, particularly those pertaining to decisions of treatment for a given client, depend upon empirical research to decide whether any deprivation could be justified by clear long-term gains in patient behavior. Current evidence does not overwhelmingly support long-term gains of psychological or behavioral treatment. The lack of evidence does not merely cast doubt on behavioral treatments but on the deprivation rendered by other treatment and rehabilitation techniques as well, i.e., routine hospitalization and incarceration. The absence of clear evidence in support of a set of techniques introduces obvious problems in carrying out the least restrictive alternative doctrine. The doctrine implies that there are *effective* treatments differing in restrictiveness. The least restrictive treatment among the alternatives should be selected. In fact, evidence for efficacy of even the most restrictive treatments is not especially convincing after treatment is terminated. Thus, selection of the least restrictive effective treatment awaits extensive empirical research.

Selection of Target Behaviors

Most of the behaviors focused upon in token economies have not raised legal questions. Behaviors related to self-care and personal hygiene, participation in therapy groups, and other have not been questioned. Some programs in psychiatric hospitals have focused upon work or job performance. Utilizing patients to perform work for the institution without payment of the minimum wage is a widespread practice independently of behavior modification programs (Ennis & Friedman, 1973). There has been debate about whether the client or institution reaps the greater benefit of work which maintains the institution. Some authors have suggested that a patient's work can even be counter to therapeutic goals because work may make retention of the patient essential to the institution and interfere with discharge (Bartlett, 1964; Mental Health Law Project, 1973).

In a large number of token economies, patients engage in work that maintains the effective operation of the institution (e.g., Aitchison & Green, 1974, Exp. 2; Arann & Horner, 1972; Ayllon & Azrin, 1965; Glickman *et al.,* 1973). Although work is frequently used as one of the behaviors to overcome inactivity and the effects of institutionalization on the part of patients, there has been concern expressed about patient exploitation. By making work part of a system of earning privileges, the possibility exists of coercing the patient to engage in work.

An important decision on patient labor has ruled that if the only purpose of patient work is saving labor for the institution, with no therapeutic value, the individual's Constitutional rights may be violated (*Jobson v. Henne*). (The violation entails involuntary servitude which is prohibited under the Thirteenth Amendment of the Constitution.) In the *Jobson* decision, the court ruled that patients could be assigned some work if the tasks were reasonably related to a therapeutic program.

The *Wyatt v. Stickney* decision extended this by ruling that involuntary patient labor related to the operation or maintenance of the hospital was not permitted whether or not the work was deemed therapeutic. Voluntary institutional work could be performed by the patient only if work was compensated with at least the minimum wage. This applied to any work, whether or not it was defined as therapeutic. The court specified that incentives other than wages (e.g., privileges or release from the hospital) could not be contingent upon work performance. Thus, the use of work as a target behavior, at least as usually focused upon in token programs, was effectively ruled out as a violation of the individual patient's rights. Even if the patient volunteered to perform work, performance of those jobs could not serve as a basis of obtaining privileges as is usually the case in token programs. The court did note some exceptions to the rule against involuntary work. Patients could be assigned select tasks that were considered therapeutic (e.g., vocational training) and unrelated to hospital functioning (e.g., personal tasks such as making one's own bed). This is consistent with part of the *Jobson* ruling that patients could be assigned normal housekeeping chores.

Right to Treatment

The client's right to treatment is directed at affording adequate treatment to individuals in institutional settings (cf. Birnbaum, 1960, 1969, 1972; Burris, 1969). Concern over this right stems from involuntary confinement. If an individual's freedom is denied through confinement to treatment, there must be some intervention provided that would be expected to return him to the community. The right to treatment also stems in part from due process of law. A patient should not be deprived indefinitely of his freedom in a "mental prison" if he is not receiving adequate care and treatment (Birnbaum, 1969).

In a landmark case pertaining to the right to treatment (*Rouse v. Cameron*), a patient was committed to a mental hospital after determination that he was not guilty (of carrying a dangerous weapon) for reasons of insanity. The patient was hospitalized for a period longer than the term that would have resulted from criminal conviction. He attempted to obtain release alleging that he received no psychiatric treatment. The eventual ruling in the case was that a patient committed to a mental hospital has a right to receive adequate treatment for the mental illness that justified his commitment. Interestingly, the court ruled that the hospital need not show that the treatment will cure or improve the patient but only that it is a bona fide attempt to do so. A treatment has to be provided which is adequate in light of present knowledge. The court cannot determine whether a treatment is an optimal one but only whether it is a reasonable plan in light of the patient's circumstances (cf. *Jones v. Robinson; Tribby v. Cameron*).

In another case, the court indicated that a patient was not receiving adequate psychiatric treatment, as defined by expert testimony (*Nason v. Superintendent of Bridgewater State Hospital*). The court ordered that appropriate treatment be administered rather than custodial care and retained jurisdiction over the case to ensure that its mandate was followed. Similarly, the *Wyatt v. Stickney* ruling stipulated that involuntarily committed patients have a constitutional right to receive such individual treatment that provides an opportunity to be cured or to improve one's mental condition.

Extensions of the right to treatment have supported these rulings and even assert that patients cannot be confined without treatment if they present no danger to themselves or others (*Donaldson v. O'Connor*). Confinement without treatment is tantamount to unlawful imprisonment (*Renelli v. Department of Mental Hygiene*). The right to treatment has been extended to cases of sexual psychopaths, drug addicts, and juveniles (*Creek v. Stone; Millard v. Cameron; Morales v. Turman; People ex rel. Blunt v. Narcotic Addiction Control Commission*).

Guaranteeing a right to adequate treatment raises a host of problems. The definition of "adequate treatment," the methods by which that adequacy will be determined for a given patient, and the measures of treatment effects are all areas of potential dispute. Procedurally, the right may entail designing programs to improve the staff, individualizing treatment plans, systematically assessing program effects, and in general, ensuring accountability on the part of the treatment facility. Even though the patient's right has been defined, the majority of treatment facilities might not be prepared to begin to seriously evaluate the efficacy of treatment empirically.

Refusal of Treatment

Although the individual who is confined may have a right to treatment, it is possible that treatment may abridge individual rights. Treatment may include both procedures and goals that conflict with an individual's rights granted by the Constitution. Various legal decisions have provided precedent for individual rights that conflict with treatment although there are some inconsistencies.

In some cases, the individual's right may be temporarily abridged due to treatment considerations. For example, in *Peek v. Ciccone*, a prisoner claimed that enforced administration of a tranquilizer was unlawful on the grounds that it was cruel and unusual punishment (in violation of the Eighth Amendment). The claim was rejected by the court. Similarly, in *Haynes v. Harris*, an inmate protested that a medical treatment was being forced upon him. His complaint was directed primarily at the methods of prison discipline and supervision. The court ruled that since the purpose of commitment was treatment, all available treatments could be provided. In other cases, for example, in the treatment of drug addicts, the courts have ruled that compulsory treatment does not violate Constitutional rights (*In re Spadafora*). Indeed, the court has stated that individuals may have to sacrifice their liberty to benefit from treatment (*People ex rel. Stutz v. Conboy*).

There are some inconsistencies in abridging individual rights with regard to enforced treatment. For example, in *Winters v. Miller,* a psychiatric patient protested enforced chemotherapy because the use of drugs violated her religious convictions (as a Christian Scientist). The court noted that the interest of society was not clearly served by forcing unwanted medication, so her individual rights were upheld. Similarly, in other cases (*Knecht v. Gillman; Mackey v. Procunier*), the court has ruled against the use of drugs as part of aversion therapy (i.e., apomorphine which induces vomiting and anectine which induces sensations of suffocating, drowning, and dying). The court has also denied authorization of electroshock with nonconsenting psychiatric patients (*New York City Health and Hospital Corporation v. Stein*). In general, the decision to override a patient's rights depends upon balancing the interest of the individual and society. Personal rights are most likely to be overriden when there is some grave and immediate danger the state wishes to protect (cf. *Holmes v. Silver Cross Hospital of Joliet, Illinois*).

As shown in some of the cases reviewed above, individuals may refuse a treatment to which they are assigned. Interestingly, the refusal of treatment does not necessarily absolve the treatment facility from responsibility for the client's welfare. For example, in one case, damages were awarded to a patient who allegedly refused treatment (*Whitree v. State*). The court implied that the hospital was obligated to provide treatment even though the patient objected. Hence, the right to refuse treatment is complicated by the responsibilities that treatment facilities have to attempt to improve the client and to protect themselves against charges that they have not fulfilled their function.

The assignment of individuals to treatment may raise issues of due process. One of the more prominent cases pertaining to the assignment of individuals to treatment was raised in litigation against the START program in which intractible prisoners were transferred to a special facility for a behavioral program (*Clonce v. Richardson*). There were several issues raised some of which were not considered because the program was terminated. The court did address the challenge from prisoners that transfer to the new facility violated the inmates' due process of law. The court ruled that the transfer to the START program involved a major change in their conditions of confinement including a substantial loss of privileges such as loss of personal articles, the ability to receive and possess religious and legal materials, the opportunity to eat a satisfactory amount of food, and other conditions the inmates characterized as unbearable. The court upheld the challenge of the inmates.

In general, the assignment of individuals to treatment presents a wide range of problems for treatment facilities and the courts. All of the issues addressed including providing the least restrictive environment, the individual's right to treatment, and right to refuse treatment combine in litigation.

Informed Consent

Some of the problems raised by implementing programs with confined clients can be alleviated if the client provides his or her consent about receiving treatment. For example, the client can agree to waive rights of access to various aspects of hospital life (e.g., taking walks on the grounds) as part of a reinforcement program in order to achieve some therapeutic goal. Essentially, the client can consent to the restrictions imposed by the intervention. Obtaining informed consent would seem to resolve many of the legal problems about infringements of client rights. In fact, there are many ambiguities about informed consent and the precise role it can occupy in treatment of involuntarily confined populations (Kassirer, 1974; Stolz, 1977; Wexler, 1975a).

Consent includes at least three elements: competence, knowledge, and voluntariness (Friedman, 1975; Martin, 1975; Wexler, 1975a). Competence refers to the individual's ability to make a well-reasoned decision, to understand the nature of the choice presented, and to meaningfully give consent. There is some question whether many individuals exposed to behavior-modification programs (e.g., some psychiatric patients, children, and retardates) have the capacity to provide truly informed consent. In these cases, of course, parents or guardians can provide consent.

Even in cases where client consent is sought and obtained, it is unclear whether this provides an adequate means of client protection. This was illustrated in a study of the adequacy of informed consent which queried psychiatric patients who signed voluntary admissions forms, thereby admitting themselves to a psychiatric hospital (Palmer & Wohl, 1972). Sixty percent of the patients who were queried about their status were unable to recall signing the admission form within 10 days after admission. Thirty-three percent of the patients did not recall the content of the form or could only inaccurately recall the contents. Some of the patients even denied having signed the form. These results call into question either the competence of the patients or the procedures employed to secure consent.

The second element of consent is knowledge which includes understanding the nature of treatment, the alternatives available, and the potential benefits and risks involved. It is difficult, if not impossible, to provide complete information to meet the requirements for a knowledgeable decision given that so little is known about many of the available treatments. An extremely important feature of consent is that the individual is aware that he or she does not have to give consent and that once given, consent can be revoked. The implications of this aspect of consent are discussed below.

The third element of consent is voluntariness or that an individual must agree to participate in treatment. Of course, agreement to participate must not be given under duress. Thus, giving a patient a "choice" between undergoing a particular treatment or suffering some sort of deprivation if he does not choose treatment is not an adequate base from which consent can be provided. It is difficult to ensure that individuals involuntarily confined to an institutional setting can provide consent to participate in a program without some duress. In prisons and psychiatric hospitals, inmates may feel compelled to participate in a program because of anticipated long-term gains from favorable evaluation by staff and administration whose opinions play an important role in release. For example, in one program in which mental patients consented to take a drug (anectine) which induces an adverse psychological state, several individuals claimed they participated in the program because they felt pressure to do so by the doctor's request (Mattocks & Jew, undated).

Voluntary consent may be impossible because the institutional environment for an involuntary patient may be inherently coercive, i.e., privileges and release may depend upon the individual's cooperativeness. Indeed, this conclusion was reached in a landmark decision (*Kaimowitz v. Michigan Department of Mental Health*). This case involved the use of psychosurgery to control the aggressive behavior of a psychiatric patient. The court ruled that truly voluntary and informed consent was not possible. The involuntary status of the patient militated against voluntary consent. Also, the nature of the experimental intervention (i.e., a dangerous and irreversible treatment) and lack of available information about its benefits and risks militated against "informed" consent. Indeed, the treatment was regarded as unconstitutional independently of consent. The crucial aspect of this decision with far-reaching implications is the view that involuntarily confined patients are not in a position to give voluntary consent because of the inherently coercive nature of the institutional environment. Wexler (1975a, 1975b) has challenged the notion that involuntarily committed patients are necessarily coerced by their status in the institution. If the lure of release is regarded as inherently coercive, all therapy for involuntarily institutionalized persons would be coercive despite their expressed desire or consent.

The issue of consent raises a host of problems for treatment in general and for reinforcement practices in particular. Even if the client initially gives consent for a particular treatment, it appears that he may be allowed to withdraw consent at will (*Knecht v. Gillman*). For example, in a token economy, a hospitalized patient may waive the right for various events such as meals, adequate sleeping quarters, and ground privileges. With the patient's consent, these may be delivered contingently. Yet, if the patient does not earn sufficient tokens to purchase the events which he consented to waive as rights, he may withdraw consent and terminate the program. Withdrawing consent may be easier than performing the token-earning behaviors that would allow purchase of the back-up events (Wexler, 1975b). From the standpoint of implementing effective programs, obtaining consent does not guarantee that the contingencies could be adequately applied given that consent can be revoked. Because of the nature of the events that are absolute rather than contingent rights and the patient's ability to withdraw consent, programs will have to rely heavily on highly attractive incentives throughout the program.

PROTECTION OF CLIENT RIGHTS

The guidelines provided by the courts have begun to outline individual rights, particularly for patients who are involuntarily committed and prisoners. The conditions that will safeguard individual rights are far from complete and will require additional litigation and legislation as well. Aside from these sources of policy for individual treatment, other safeguards have been suggested. For example, Kittrie (1971) has proposed a *Therapeutic Bill of Rights* which could be codified as law or merely serve as guidelines for institutions. The rights, listed in Table 9, encompass and extend many of the existing court rulings. The rights are designed to protect the fundamental rights and liberties of the individual and at the same time to allow therapeutic interventions to function. To achieve this balance, the rights are formulated in general terms so that they can be applied across different patients, goals, and methods of treatment (Kittrie, 1971).

Of course, the generality of the proposed rights will require elaborate interpretation in any given case. Indeed, some of the ambiguity will raise major questions about entire treatment programs. For example, the proposal includes a recommendation for a right of an involuntarily confined patient to receive treatment. Yet, the definition of "treatment" is not without problems. Is "treatment" some interven-

Table 9.   The Therapeutic Bill of Rights (Kittrie, 1971, pp. 402–404)

1. No person shall be compelled to undergo treatment except for the defense of society.
2. Man's innate right to remain free of excessive forms of human modification shall be inviolable.
3. No social sanctions may be invoked unless the person subjected to treatment has demonstrated a clear and present danger through truly harmful behavior which is immediately forthcoming or has already occurred.
4. No person shall be subjected to involuntary incarceration or treatment on the basis of finding of a general condition or status alone. Nor shall the mere conviction of a crime or a finding of not guilty by reason of insanity suffice to have a person automatically committed or treated.
5. No social sanctions, whether designated criminal, civil, or therapeutic, may be invoked in the absence of the previous right to a judicial or other independent hearing, appointed counsel, and an opportunity to confront those testifying about one's past conduct or therapeutic needs.
6. Dual interference by both the criminal and therapeutic process is prohibited.
7. An involuntary patient shall have the right to receive treatment.
8. Any compulsory treatment must be the least required reasonably to protect society.
9. All committed persons should have direct access to appointed counsel and the right, without any interference, to petition the courts for relief.
10. Those submitting to voluntary treatment should be guaranteed that they will not be subsequently transferred to a compulsory program through administrative action.

tion that a professional so labels or one which in fact has been shown to effect therapeutic change? It is generally accepted that a large number of treatment procedures are defined as therapeutic and advanced by professionals for diverse disorders although the procedures have little supporting evidence.

As a second question, will treatment defined by a professional invariably meet legal requirements for treatment set by the courts? For example, normalized interactions between staff and patients may be justified by professionals under the rubric of "milieu therapy." However, this form of treatment is not sufficiently specific or individualized to be satisfactory to the courts (cf. Martin, 1975). Overall, general recommendations in the form of a bill of rights should contribute markedly in their own right but more specific guarantees and guidelines will be required to handle specific cases.

Aside from general recommendations to develop guidelines for treatment, some specific solutions for protecting client rights have been proposed. One solution proposed that the relationship of the patient and the therapist or institution be altered vis-à-vis treatment (Schwitzgebel, 1975). The proposal recommends that treatment be conceptualized as a contractual activity where patients negotiate the conditions of treatment with a therapist. Specific treatment contracts could be written between a therapist and patient. The contract could make explicit the goals, methods, risks, and benefits of treatment. This extends beyond informed consent because the goals are explicit, and more importantly, the patient has a role in negotiating the final goals of treatment.

While making the goals explicit, the contract might specify the contingencies for therapeutic success and failure so that the therapist or institution is accountable. By placing treatment in terms of a contract, the patient has some legal power to sue for a breach of contract, to be compensated for injury, and to demand effective treatments. The contract may or may not guarantee a therapeutic outcome. Given a particular behavior problem, the qualified efficacy of various techniques, and the vicissitudes of clinical practice such a contract may be unreasonable. However, the therapist would specify the procedures to be used, their probable result, and alternatives in the absence of effective outcome. Perhaps, one of the greatest advantages of a contract from the standpoint of a patient's rights is that it could specify the conditions for cure or sufficient improvement to obtain release from confinement.

There are few examples of explicit contracts in treatment programs in which the goals of treatment are clearly formulated and in some way guaranteed so that the therapist is accountable. An example, mentioned earlier, involved a treatment program designed to alter the behavior of an 8-year-old boy who engaged in tantrums and negativistic behavior (Ayllon & Skuban, 1973). The program involved training the boy to comply with instructions and to not engage in tantrums across a wide range of situations in everyday life. The explicitness with which the nature and goals of treatment were presented is illustrated in the contract shown in Table 10. The contract specifies the goals of treatment, the nature of the procedures used, and the criteria for judging whether treatment effected adequate change. An extraordinary feature of the contract, given the usual practice in therapy, is that the

Table 10.   A Contingency Contract for Therapy (Ayllon & Skuban, 1973)

I. Overview of Problem and Therapeutic Program

The overall objective of this therapeutic program is to develop and stabilize Mike's behavior patterns so that he may be considered for admission to school this fall. In general, this will involve strengthening some requisite behaviors such as following commands from an adult, and eliminating others, such as the screaming and tantrumming that accompany most of his refusals to follow instructions.

Mike has a discouraging behavior history for most teachers to consider working with. Because his characteristic reaction to requests is to throw tantrums, he is considered "untestable" by standard psychological means. This does not necessarily mean that he cannot do the items on a test, but rather that he has little or no control over his own behavior. His uncooperativeness quickly discourages most people from making much of an effort to work with him. What is clearly needed is an intensive rehabilitation program designed to enable Mike to build patterns of self-control which would lead to the elimination or drastic reduction of his disruptive behavior. This, in turn, would open other possibilities for developing Mike's potential, that is, the avenues which are blocked by his unmanageable behavior.

The overall goal of this 8-week program will be the development of self-control with its reciprocal outcome of decreasing or eliminating tantrums and disruptive behaviors. Implementation of this program will require that the child and his trainer engage in such activities as trips to the zoo, museums, parks, movies, swimming pools, shopping centers, supermarkets, and so on as well as having lunch and snacks together. These settings are included to expose Mike to a maximal number of normal situations where expectations of a standard of conduct are imposed by the setting itself.

As much as possible, the techniques used in the day program will be designed with the ultimate objective of utilization in the home. An attempt will be made to see that procedures used in the program are transferred to home management at the termination of treatment. The therapist will give instructions weekly to the parents by phone to insure that efforts both at home and in rehabilitation do not conflict.

II. Behavioral Objectives of Therapy

1. The objective of the therapeutic program is to teach Mike to comply with between 80–100 per cent of the verbal commands given to him by an adult(s). Compliance will be defined as Mike's beginning to perform the behavior specified by the command within 15 sec after it has been stated and then completing the specified task.

2. In addition, we intend to eliminate or drastically reduce Mike's excessive screaming and tantrumming. The goal is not to tantrum more frequently than once out of 30 commands and for no longer than 1 min at a time.

3. Evaluation of treatment outcome: The decision as to the attainment of these specific objectives will rest upon Mike's performance during a 30 min test session to be conducted in a classroom situation. At this session the therapist, the parents, and an additional person will make 10 verbal requests each of Mike, for a total of 30 verbal requests. Mike must comply with 80–100 per cent of these requests for the program to be considered a success. In addition, he must have tantrummed not more than once, and for not more than 1 min, during this final evaluation.

III. Time and Place of Therapeutic Intervention

1. The therapeutic program will start on _____ and terminate on _____. Evaluation of the effectiveness of treatment will be held on or about the termination date of the therapeutic program.

2. Location: The meeting place will be at the _____. Session activities, however, will involve time spent elsewhere, for example, having lunch, trips to shopping centers, amusements, and other special events. If the facility is not available, some other place agreeable can be designated as meeting and base center.

3. Days of training: Therapy sessions will be scheduled 5 days per week. The specific days may vary from week to week to comply with the objectives of the program. The family will be advised of the therapy schedule 1 week in advance.

4. Hours per day: Therapeutic sessions will be scheduled for 7 hr. a day. Session time may be extended when therapeutically necessary as decided by the therapist.

5. Absences: There will be 4 notified absenced allowed. The mother is expected to notify the therapist at least 1 hr before the scheduled therapy sessions. Any additional absences will require an additional fee of $10 per absence.

IV. Fees

Achievement of the behavioral objectives is expected to take 7 weeks of training from _____. This training will cost a total of _____. The monies will be disbursed in the following manner.

1. A check for 2/3 of the total amount will be given to the therapist at the beginning of therapy.

2. The balance of 1/3 will be paid to the therapist upon the achievement of the program objectives as specified above on about the date of termination of the program. In the event that the above objectives are not reached by this date, therapy will be discontinued and the balance will be forfeited by the therapist.

3. All expense incurred during training will be defrayed by the therapist. This will include admission to baseball games, the city zoo, swimming pools, and so on, as well as the cost of field trips, lunch, and snacks.

* * * *

By my signature I do hereby attest that I have read the above proposal and agree to the conditions stated therein.

Parent

Supervising Therapist

Co-Therapist

Date

therapist's fee was based upon the extent to which the original objectives were achieved. Usually, fees are made contingent upon providing services independently of their effects on behavior. Yet, making fees at least partially contingent upon outcome increases accountability of the treatment agent.

The contractual model is desirable because it allows the patient direct access to the therapist and provides the opportunity to negotiate the goals of treatment. It is unclear whether the contractual agreement will by itself provide sufficient protection for a patient. The difference in status, power, and information about treatment as well as the patient's or inmate's confined and involuntary status may limit the legitimacy of the contract arrangement. For the contract to be upheld by the courts, the usual conditions of informed consent may have to be met (Friedman, 1975). That is, the patient must be knowledgeable, competent, and submit voluntarily to the conditions of the contract.

Guidelines for the use of behavior modification procedures and the protection of client rights are being developed by various states and professional organizations (cf. Stolz, 1976; Wexler, 1975a). One of the more well-known proposals was developed in Florida in response to a program that engaged in several abuses in a token-economy program for mentally retarded and delinquent and disturbed boys. The abuses, many of which were inappropriately viewed as behavior modification, included severe physical punishment, forced sexual acts, and deprivation, all in the form of several specific consequences for undesirable behavior.

These abuses led to formation of a Task Force to develop guidelines based upon psychological and legal principles against which subsequent programs could be evaluated. The guidelines included recommendations pertaining to competence, informed consent, and the least restrictive alternative doctrine (Friedman, 1975; Wexler, 1975a). Importantly, they outlined procedures for selecting methods of treatment. In general, the guidelines proposed the development of review committees to oversee any proposed treatment program. The committees could include a peer review committee which consists of experts in the area of behavior modification and a legal and ethical protection review committee which consists of, among others, a lawyer, a behavioral scientist, and lay person such as a parent of a handicapped client, all of whom would represent the interests and civil liberties of the client. These committees could judge the adequacy of treatment from different perspectives and ensure that the program combined treatment interests with the client's rights.

To facilitate safeguarding of client rights, a three-level scheme was proposed as part of the Florida Guidelines (Wexler, 1975a). The scheme classifies the behaviors to be modified and treatment techniques. The levels represent treatments that are increasingly intrusive to the client and, therefore, require increasingly greater scrutiny by advocacy and review panels. At the first level, the behaviors that are changed and the techniques that are used are those generally regarded as standard, resonable, and conventional. The behaviors included here might be self-help responses, language acquisition, tantrums, self-stimulation, and similar areas where there is wide agreement about the focus. The procedures included here might be positive reinforcement using praise or other events which do not infringe upon the absolute rights of the client.

At the second level, somewhat more intrusive procedures could be used to alter behaviors in the previous level. These procedures would only be used if necessary and also represent relatively standard, reasonable, and conventional techniques. The techniques might include mildly aversive procedures such as time-out or response cost. At the third level, behaviors and procedures not specified in the previous levels would be included. Behaviors included here might be relatively controversial areas such as patterns of sexual behavior where the direction of behavior change or whether change should be sought is questionable. The procedures included at this level might consist of last-resort interventions such as electric shock or drugs or perhaps less well-established interventions whose status is unclear.

Classification of behaviors and techniques into three levels dictates the amount of scrutiny required prior to program approval. Level-one behaviors and techniques might not need approval from a committee. Higher levels might require specific review procedures. Behaviors and techniques in a third level would be the most controversial and the most likely to infringe upon client rights. The committee review process and the right of the client to representation by a lawyer might be utilized on a case-by-case basis. Additionally, informed consent as well as the least restrictive alternative doctrine could be enforced by the review process. An advantage of the hierarchical arrangement of behaviors and techniques is that it fits well with the least restrictive alternative doctrine. Procedures at a lower level are likely to be less restrictive than those at a higher level. Also, justification of using an intervention at a higher level would require demonstration that less controversial procedures were ineffective.

In general, the precise method by which patient rights can be best guaranteed remains to be determined. None of the methods proposed is flawless by any means. Indeed, variations of some of these methods already have been employed in cases where the court still decided that the client's best interests were not protected. For example, in the *Kaimowitz* case, psychosurgery aimed at ameliorating destructive behavior of a sexual psychopath had been approved by two review committees and a scientific and a human rights review committee. Additionally, consent was provided by the patient. As noted earlier, the court ruled that voluntary and informed consent was not possible given the involuntarily confined status of the patient and the nature of the intervention.

Characteristics of different treatment populations, the conditions which bring them into treatment, the behaviors focused upon, and, perhaps most importantly, the intrusiveness of the intervention will dictate different solutions to protect client rights. An important implicit source of protection may arise from the general concern over individual rights. Recent litigation has increased the sensitivity of individuals responsible for designing and implementing treatment to the rights of their clients. The increased sensitivity may reduce the likelihood that controversial treatment techniques that threaten individual liberties will be proposed. The litigation, accountability, and institutional and personal responsibility extends beyond the application of behavior modification to treatment in general. Thus, the consequences of court involvement in treatment could alter the scope of treatment and rehabilitation and the criteria used to evaluate diverse therapeutic techniques.

CONCLUSION

The application of behavior modification in institutional settings and, indeed, on a larger social scale, has raised concern over ethical and legal issues. The ethical issues include concern over the technology of behavioral control and the purposes for which such control would be used. The appearance of a potent technology of behavior change threatens to extend the power of those who might not act in the best interests of society. Concern with the misuses of technology, in general, probably is well founded given the potential and occasional misuse of other technologies. The threat of a technology to abridge individual freedom has been actively discussed.

Authors sympathetic to behavior modification have pointed out that the behavioral technology itself should not be criticized because of its implications. In fact, coercive techniques of controlling people have been available and utilized in society and are not spawned by the development of behavior modification. Also, a behavioral technology may offer a means of averting control by others in that the people can apply behavioral techniques either to control the controllers or themselves.

Although the ethical questions have raised important issues, many of the discussions have remained on the abstract plane because of their basis on judgments of the good life and hypothetical circumstances in which specific practices would be odious. Legal issues have in many ways operationalized some of the ethical concerns. The increased involvement of the courts in treatment and rehabilitation, particularly with involuntarily confined patients and prison inmates, has had direct implications for behavior modification programs. Many of the decisions have been directed at the use of highly controversial aversive techniques and nonbehavioral techniques such as psychosurgery.

Although decisions directly pertaining to the use of reinforcement programs are relatively infrequent, a number of related decisions have altered the type of programs that can be conducted. For example, in token economies back-up events that have been commonly used are no longer routinely available. Events that behavior modifiers have viewed as *privileges* to be allocated contingent upon behavior have been ruled as *absolute rights* of patients and inmates. Basic amenities of living including living quarters, clothes, meals, access to group privileges, interaction with others, religious services, and other potential reinforcers must be given noncontingently, except in unusual circumstances.

The behaviors focused upon in token programs have been less of a source of concern. One area that has now been limited by the courts is the use of patients to perform jobs to maintain the institution. Patient labor has been an issue prior to the development of reinforcement programs. In light of recent decisions, use of patients to perform jobs, even if considered therapeutic, has been restricted. There still are tasks that patients can perform, particularly those related to self-care. However, the routine reinforcement of work behavior with tokens is not legally available.

Increased attention is being given to a patient's right to receive and to refuse treatment. Many ambiguities remain such as the occasions in which enforced treatment can abridge individual rights and when refusal of treatment is or is not an alternative. The nature of appropriate treatment *per se* might be influenced by the decisions reached by the courts. Treatments that are the least restrictive must be employed although there is great ambiguity about what a bona fide treatment might be and the dimensions along which restrictiveness might be evaluated.

The issue of informed consent raises particular problems because many of the individuals for whom treatment is provided cannot provide consent due to incompetence. Moreover, the status of involuntarily confined individuals and the duress that confined individuals may experience to provide consent make the issue extremely complex. For many treatments that are proposed, consent cannot be informed simply because of the paucity of empirical evidence about their direct and inadvertent effects.

Concern over legal issues has spawned professional groups, legislative bodies, and special committees to devise guidelines for treatment. The guidelines are directed at taking into account both the requirements for treatment and the rights of the clients. Recommendations have been advanced in the form of general guidelines for all treatments and the use of contractual arrangements where clients or individuals who represent the clients can negotiate treatment means and ends directly. Systems of classifying behavioral techniques and therapeutic goals have been suggested to help simplify those areas that must be closely monitored without impeding application of techniques that are not intrusive. Review committees have been suggested to oversee treatment programs so that professionals and lay individuals with diverse interests can pass judgment on the adequacy of treatment and the means to protect client rights.

SUMMING UP

11

The previous chapters have attempted to integrate the token-economy research. The complexity and diversity of token programs and the large number of issues raised throughout previous chapters may obfuscate the general conclusions that were drawn about the token economy. As any area, depth into the research leads to qualifications and scholarly hedging. Although this was not intended, it may be a natural consequence of trying to integrate an extensive literature. This chapter provides a conspectus of a few of the major conclusions that can be drawn and the issues that were raised.

THE ACCOMPLISHMENTS OF THE TOKEN ECONOMY

The outcome literature for the token economy is vast. From all of the studies, what can be said about the efficacy of the token economy? Actually, the research overall is somewhat overwhelming because the token economy has been applied so widely. The populations studied include psychiatric patients, retardates, individuals in classroom settings, delinquents, adult offenders, addicts and alcoholics, and several others. The settings studied encompass traditional institutions for diverse populations, classrooms for preschool through university students, small home-style facilities, day-care centers, the home, and others. The extension of token reinforcement to naturalistic social settings dramatizes the breadth of applications.

Diverse behaviors have been successfully altered related to areas such as self-care, psychiatric impairment, social interaction, education, language development, vocational training, and others. Aside from specific discrete responses within each of these categories, changes have been effected in more global measures such as attitudes (e.g., self-concept) or social behaviors not directly focused upon in treatment (e.g., posttreatment adjustment in the community for delinquents or psychiatric patients). Finally, responses that transcend the traditional concerns of treatment, rehabilitation, and education have been altered. Thus, the recent focus on socially and environmentally relevant responses in the community demonstrates the efficacy of reinforcement techniques on a variety of behaviors in everyday life. Given the extensive literature, it cannot be stated that the token economy has achieved its changes across a narrow range of conditions. Indeed, it is unclear what other treatment modality would even begin to approach the number and breadth of empirical evaluations of behavior change as the token economy.

The above evaluation of the token economy may appear to be too strong an endorsement. Thus, it is important to evaluate the accomplishments of the token economy from different perspectives. For convenience, the accomplishments of the token economy might be evaluated against absolute and relative criteria. The absolute criterion here refers to the accomplishments that have been achieved in the current development of the token economy compared with those that need to be achieved or are possible. The relative criterion here refers to the accomplishments of the token economy compared with the accomplishments achieved by other treatment techniques.

From the standpoint of some absolute criterion, the token economy has by no means solved problems in diverse areas of treatment, rehabilitation, and education. It is not clear what the realistic and ideal treatment goals are with diverse populations. Actually, the goals need to be individualized across populations and, of course, even within populations. Yet, some general goals can be discussed against which the accomplishments of the token economy could be evaluated.

Consider how the token economy has fared with ambitious and not necessarily achievable goals. For example, the token economy has not routinely or perhaps even occasionally made the regressed chronic psychotic patient indistinguishable from normal. Similarly, retarded children have not routinely acquired skills to advance their diagnostic status. Even educably retarded children are not routinely advanced to normal levels of intelligence where they no longer can be distinguished from their nonretarded peers. Nor are prisoners trained to give up crime after going through a token economy. In short, treatment has not achieved what many might regard as ultimate goals within specific areas of treatment.

Of course, the difficulty in evaluating treatment from the standpoint of absolute criteria, as defined here, is that the criteria themselves are unclear. It is not clear what the limits of therapeutic change are with many treatment populations. Thus, the goals that are attainable are unknown. These goals themselves depend upon developing effective treatments which extend the accomplishments of previous treatments. Nevertheless, evaluating the token economy from the standpoint of achieving a final or totally satisfactory resolution to problems to which it has been applied requires restraint and humility. The enthusiasm evident in the literature reflects excitement over progress rather than celebration after completion of the task set for the mental health professions.

From the standpoint of a relative criterion, the accomplishments of the token economy can be compared with those of other techniques directed toward similar goals. The relative standing of the token economy vis-à-vis other treatment, re-habilitation, and education techniques can be examined by looking at the mass of literature that has been established in token-economy research relative to other literatures and at direct comparative studies.

In terms of the amount of literature available, few if any methods begin to approach the demonstrated efficacy of the token economy. Fields encompassed by mental health are not at a loss for new treatment techniques. A large number of techniques and methods of behavior change in treatment and education are strongly advocated and adhered to among professionals. Unfortunately, the number of therapy or educational techniques advanced and the tenacity with which these are embraced are not at all commensurate with the empirical data that can be brought to bear in their behalf. Yet, the token economy has as its strength a relatively large body of evidence in support of its claims. The evidence itself does not resolve a number of questions. However, there is markedly less evidence for a variety of other approaches which might be viewed as competitors in deciding treatment interventions.

As a second point, comparative studies have tested the token economy against other methods. Although there are several comparative studies, they encompass diverse techniques, populations, and behaviors. Thus, there is relatively little evidence comparing the token economy and one particular technique upon which firm conclusions could be based. Yet, the comparative literature with all of its problems does seem to converge on the general conclusion that the token economy often is more effective than traditional techniques in treatment and education as well as specific forms of psychotherapy and pharmacotherapy. Of course, additional comparative studies are needed but the available evidence does place the token economy in a relatively favorable light.

Overall, the token economy has not resolved the outstanding treatment problems in an absolute sense. Indeed, few probably would be satisfied if the current accomplishments represent the limits of treatment. However, the token economy has a great deal to support it relative to other approaches. Criticism of the token economy has not always distinguished the sense in which the accomplishments are evaluated. Commonly, absolute criteria are invoked to discredit the accomplishments of a technique to which one is unfavorably disposed. Thus, criticism can be levied by saying that psychiatric patients are not indistinguishable from normals after going through a token economy. However, this is not a very telling criticism of the token economy because no treatment technique clearly has such an accomplishment to its credit. The criticism based upon the discrepancy between the actual accomplishments and the ideal accomplishments apply to most if not all techniques. In an advancing science, the major question is what is the most effective method available at this time to accomplish the goals. The answer depends upon empirical research, a criterion still infrequently invoked in the mental-health–related sciences.

OUTSTANDING ISSUES

Commenting on the accomplishments of the token economy should in no way imply that most of the problems associated with this technique are resolved. Many areas remain to be researched. A major area where research is needed is in developing strategies to ensure that changes made during token-reinforcement programs are maintained and transfer to extratreatment settings. Critics frequently point out that reinforcement programs produce reliable changes only when the client is directly under the control of the contingencies. Of course, this is not necessarily true, as reviewed in earlier chapters where follow-up data have been gathered across diverse populations. As with any criticism, there is some basis for noting the reliance of behavior upon the contingencies of reinforcement. The consistency of the demonstrations showing the importance of the contingencies in studies using the familiar ABAB design suggests the transience of behavior change. However, achieving behavior change during treatment should not be taken lightly in its own right. For many populations who live in permanent institutional settings such as the severely and profoundly retarded, programs may be implemented on a permanent or quasi-permanent basis and the concerns of transfer and maintenance are attenuated.

The usual goal of treatment is to ensure that behavior changes are maintained and transfer once the individual leaves the treatment setting. Currently, the token economy represents one development in the technology of behavior change. The technology of behavior change is more advanced than the technology for maintaining behavior and ensuring transfer. The literature suggests that procedures used to change behavior differ from those used to maintain that change. Now that the technology of behavior change is reasonably researched, the technology of maintenance and transfer is receiving increased attention. The number of recent publications that review the emerging technology of maintenance and transfer attest to the contemporary attention in this area (Baer & Stokes, 1976; Marholin et al., 1976; Stokes & Baer, 1977; Wildman & Wildman, 1975). Procedures have been developed to ensure that behavior changes transfer across settings and agents. Already a select number of procedures can be recommended for use once behavior change has been accomplished. An effective program requires implementing both behavior change and maintenance and transfer techniques. Overall, the criticism about transient and setting specific changes associated with reinforcement techniques will abate as the technology develops. Sufficient accomplishments already have been made to regard this development as inevitable rather than as a mere promissory note.

The issue of maintaining behavior changes embraces another important area for research. Training behavior-change agents to implement reinforcement techniques represents an area where response maintenance is particularly important. Methods to train aides, teachers, parents, peers, and others to implement reinforcement techniques have been developed. Yet, maintaining these changes appears to be a neglected area. A problem with the behavior of staff is that newly acquired behaviors often are not reinforced in the settings or as part of the natural contingencies. An effective teacher or psychiatric aide does not necessarily maintain these behaviors because of increased effectiveness. Indeed, as suggested in an earlier chapter, trained behaviors of staff routinely decline even though staff have effected marked changes in their clients during a behavioral program.

In most institutional settings, the consequences delivered to the staff do not depend upon the accomplishments that have been made with the clients. For example, academic performance of students usually is not used to decide the meritorious consequences to be delivered to a teacher. A teacher who is "loved by his or her students" might be conspicuously rewarded. Rarely does one hear of rewards delivered on the basis of what a teacher has effectively taught. Similarly, across diverse institutional settings, the natural contingencies in which staff operate do not seem to consistently support advancing the clients along the dimensions important to the setting. Thus, it is no surprise that staff fail to maintain behavior changes developed in staff training programs. Perhaps, an area that warrants attention is the alteration of contingencies within existing systems or hierarchies of staff so that each level is accountable to another and that reinforcers normally available in the setting are based upon staff accomplishments with the clients.

WHEN AND HOW TO USE THE TOKEN ECONOMY

There are few defining characteristics of the token economy outside of the delivery of tangible conditioned reinforcers contingent upon specific behaviors. In many programs, the reinforcers are not so tangible and the definition is even more vague. Perhaps, the token economy is best conceptualized as a very general procedure of programming contingent consequences. The procedure allows for diverse variations and program options. Programs differ in the types of contingencies that are used, how consequences are administered, the range of back-up events that are used or whether back-up events are used at all, whether supplementary techniques are used, and so on. Administration of the token program can vary so that staff, peers, or the clients themselves can be responsible for defining the contingencies and administering tokens. The large number of dimensions along which token programs vary accounts for the ease with which this treatment technique has been applied across different populations.

Overall, the impressive feature of the token economy is its adaptability to diverse populations differing widely in characteristics. Many treatment methods are suitable for a relatively restricted range of clients. For example, rehabilitation techniques for psychiatric patients, the mentally retarded, and prison inmates are rarely seen as the same. Populations usually are thought to require different techniques. In contrast, the token economy can be applied to diverse populations by making adjustments in specific features of the program such as the events used to back up the system, the behaviors focused upon, perhaps, even the tokens used, and other changes. The flexibility of the token system is an advantage even when designing programs for homogeneous population. For example, for a given group of patients, the token economy provides a general system to change behavior but can take into account individual differences in treatment.

The adaptability of the token economy is an advantage because it means that the token economy can be applied widely. The adaptability may be a disadvantage, as well. The ease with which a token economy can be applied could result in its routine use whenever behavior change is needed. Merely because the token economy can be implemented across populations and situations does not mean that it should be in any given circumstance. The conditions under which a token economy should be used remain to be resolved.

There are reasons for not routinely applying a token program when behavior change is needed. First, a token economy may introduce into a situation contingencies and reinforcers that are not routinely available. When this is the case, a program essentially introduces "artificial" events that at some point have to be withdrawn if behavior change in the natural environment is desired. For example, it may not be desirable to routinely turn classrooms into token economies. Tokens and complex exchange systems for behaviors performed in the classroom are not routinely employed in everyday life. For example, one does not usually receive tokens backed by snacks for reading books. Turning a classroom into a token program means that specific techniques might be needed to withdraw the system and to develop responsiveness to reinforcing events that are readily available. Of course, token economies can vary widely in the degree to which "artificial" reinforcers are utilized. In many classroom programs, tokens are merely exchanged for existing events such as recess, selection of a particular academic task, sitting with friends, and so on. In cases where naturally available events are used as back-up events, the program may be more easily faded than where events extraneous to the setting are introduced.

Second, introducing a token program may create problems associated with practical aspects of monitoring and administering tokens. A token economy usually requires monitoring earnings and expenditures of the clients, minimizing the opportunities for theft or illicit earnings of tokens, and so on. Records normally collected on target behaviors often are supplemented with records of savings. From the standpoint of staff time, the administrative demands of a token economy may be greater than for reinforcement programs that do not rely on tokens.

In general, the token economy is not always needed to effect behavior change. Other reinforcers such as praise, feedback, or access to privileges and activities available in the setting have effectively altered diverse behaviors. The token economy seems especially useful in particular situations. First, and most obvious, tokens usually are more effective than other reinforcers employed in a treatment setting. Thus, when other events prove ineffective or produce less behavior change than needed, a program that uses tokens backed by other events is likely to be more effective.

Second, in many situations, it is difficult to convey the goals of treatment to the clients and to provide feedback for specific accomplishments toward the goals. Tokens provide an extremely concrete way of conveying the relationship between behavior and specific consequences. This often is not accomplished with other reinforcing events where magnitude of the event is not as concretely tied to specific levels of performance.

Third, the token economy is especially useful where it is difficult to ensure that staff consistently deliver reinforcing events to the clients. For example, in many programs behavior change of the clients might be achieved with contingent staff or parental praise. Yet, it is difficult to monitor and to ensure that praise will be regularly administered. The tokens provide a convenient way to structure staff-client interactions. The delivery of a token serves as much as a cue for a staff member to engage in certain behaviors as it does for the client to perform the target response. The exchange of tokens between staff and clients provides a way to monitor the behavior of those who are responsible for change as well as those whose behavior is to be changed.

Finally, the token economy is useful where staff attend to a group of individuals. As noted repeatedly, the token economy provides a general way to administer individual programs. Tokens can be conveniently dispensed to several individuals. The individuals decide upon their own back-up events when tokens are exchanged. Thus, the use of tokens by the staff reduces the need to provide idiosyncratic reinforcing events delivered at the moment that the target behavior is performed. No doubt there are other situations in which the token economy provides an advance in convenience or staff economy of time.

THE ETHICAL AND LEGAL CHALLENGE

Recent discussions of the ethics of behavior modification and legal decisions pertaining to treatment have raised the question whether the token economy will continue to exist at all or in the form that has been commonly used (Berwick & Morris, 1974; Wexler, 1973). Certainly, recent litigation has altered the legality of specific features that have been commonly employed in token economies including the contingent delivery of events and alteration of select behaviors whose therapeutic value has been questioned. Also, a liberal interpretation of general policies such as the least restrictive alternative doctrine could limit the applications of contingencies in general with confined populations. Although specific features of the token economy will change, it is unlikely that the token economy will be discontinued.

There may be careful scrutiny of the use of token programs with involuntarily confined populations but these are only a small part of the existing applications. Token economies are firmly entrenched in areas where relatively few questions arise such as in programs in the classroom. In these applications, the traditional goals (e.g., academic performance) are achieved in a more accelerated fashion than usually is the case. Of course, there continues to be concern in education about the use of punishment, and rightfully so. However, the use of token reinforcement to alter behavior is not likely to receive great attention as long as it stays within the goals of traditional education. Similarly, applications in the natural environment increasingly rely on token reinforcement. These applications include outpatient extensions in the treatment of problems in individual behavior therapy as well as parent- and spouse-controlled programs in the home. On a larger scale, token reinforcement has been extended to social behaviors in the community. Applications in the natural environment are less likely to come under the scrutiny of the courts because of the voluntary nature of participation in treatment. In short, recent litigation simply does not apply to many applications of token reinforcement.

In areas where the recent litigation does apply, court decisions are likely to exert positive influences toward the advancement of reinforcement techniques rather than their demise. This already can be seen in specific programs. For example, recent advances in token economies in prisons show sensitivity to the types of events that can be delivered contingently. Many prison programs have added creative reinforcers that can be used legally and permit inmates to withdraw from the program at will.

On a more general level, the legal issues should increasingly bring into awareness the importance of demonstrably effective treatment strategies. Although behavior modification is coming under close scrutiny, so may all techniques. The basis for using a given technique will be increasingly challenged. The evidence that can be advanced in support of a given technique may become more visible. More likely, the lack of evidence for many existing treatments will become apparent. In a sense, the court's participation in questions of treatment may serve as a monitoring agent of sorts that may exert a constructive implicit influence on the basis for specific treatment programs.

The legal challenges also may accelerate the search for treatments that are acceptable to clients. Although there may be alternative treatments for a particular problem, these may differ markedly in their preference value from the standpoint of the clients. Because clients are increasingly involved in their own treatment decisions, research will have to address the suitability of treatment from the standpoint of the recipient. Some preliminary attempts in education have been made where teachers and students are asked to indicate their preference for the type of reinforcement program that seems most favorable among different interventions.

The legal challenges also sensitize investigators to the restrictiveness of the treatments that can be recommended. If the least restrictive alternative doctrine is enforced, this might alter the spectrum of institutional treatments. It is unclear whether most treatments routinely employed represent the least restrictive of the available treatments. Also, available treatments are not necessarily effective treatments. If this distinction is recognized by the courts, the balance of restrictiveness and effectiveness will need to be elaborated.

Overall, ethical and legal considerations should not be viewed as a threat to token economies or to behavioral techniques in general. Indeed, contemporary challenges probably will contribute to the development of behavioral techniques. Very few existing practices appear to fall under the umbrella of the courts, at least at this point in time. Also, most programs do not involve controversial applications, particularly those in naturalistic rather than institutional settings. The challenges of the court stress the accountability of specific treatment techniques. Accountability should prove to be a strength of behavioral research because the goals of treatment and means to achieve those goals are explicit. Also, data assessing the extent to which the goals are achieved are routinely collected. In short, the methodology of behavioral treatments should provide a strength in light of contemporary concerns over developing acceptable and effective treatments.

REFERENCES

Abrahms, J. L., & Allen, G. J. Comparative effectiveness of situational programming, financial pay-offs, and group pressure in weight reduction. *Behavior Therapy,* 1974, *5,* 391–400.

Abrams, L., Hines, D., Pollack, D., Ross, M., Stubbs, D. A., & Polyot, C. J. Transferable tokens: Increasing social interaction in a token economy. *Psychological Reports,* 1974, *35,* 447–452.

Aitchison, R. A. A low cost rapid delivery point system with "automatic" recording. *Journal of Applied Behavior Analysis,* 1972, *5,* 527–528.

Aitchison, R. A., & Green, D. R. A token reinforcement system for large wards of institutionalized adolescents. *Behaviour Research and Therapy,* 1974, *12,* 181–190.

Alba, E., & Pennypacker, H. S. A multiple choice change score comparison of traditional and behavioral college teaching procedures. *Journal of Applied Behavior Analysis,* 1972, *5,* 121–124.

Alexander, J. F., & Parsons, B. V. Short-term behavioral intervention with delinquent families: Impact on family process and recidivism. *Journal of Abnormal Psychology,* 1973, *81,* 219–225.

Allen, D. J., & Magaro, P. A. Measures of change in token-economy programs. *Behaviour Research and Therapy,* 1971, *9,* 311–318.

Allen, G. J. Case study: Implementation of behavior modification techniques in summer camp settings. *Behavior Therapy,* 1973, *4,* 570–575.

Allen, K. E., Hart, B. M., Buell, J. S., Harris, F. R., & Wolf, M. M. Effects of social reinforcement on isolate behavior of a nursery school child. *Child Development,* 1964, *35,* 511–518.

Alvord, J. R. The home token economy: A motivational system for the home. *Corrective Psychiatry and Journal of Social Therapy,* 1971, *17,* 6–13.

Anderson, L. T., & Alpert, M. Operant analysis of hallucination frequency in a hospitalized schizophrenic. *Journal of Behavior Therapy and Experimental Psychiatry,* 1974, *5,* 13–18.

Anderson, D., Morrow, J. E., & Schleisinger, R. The effects of token reinforcers on the behavior problems of institutionalized female retardates. Paper presented at Western Psychological Association Convention, San Francisco, May, 1967.

Andrews, G., & Ingham, R. J. Stuttering: An evaluation of follow-up procedures for syllable-timed speech/token system therapy. *Journal of Communication Disorders,* 1972, *5,* 307–319.

Arann, L., & Horner, V. M. Contingency management in an open psychiatric ward. *Journal of Behavior Therapy and Experimental Psychiatry,* 1972, *3,* 31–37.

Arnett, M. S., & Ulrich, R. E. Behavioral control in a home setting. *Psychological Record,* 1975, *25,* 395–413.

Ascare, D., & Axelrod, S. Use of a behavior modification procedure in four "open" classrooms. *Psychology in the Schools,* 1973, *10,* 243–248.

Atthowe, J. M., Jr. Token economies come of age. *Behavior Therapy,* 1973, *4,* 646–654.

Atthowe, J. M., Jr., & Krasner, L. Preliminary report on the application of contingent reinforcement procedures (token economy) on a "chronic" psychiatric ward. *Journal of Abnormal Psychology,* 1968, *73,* 37–43.

Axelrod, S. Comparison of individual and group contingencies in two special classes. *Behavior Therapy*, 1973,*4*, 83–90.

Axelrod, S., Hall, R. V., & Maxwell, A. Use of peer attention to increase study behavior. *Behavior Therapy*, 1972, *3*, 349–351.

Ayers, S. K. B., Potter, R. E., & McDearmon, J. R. Using reinforcement therapy and precision teaching techniques with adult aphasics. *Journal of Behavior Therapy and Experimental Psychiatry*, 1975, *6*, 301–305.

Ayllon, T. Intensive treatment of psychotic behavior by stimulus satiation and food reinforcement. *Behaviour Research and Therapy*, 1963, *1*, 53–61.

Ayllon, T. Some behavioral problems associated with eating in chronic schizophrenic patients. In L. P. Ullmann & L. Krasner (Eds.), *Case studies in behavior modification*. New York: Holt, Rinehart & Winston, 1965.

Ayllon, T., & Azrin, N. H. Reinforcement and instructions with mental patients. *Journal of the Experimental Analysis of Behavior*, 1964, *7*, 327–331.

Ayllon, T., & Azrin, N. H. The measurement and reinforcement of behavior of psychotics. *Journal of the Experimental Analysis of Behavior*, 1965, *8*, 357–383.

Ayllon, T., & Azrin, N. H. Reinforcer sampling: A technique for increasing the behavior of mental patients. *Journal of Applied Behavior Analysis*, 1968, *1*, 13–20. (a)

Ayllon, T., & Azrin, N. H. *The token economy: A motivational system for therapy and rehabilitation.* New York: Appleton-Century-Crofts, 1968. (b)

Ayllon, T., Garber, S., & Pisor, K. The elimination of discipline problems through a combined school–home motivational system. *Behavior Therapy*, 1975, *6*, 616–626.

Ayllon, T., & Haughton, E. Control of the behavior of schizophrenic patients by food. *Journal of the Experimental Analysis of Behavior*, 1962, *5*, 343–352.

Ayllon, T., & Haughton, E. Modification of symptomatic verbal behaviour of mental patients. *Behaviour Research and Therapy*, 1964, *2*, 87–97.

Ayllon, T., & Kelly, K. Effects of reinforcement on standardized test performance. *Journal of Applied Behavior Analysis*, 1972, *5*, 477–484.

Ayllon, T., & Kelly, K. Reinstating verbal behavior in a functionally mute retardate. *Professional Psychology*, 1974, *5*, 385–393.

Ayllon, T., Layman, D., & Burke, S. Disruptive behavior and reinforcement of academic performance. *Psychological Record*, 1972, *22*, 315–323.

Ayllon, T., Layman, D., & Kandel, H. J. A behavioral-educational alternative to drug control of hyperactive children. *Journal of Applied Behavior Analysis*, 1975, *8*, 137–146.

Ayllon, T. & Michael, J. The psychiatric nurse as a behavioral engineer. *Journal of the Experimental Analysis of Behavior*, 1959, *2*, 323–334.

Ayllon, T., & Roberts, M. D. Eliminating discipline problems by strengthening academic performance. *Journal of Applied Behavior Analysis*, 1974, *7*, 71–76.

Ayllon, T., Roberts, M., & Milan, M. *Behavior modification and prison rehabilitation: Toward an effective humanism.* New York: Wiley & Sons, in press.

Ayllon, T., & Skuban, W. Accountability in psychotherapy: A test case. *Journal of Behavior Therapy and Experimental Psychiatry*, 1973, *4*, 19–30.

Ayllon, T., Smith, D., & Rogers, M. Behavioral management of school phobia. *Journal of Behavior Therapy and Experimental Psychiatry*, 1970, *1*, 125–138.

Azrin, N. H., Flores, T., & Kaplan, S. J. Job-finding club: A group-assisted program for obtaining employment. *Behaviour Research and Therapy*, 1975, *13*, 17–27.

Azrin, N. H., & Holz, W. C. Punishment. In W. K. Honig (Ed.), *Operant behavior: Areas of research and application.* New York: Appleton-Century-Crofts, 1966.

Azrin, N. H., & Lindsley, O. R. The reinforcement of cooperation between children. *Journal of Abnormal and Social Psychology*, 1956, *52*, 100–102.

Baer, D. M. Escape and avoidance response of preschool children to two schedules of reinforcement withdrawal. *Journal of the Experimental Analysis of Behavior*, 1960, *3*, 155–159.

Baer. D. M. Some remedial uses of the reinforcement contingency. In J. Shlien (Ed.), *Research in psychotherapy, Volume III.* Washington, D.C.: American Psychological Association, 1968.

Baer, D. M. A case for the selective reinforcement of punishment. In C. Neuringer & J. L. Michael (Eds.), *Behavior modification in clinical psychology.* New York: Appleton-Century-Crofts, 1970.

Baer, D. M., & Guess, D. Receptive training of adjectival inflections in mental retardates. *Journal of Applied Behavior Analysis,* 1971, *4,* 129–139.

Baer, D. M., & Guess, D. Teaching productive noun suffixes to severely retarded children. *American Journal of Mental Deficiency,* 1973, *77,* 498–505.

Baer, A. M., Rowbury, T., & Baer, D. M. The development of instructional control over classroom activities of deviant preschool children. *Journal of Applied Behavior Analysis,* 1973, *6,* 289–298.

Baer, D. M., & Stokes, T. F. Discriminating a generalization technology: Recommendations for research in mental retardation. Presented at the Fourth International Congress of the International Association for the Scientific Study of Mental Deficiency, Washington, D. C., August, 1976.

Baer, D. M., & Wolf, M. M. The entry into natural communities of reinforcement. In R. Ulrich, T. Stachnik, & J. Mabry, *Control of human behavior, Volume 2.* Glenview, Illinois: Scott, Foresman and Company, 1970.

Baer, D. M., Wolf, M. M., & Risley, T. R. Some current dimensions of applied behavior analysis. *Journal of Applied Behavior Analysis,* 1968, *1,* 91–97.

Bailey, J. S., Timbers, G. D., Phillips, E. L., & Wolf, M. M. Modification of articulation errors of pre-delinquents by their peers. *Journal of Applied Behavior Analysis,* 1971, *4,* 265–281.

Bailey, J. S., Wolf, M. M., & Phillips, E. L. Home-based reinforcement and the modification of pre-delinquents' classroom behavior. *Journal of Applied Behavior Analysis,* 1970, *3,* 223–233.

Ball, R. S. Reinforcement conditioning of verbal behavior by verbal and nonverbal stimuli in a situation resembling a clinical interview. Unpublished Dissertation, Indiana University, 1953.

Ball, T. S. Issues and implications of operant conditioning: The reestablishment of social behavior. *Hospital & Community Psychiatry,* 1968, *19,* 230–232.

Baltes, M. M. Operant principles applied to acquisition and generalization of nonlittering behavior in children. *Proceedings, 81st Annual Convention, American Psychological Association,* 1973, *8,* 889–890.

Baltes, M. M., & Hayward, S. C. Behavioral control of littering in a football stadium. Paper presented at the 82nd annual convention of the American Psychological Association, New Orleans, September, 1974.

Bandura, A. *Principles of behavior modification.* New York: Holt, Rinehart and Winston, 1969.

Bandura, A. Modeling theory. In W. S. Sahakian (Ed.), *Psychology of learning: Systems, models, and theories.* Chicago: Markham, 1970.

Bandura, A. Behavior theory and the models of man. *American Psychologist,* 1974, *29,* 859–869.

Barnard, J. D., Christophersen, E. R., & Wolf, M. M. Supervising paraprofessional tutors in a remedial reading program. *Journal of Applied Behavior Analysis,* 1974, *7,* 481.

Baron, A., Kaufman, A., & Stauber, K. A. Effects of instructions and reinforcement-feedback on human operant behavior maintained by fixed-interval reinforcement. *Journal of the Experimental Analysis of Behavior,* 1969, *12,* 701–712.

Barrett, B. H. Reduction in rate of multiple tics by free operant conditioning methods. *Journal of Nervous and Mental Disease,* 1962, *135,* 187–195.

Barrett, B. H., & Lindsley, O. R. Deficits in acquisition of operant discrimination in institutionalized retarded children. *American Journal of Mental Deficiency,* 1962, *67,* 424–436.

Barrish, H. H., Saunders, M., & Wolf, M. M. Good behavior game: Effects of individual contingencies for group consequences on disruptive behavior in a classroom. *Journal of Applied Behavior Analysis*, 1969, *2*, 119–124.

Barry, J. V. *Alexander Maconochie of Norfolk Island: A study of a pioneer in penal reform.* London: Oxford University Press, 1958.

Bartlett, F. L. Institutional peonage: Our exploitation of mental patients. *Atlantic Monthly*, 1964, *214*, 116–119.

Bassett, J. E., Blanchard, E. B., & Koshland, E. Applied behavior analysis in a penal setting: Targeting "free world" behaviors, *Behavior Therapy*, 1975, *6*, 639–648.

Bath, K. E., & Smith, S. A. An effective token economy program for mentally retarded adults. *Mental Retardation*, 1974, *12*, 41–44.

Battalio, R. C., Kagel, J. H., Winkler, R. C., Fisher, E. B., Jr., Basmann, R. L., & Krasner, L. A test of consumer demand theory using observations of individual consumer purchases. *Western Economic Journal*, 1973, *11*, 411–428.

Battalio, R. C., Kagel, J. H., Winkler, R. C., Fisher, E. B., Jr., Basmann, R. L., & Krasner, L. An experimental investigation of consumer behavior in a controlled environment. *Journal of Consumer Research*, 1974, *1*, 52–60.

Bauermeister, J. J., & Jemail, J. A. Modification of "elective mutism" in the classroom setting: A case study. *Behavior Therapy*, 1975, *6*, 246–250.

Bechterev, V. K. *General principles of human reflexology.* New York: International Publishers, 1932.

Becker, W. C., Madsen, C. H., Arnold, C. R., & Thomas, D. R. The contingent use of teacher attention and praising in reducing classroom behavior problems. *Journal of Special Education*, 1967, *1*, 287–307.

Bednar, R. L. Zelhart, P. F., Greathouse, L., & Weinberg, S. Operant conditioning principles in the treatment of learning and behavior problems with delinquent boys. *Journal of Counseling Psychology*, 1970, *17*, 492–497.

Begelman, D. A. Ethical and legal issues of behavior modification. In M. Hersen, R. M. Eisler, & P. M. Miller (Eds.), *Progress in behavior modification, Volume 1.* New York: Academic, 1975.

Bem, D. J. Self-perception: An alternative interpretation of cognitive dissonance phenomena. *Psychological Review*, 1967, *74*, 183–200.

Bennett, P. S., & Maley, R. F. Modification of interactive behaviors in chronic mental patients. *Journal of Applied Behavior Analysis*, 1973, *6*, 609–620.

Berkowitz, B. P., & Graziano, A. M. Training parents as behavior therapists: A review. *Behaviour Research and Therapy*, 1972, *10*, 297–317.

Berkowitz, S., Sherry, P. J., & Davis, B. A. Teaching self-feeding skills to profound retardates using reinforcement and fading procedures. *Behavior Therapy*, 1971, *2*, 62–67.

Berwick, P. T., & Morris, L. A. Token economies: Are they doomed? *Professional Psychology*, 1974, *5*, 434–439.

Best, D. L., Smith, S. C., Graves, D. J., & Williams, J. E. The modification of racial bias in preschool children. *Journal of Experimental Child Psychology*, 1975, *20*, 193–205.

Betancourt, F. W., & Zeiler, M. D. The choices and preferences of nursery school children. *Journal of Applied Behavior Analysis*, 1971, *4*, 299–304.

Bigelow, G., & Liebson, I. Cost factors controlling alcoholic drinking. *Psychological Record*, 1972, *22*, 305–314.

Bijou, S. W. A systematic approach to an experimental analysis of young children. *Child Development*, 1955, *26*, 161–168.

Bijou, S. W. Methodology for an experimental analysis of child behavior. *Psychological Reports*, 1957, *3*, 243–250. (a)

Bijou, S. W. Patterns of reinforcement and resistance to extinction in young children. *Child Development*, 1957, *28*, 47–54. (b)

Bijou, S. W. Operant extinction after fixed-interval schedules with young children. *Journal of the Experimental Analysis of Behavior*, 1958, *1*, 25–29.

Bijou, S. W. Learning in children. *Monographs of the Society for Research in Child Development*, 1959, *24*, No. 5 (Whole No. 74).

Bijou, S. W. Theory and research in mental (developmental) retardation. *Psychological Record*, 1963, *13*, 95–110.

Bijou, S. W. A functional analysis of retarded development. In N. R. Ellis (Ed.), *International review of research in mental retardation, Vol. I*. Academic Press, 1966.

Bijou, S. W., & Baer, D. M. *Child development Volume I: A systematic and empirical theory*. New York: Appleton-Century-Crofts, 1961.

Bijou, S. W., & Baer, D. M. Operant methods in child behavior and development. In W. K. Honig (Ed.), *Operant behavior: Areas of research and application*. New York: Appleton-Century-Crofts, 1966.

Bijou, S. W., Peterson, R. F., Harris, F. R., Allen, K. E., & Johnston, M. S. Methodology for experimental studies of young children in natural settings. *Psychological Record*, 1969, *19*, 177–210.

Bijou, S. W., & Sturges, P. T. Positive reinforcers for experimental studies with children—consumables and manipulatables. *Child Development*, 1959, *30*, 151–170.

Birky, H. J., Chambliss, J. E., & Wasden, R. A comparison of residents discharged from a token economy and two traditional psychiatric programs. *Behavior Therapy*, 1971, *2*, 46–51.

Birnbaum, M. The right to treatment. *American Bar Association Journal*, 1960, *10*, 499–505.

Birnbaum, M. A rationale for the right. In D. S. Burris (Ed.), *The right to treatment: A symposium*. New York: Springer Publishing Company, 1969.

Birnbaum, M. The right to treatment—some comments on implementation. *Duquesne Law Review*, 1972, *10*, 579–608.

Birnbaum, P. (Ed.), *A treasury of Judaism*. New York: Hebrew Publishing Co., 1962.

Birnbrauer, J. S., Bijou, S. W., Wolf, M. M., & Kidder, J. D. Programmed instructions in the classroom. In L. P. Ullmann & L. Krasner (Eds.), *Case studies in behavior modification*. New York: Holt, Rinehart and Winston, 1965. (a)

Birnbrauer, J. S., & Lawler, J. Token reinforcement for learning. *Mental Retardation*, 1964, *2*, 275–279.

Birnbrauer, J. S., Wolf, M. M., Kidder, J. D., & Tague, C. E. Classroom behavior of retarded pupils with token reinforcement. *Journal of Experimental Child Psychology*, 1965, *2*, 219–235. (b)

Blanchard, E. B., & Young, L. D. Clinical applications of biofeedback. *Archives of General Psychiatry*, 1974, *30*, 573–589.

Bolstad, O. D., & Johnson, S. M. Self-regulation in the modification of disruptive behavior. *Journal of Applied Behavior Analysis*, 1972, *5*, 443–454.

Boren, J. J., & Colman, A. D. Some experiments on reinforcement principles within a psychiatric ward for delinquent soldiers. *Journal of Applied Behavior Analysis*, 1970, *3*, 29–37.

Born, D. G., Gledhill, S. M., & Davis, M. L. Examination performance in lecture-discussion and personalized instruction courses. *Journal of Applied Behavior Analysis*, 1972, *5*, 33–43.

Bornstein, P. H., & Hamilton, S. B. Token rewards and straw men. *American Psychologist*, 1975, *7*, 780–781.

Boudin, H. M. Contingency contracting as a therapeutic tool in the deceleration of amphetamine use. *Behavior Therapy*, 1972, *3*, 604–608.

Brady, J. P., & Lind, D. L. Experimental analysis of hysterical blindness. *Archives of General Psychiatry*, 1961, *4*, 331–339.

Braukmann, C. J., & Fixsen, D. L. Behavior modification with delinquents. In M. Hersen, R. M. Eisler, & P. M. Miller (Eds.), *Progress in behavior modification, Volume 1*. New York: Academic Press, 1975.

Braukmann, C. J., Fixsen, D. L., Phillips, E. L., & Wolf, M. M. Behavioral approaches to treatment in the crime and delinquency field. *Criminology*, 1975, *13*, 199–331.

Breyer, N. L., & Allen, G. J. Effects of implementing a token economy on teacher attending behavior. *Journal of Applied Behavior Analysis*, 1975, *8*, 373–380.

Bricker, W. A., Morgan, D. G., & Grabowski, J. G. Development and maintenance of a behavior modification repertoire of cottage attendants through T. V. feedback. *American Journal of Mental Deficiency*, 1972, *77*, 128–136.

Brickes, W. A., & Brickes, D. D. Development of receptive vocabulary in severely retarded children. *American Journal of Mental Deficiency*, 1970, *74*, 599–607.

Brierton, G., Garms, R., & Metzger, R. Practical problems encountered in an aide administered token reward cottage program. *Mental Retardation*, 1969, *7*, 40–43.

Brigham, T. A. Some speculations about self-control. In T. A. Brigham & A. C. Catania (Eds.), *The handbook of applied behavior research: Social and instructional processes.* New York: Irvington Press/Halstead Press, in press.

Brigham, T. A., Finfrock, S. R., Breunig, M. K., & Bushell, D. The use of programmed materials in the analysis of academic contingencies. *Journal of Applied Behavior Analysis*, 1972, *5*, 177–182.

Brigham, T. A., Graubard, P. S., & Stans, A. Analysis of the effects of sequential reinforcement contingencies on aspects of composition. *Journal of Applied Behavior Analysis*, 1972, *5*, 421–429.

Briscoe, R. V., Hoffman, D. B., & Bailey, J. S. Behavioral community psychology: Training a community board to problem solve. *Journal of Applied Behavior Analysis*, 1975, *8*, 157–168.

Broden, M., Bruce, C., Mitchell, M. A., Carter, V., & Hall, R. V. Effects of teacher attention on attending behavior of two boys at adjacent desks. *Journal of Applied Behavior Analysis*, 1970, *3*, 199–203.

Broden, M., Hall, R. V., Dunlap, A., & Clark, R. Effects of teacher attention and a token reinforcement system in a junior high school special education class. *Exceptional Children*, 1970, *36*, 341–349.

Broden, M., Hall, R. V., & Mitts, B. The effect of self-recording on the classroom behavior of two eighth-grade students. *Journal of Applied Behavior Analysis*, 1971, *4*, 191–199.

Brown, D., Reschly, D., & Sabers, D. Using group contingencies with punishment and positive reinforcement to modify aggressive behaviors in a Head Start Classroom. *Psychological Record*, 1974, *24*, 491–496.

Brown, J., Montgomery, R., & Barclay, J. An example of psychologist management of teacher reinforcement procedures in the elementary classroom. *Psychology in the Schools*, 1969, *6*, 336–340.

Brown, L., & Pearce, E. Increasing the production rates of trainable retarded students in a public school simulated workshop. *Education and Training of the Mentally Retarded*, 1970, *5*, 15–22.

Bucher, B., & Hawkins, J. Comparison of response cost and token reinforcement systems in a class for academic underachievers. In R. D. Rubin, J. P. Brady, & J. D. Henderson (Eds.), *Advances in behavior therapy, Volume 4.* New York: Academic Press, 1973.

Budd, K., & Baer, D. M. Behavior modification and the law: Implications of recent judicial decisions. Unpublished manuscript, University of Kansas, 1976.

Buehler, R. E., Patterson, G. R., & Furniss, J. M. The reinforcement of behaviour in institutional settings. *Behaviour Research and Therapy*, 1966, *4*, 157–167.

Buell, J., Stoddard, P., Harris, F., & Baer, D. M. Collateral social development accompanying reinforcement of outdoor play in a preschool child. *Journal of Applied Behavior Analysis*, 1968, *1*, 167–173.

Burchard, J. D. Systematic socialization: A programmed environment for the habilitation of antisocial retardates. *Psychological Record*, 1967, *17*, 461–476.

Burchard, J. D., & Barrera, F. An analysis of timeout and response cost in a programmed environment. *Journal of Applied Behavior Analysis*, 1972, *5*, 271–282.

Burchard, J. D., & Tyler, V. O. The modification of delinquent behaviour through operant conditioning. *Behaviour Research and Therapy*, 1965, *2*, 245–250.

Burgess, A. *A clockwork orange.* New York: W. W. Norton, 1963.

Burgess, R. L., Clark, R. N., & Hendee, J. C. An experimental analysis of anti-litter procedures. *Journal of Applied Behavior Analysis*, 1971, *4*, 71–75.

Burris, D. S. (Ed.), *The right to treatment: A symposium.* New York: Springer Publishing Company, 1969.

Bushell, D., Jr. The design of classroom contingencies. In F. S. Keller and E. Ribes-Inesta (Eds.), *Behavior modification: Applications to education.* New York: Academic Press, 1974.

Bushell, D., Wrobel, P. A., & Michaelis, M. L. Applying "group" contingencies to the classroom study behavior of preschool children. *Journal of Applied Behavior Analysis,* 1968, *1,* 55–61.

Business Week, New tool: "Reinforcement" for good work. December 18, 1971, 76–77.

Cahoon, D. D. Issues and implications of operant conditioning: Balancing procedures against outcomes. *Hospital & Community Psychiatry,* 1968, *19,* 228–229.

Calhoun, J. F. Modifying the academic performance of the chronic psychiatric inpatient. *Journal of Consulting and Clinical Psychology,* 1974, *42,* 621.

Campbell, J. Improving the physical fitness of retarded boys. *Mental Retardation,* 1974, *12,* 31–35.

Campbell, D. T., & Stanley, J. C. *Experimental and quasi-experimental designs for research.* Chicago: Rand-McNally, 1963.

Carcopino, J. *Daily life in ancient Rome.* New Haven: Yale University Press, 1940.

Carlson, C. G., Hersen, M., & Eisler, R. M. Token economy programs in the treatment of hospitalized adult psychiatric patients. *Journal of Nervous and Mental Disease,* 1972, *155,* 192–204.

Carpenter, P., & Carom, R. Green stamp therapy: Modification of delinquent behavior through food trading stamps. *Proceedings, 76th Annual Convention, American Psychological Association,* 1968, *3,* 531–532.

Carroccio, D. F., Latham, S., & Carroccio, B. B. Rate-contingent guitar rental to decelerate stereotyped head/face-touching of an adult male psychiatric patient. *Behavior Therapy,* 1976, *7,* 104–109.

Cash, W. M., & Evans, I. M. Training pre-school children to modify their retarded siblings' behavior. *Journal of Behavior Therapy and Experimental Psychiatry,* 1975, *6,* 13–16.

Catania, A. C. *Contemporary research in operant behavior.* Glenview, Ill.: Scott, Foresman, and Company, 1968.

Catania, A. C. The myth of self-reinforcement. *Behaviorism,* 1975, *3,* 192–199.

Chadwick, B. A., & Day, R. C. Systematic reinforcement: Academic performance of under-achieving students. *Journal of Applied Behavior Analysis,* 1971, *4,* 311–319.

Chapman, C., C., Risley, T. R. Anti-litter procedures in an urban high-density area. *Journal of Applied Behavior Analysis,* 1974, *7,* 377–384.

Chase, J. D. Token economy programs in the Veterans Administration. Unpublished manuscript, VA Department of Medicine and Surgery, Washington, D. C., 1970.

Choban, M. C., Cavior, N., & Bennett, P. Effects of physical attractiveness of patients on outcome in a token economy. Paper presented at 82nd Annual Convention of the American Psychological Association, New Orleans, August, 1974.

Chomsky, N. The case against B. F. Skinner. In F. W. Matson (Ed.), *Without/within: Behaviorism and humanism.* Monterey, California: Brooks/Cole, 1973.

Christensen, D. E. Effects of combining methylphenidate and a classroom token system in modifying hyperactive behavior. *American Journal of Mental Deficiency,* 1975, *80,* 266–276.

Christensen, D. E., & Sprague, R. L. Reduction of hyperactive behavior by conditioning procedures alone and combined with methylphenidate (Ritalin). *Behaviour Research and Therapy,* 1973, *11,* 331–334.

Christophersen, E. R., Arnold, C. M., Hill, D. W., & Quilitch, H. R. The home point system: Token reinforcement procedures for application by parents of children with behavior problems. *Journal of Applied Behavior Analysis,* 1972, *5,* 485–497.

Christy, P. R. Does use of tangible rewards with individual children affect peer observers? *Journal of Applied Behavior Analysis,* 1975, *8,* 187–196.

Church, R. M. The varied effects of punishment on behavior. *Psychological Review,* 1963, *70,* 369–402.

Clark H. B., & Sherman, J. A. Teaching generative use of sentence answers to three forms of questions. *Journal of Applied Behavior Analysis,* 1975, *8,* 321–330.

Clark, M., Lachowicz, J., & Wolf, M. M. A pilot basic education program for school dropouts incorporating a token reinforcement system. *Behaviour Research and Therapy,* 1968, *8,* 183–188.

Clark, R. N., Burgess, R. L., & Hendee, J. C. The development of anti-litter behavior in a forest campground. *Journal of Applied Behavior Analysis,* 1972, *5,* 1–5.

Clayton, T. The adolescent and the psychiatric hospital. *Hospital & Community Psychiatry,* 1973, *24,* 398–405.

Clements, C. B., & McKee, J. M. Programmed instruction for institutionalized offenders: Contingency management and performance contracts. *Psychological Reports,* 1968, *22,* 957–964.

Clonce v. Richardson, 379 F. Supp. 338 (W. D. Mo. 1974).

Coghlan, A. J., Dohrenwend, E. F., Gold, S. R., & Zimmerman, R. S. A psychobehavioral residential drug abuse program: A new adventure in adolescent psychiatry. *International Journal of Addictions,* 1973, *8,* 767–777.

Cohen, H. L. Programming alternatives to punishment: The design of competence through consequences. In S. W. Bijou & E. Ribes-Inesta (Eds.), *Behavior modification: Issues and extensions.* New York: Academic Press, 1972.

Cohen, H. L. Responses to questions asked of the panel of experts. In the United States District Court for the Western District of Missouri, Southern Division, National Prison Project, Washington, D. C., 1974.

Cohen, H. L., Filipczak, J. A., & Bis, J. S. CASE project. In J. Shlien (Ed.), *Research in psychotherapy, Volume 3.* Washington, D. C.: American Psychological Association, 1968.

Cohen, H. L., & Filipczak, J. *A new learning environment.* San Francisco: Jossey-Bass, 1971.

Cohen, M., Liebson, I., & Faillace, L. The role of reinforcement contingencies in chronic alcoholism: An experimental analysis of one case. *Behaviour Research and Therapy,* 1971, *9,* 375–379.

Cohen, M., Liebson, I., Faillace, L., & Speers, W. Alcoholism: Controlled drinking and incentives for abstinence. *Psychological Reports,* 1971, *28,* 575–580.

Cohen, R., Florin, I., Grusche, A., Meyer-Osterkamp, S., & Sell, H. The introduction of a token economy in a psychiatric ward with extremely withdrawn chronic schizophrenics. *Behaviour Research and Therapy,* 1972, *10,* 69–74.

Colman, A. D., & Baker, S. L. Utilization of an operant conditioning model for the treatment of character and behavior disorders in a military setting. *American Journal of Psychiatry,* 1969, *125,* 1395–1403.

Colman, A. D., & Boren, J. J. An information system for measuring patient behavior and its use by staff. *Journal of Applied Behavior Analysis,* 1969, *2,* 207–214.

Cooper, J. L., & Greiner, J. M. Contingency management in an introductory psychology course produces better retention. *Psychological Record,* 1971, *21,* 391–400.

Cooper, M. L., Thomson, C. L., & Baer, D. M. The experimental modification of teacher attending behavior. *Journal of Applied Behavior Analysis,* 1970, *3,* 153–157.

Corte, H. E., Wolf, M. M., & Locke, B. J. A comparison of procedures for eliminating self-injurious behavior of retarded adolescents. *Journal of Applied Behavior Analysis,* 1971, *4,* 201–213.

Cossairt, A., Hall, R. V., & Hopkins, B. L. The effects of experimenter's instructions, feedback, and praise on teacher praise and student attending behavior. *Journal of Applied Behavior Analysis,* 1973, *6,* 89–100.

Cotler, S. B., Applegate, G., King, L. W., & Kristal, S. Establishing a token economy program in a state hospital classroom: A lesson in training student and teacher. *Behavior Therapy,* 1972, *3,* 209–222.

Cotter, V. W. Effects of music on performance of manual tests with retarded adolescent females. *American Journal of Mental Deficiency,* 1971, *76,* 242–248.

Covington v. Harris, 419 F. 2d. 617 (D. C. 1969).

Cowles, J. T. Food-tokens as incentives for learning by chimpanzees. *Comparative Psychological Monographs*, 1937, *14*, No. 71.

Craig, H. B., & Holland, A. L. Reinforcement of visual attending in classrooms for deaf children. *Journal of Applied Behavior Analysis*, 1970, *3*, 97–109.

Creek vs. Stone,   79 F. 2d 106 (D C. 1967).

Crossman, E. Communication. *Journal of Applied Behavior Analysis*, 1975, *8*, 348.

Curran, J. P., Lentz, R. J., & Paul, G. L. Effectiveness of sampling-exposure procedures on facilities utilization by psychiatric hard-core chronic patients. *Journal of Behavior Therapy and Experimental Psychiatry*, 1973, *4*, 201–207.

Dalton, A. J., Rubino, C. A., & Hislop, M. W. Some effects of token rewards on school achievement of children with Down's syndrome. *Journal of Applied Behavior Analysis*, 1973, *6*, 251–259.

Danaher, B. G. Theoretical foundations and clinical applications of the Premack Principle: Review and critique. *Behavior Therapy*, 1974, *5*, 307–324.

Datel, W. E., & Legters, L. J. The psychology of the army recruit. Paper presented at American Medical Association, Chicago, June, 1970.

Davidson, W. S., II, & Seidman, E. Studies of behavior modification and juvenile delinquency: A review, methodological critique, and social perspective. *Psychological Bulletin*, 1974, *81*, 998–1011.

Davison, G. C. Appraisal of behavior modification techniques with adults in institutional settings. In C. M. Franks (Ed.), *Behavior therapy: Appraisal and status*. New York: McGraw-Hill, 1969.

Davison, G. C., & Stuart, R. B. Behavior therapy and civil liberties. *American Psychologist*, 1975, *30*, 755–763.

deCharms, R. *Personal causation: The internal affective determinants of behavior*. New York: Academic Press, 1968.

Deci, E. L. Effects of externally mediated rewards on intrinsic motivation. *Journal of Personality and Social Psychology*, 1971, *18*, 105–115.

Deci, E. L. Effects of contingent and noncontingent rewards and controls on intrinsic motivation. *Organizational Behavior and Human Performance*, 1972, *8*, 217–229. (a)

Deci, E. L. Intrinsic motivation, extrinsic reinforcement and inequity. *Journal of Personality and Social Psychology*, 1972, *22*, 113–120. (b)

Deci, E. L. *Intrinsic motivation*. New York: Plenum, 1975.

Deitz, S. M., & Repp, A. C. Differentially reinforcing low rates of misbehavior with normal elementary school children. *Journal of Applied Behavior Analysis*, 1974, *7*, 622.

DeRisi, W. J. Responses to questions asked of the panel of experts. In the United States District Court for the Western Division of Missouri, Southern Division, National Prison Project, Washington, D. C., 1974.

DeVries, D. L. Effects of environmental change and of participation on the behavior of mental patients. *Journal of Consulting and Clinical Psychology*, 1968, *32*, 532–536.

Dickinson, D. J. But what happens when you take that reinforcement away? *Psychology in the Schools*, 1974, *11*, 158–160.

DiScipio, W. J., & Trudeau, P. F. Symptom changes and self-esteem as correlates of positive conditioning of grooming in hospitalized psychotics. *Journal of Abnormal Psychology*, 1972, *80*, 244–248.

Doleys, S. M., & Slapion, M. J. The reduction of verbal repetitions by response cost controlled by a sibling. *Journal of Behavior Therapy and Experimental Psychiatry*, 1975, *6*, 61–63.

Donaldson vs. O'Connor, 493 F. 2d. 507 (5th Cir. 1974).

Doolittle, J. *Social life of the Chinese: With some account of their religious, governmental, educational, and business customs and opinions, Volume 1*. New York: Harper & Brothers, 1865.

Doty, D. W. Role playing and incentives in the modification of the social interaction of chronic psychiatric patients. *Journal of Consulting and Clinical Psychology*, 1975, *43*, 676–682.

Doty, D. W., McInnis, T., & Paul, G. L. Remediation of negative side effects of an on-going response-cost system with chronic mental patients. *Journal of Applied Behavior Analysis,* 1974, *7,* 191–198.

Drabman, R. S. Child versus teacher administered token programs in a psychiatric hospital school. *Journal of Abnormal Child Psychology,* 1973, *1,* 68–87.

Drabman, R. S. Behavior modification in the classroom. In W. E. Craighead, A. E. Kazdin, & M. J. Mahoney (Eds.), *Behavior modification: Principles, issues, and applications.* Boston: Houghton Mifflin, 1976.

Drabman, R. S., & Lahey, B. B. Feedback in classroom behavior modification: Effects on the target and her classmates. *Journal of Applied Behavior Analysis,* 1974, *7,* 591–598.

Drabman, R. S., Spitalnik, R., & O'Leary, K. D. Teaching self-control to disruptive children. *Journal of Abnormal Psychology,* 1973, *82,* 10–16.

Drabman, R., Spitalnik, R., & Spitalnik, K. Sociometric and disruptive behavior as a function of four types of token reinforcement programs. *Journal of Applied Behavior Analysis,* 1974, *7,* 93–101.

Duran, F. D. *The Aztecs.* New York: Orion Press, 1964.

Edlund, C. V. The effect on the test behavior of children, as reflected in the IQ scores, when reinforced after each correct response. *Journal of Applied Behavior Analysis,* 1972, *5,* 317–319.

Edwards, C. D., & Williams, J. E. Generalization between evaluative words associated with racial figures in preschool children. *Journal of Experimental Research in Personality,* 1970, *4,* 144–155.

Eggan, F., Gilbert, W., McAllister, J., Nash, P., Opler, M., Provinse, J., & Tax, S. *Social anthropology of North American tribes.* Chicago: University of Chicago Press, 1937.

Eitzen, D. S. The effects of behavior modification on the attitudes of delinquents. *Behaviour Research and Therapy,* 1975, *13,* 295–299.

Elliott, R., & Tighe, T. Breaking the cigarette habit: Effects of a technique involving threatened loss of money. *Psychological Record,* 1968, *18,* 503–513.

Ellis, N. R. Amount of reward and operant behavior in mental defectives. *American Journal of Mental Deficiency,* 1962, *66,* 595–599.

Ellis, N. R., Barnett, C. D., & Pryer, M. W. Operant behavior in mental defectives: Exploratory studies. *Journal of the Experimental Analysis of Behavior,* 1960, *3,* 63–69.

Ellsworth, J. R. Reinforcement therapy with chronic patients. *Hospital & Community Psychiatry,* 1969, *20,* 36–38.

Ellsworth, P. D., & Colman, A. D. The application of operant conditioning principles to work group experience. *American Journal of Occupational Therapy,* 1970, *24,* 562–568.

Emshoff, J. G., Redd, W. H., & Davidson, W. S. Generalization training and the transfer of treatment effects with delinquent adolescents. *Journal of Behavior Therapy and Experimental Psychiatry,* 1976, *7,* 141–144.

Engelin, R., Knutson, J., Laughy, L., & Garlington, W. Behaviour modification techniques applied to a family unit—a case study. *Journal of Child Psychology and Psychiatry,* 1968, *9,* 245–252.

Ennis, B. J., & Friedman, P. R. (Eds.), *Legal rights of the mentally handicapped, Volumes 1 and 2.* Practicing Law Institute, The Mental Health Law Project, 1973.

Eriksson, J. H., Gotestam, K. G., Melin, L., & Ost, L. A token economy treatment of drug addiction. *Behaviour Research and Therapy,* 1975, *13,* 113–125.

Evans, D. R., Howath, P., Sanders, S., & Dolan, J. Reinforcement of attention and academic performance in a special education class. *Psychological Reports,* 1974, *35,* 1143–1146.

Everett, P. B., Hayward, S. C., & Meyers, A. W. The effects of a token reinforcement procedure on bus ridership. *Journal of Applied Behavior Analysis,* 1974, *7,* 1–9.

Eysenck, H. J. Learning theory and behaviour therapy. *Journal of Mental Science,* 1959, *105,* 61–75.

Eysenck, H. J. *Psychology is about people.* New York: The Library Press, 1972.

Farmer, J., Lachter, G. D., Blaustein, J. J., & Cole, B. K. The role of proctoring in personalized instruction. *Journal of Applied Behavior Analysis,* 1972, *5,* 401–404.

Feallock, R. A., & Miller, L. K. The design and evaluation of a worksharing system for experimental living. Unpublished paper. University of Kansas, 1974.

Federal Bureau of Prisons. START—Revised program. Washington, D. C.: 1972.

Feingold, B. D., & Mahoney, M. J. Reinforcement effects on intrinsic interest: Undermining the overjustification hypothesis. *Behavior Therapy,* 1975, *6,* 367–377.

Feingold, L., & Migler, B. The use of experimental dependency relationships as a motivating procedure on a token economy ward. In R. D. Rubin, H. Fensterheim, J. D. Henderson, & L. P. Ullmann (Eds.), *Advances in behavior therapy.* New York: Academic Press, 1972.

Felixbrod, J. J., & O'Leary, K. D. Effects of reinforcement of children's academic behavior as a function of self determined and externally imposed contingencies. *Journal of Applied Behavior Analysis,* 1973, *6,* 241–250.

Felixbrod, J. J., & O'Leary, K. D. Self-determination of academic standards by children: Toward freedom from external control. *Journal of Educational Psychology,* 1974, *66,* 845–850.

Ferritor, D. E., Buckholdt, D., Hamblin, R. L., & Smith, L. The noneffects of contingent reinforcement for attending behavior on work accomplished. *Journal of Applied Behavior Analysis,* 1972, *5,* 7–17.

Ferster, C. B., Culbertson, S., & Boren, M. C. P. *Behavior principles* (Second Edition). Englewood Cliffs, New Jersey: Prentice-Hall, 1975.

Ferster, C. B., & DeMyer, M. K. The development of performances in autistic children in an automatically controlled environment. *Journal of Chronic Diseases,* 1961, *13,* 312–345.

Ferster, C. B., & DeMyer, M. K. A method for the experimental analysis of the behavior of autistic children. *American Journal of Orthopsychiatry,* 1962, *1,* 87–110.

Ferster, C. B., & Skinner, B. F. *Schedules of reinforcement.* New York: Appleton-Century-Crofts, 1957.

Fethke, G. C. The relevance of economic theory and technology to token reinforcement systems: A comment. *Behaviour Research and Therapy,* 1972, *10,* 191–192.

Fethke, G. C. Token economies: A further comment. *Behaviour Research and Therapy,* 1973, *11,* 225–226.

Fielding, L. T., Errickson, E., & Bettin, B. Modification of staff behavior: A brief note. *Behavior Therapy,* 1971, *2,* 550–553.

Filipczak, J., & Cohen, H. L. The Case II contingency system and where it is going. Paper presented at the American Psychological Association, Honolulu, Hawaii, September, 1972.

Fineman, K. R. An operant conditioning program in a juvenile detention facility. *Psychological Reports,* 1968, *22,* 1119–1120.

Fixsen, D. L., Phillips, E. L., & Wolf, M. M. Achievement Place: The reliability of self-reporting and peer-reporting and their effects on behavior. *Journal of Applied Behavior Analysis,* 1972, *5,* 19–30.

Fixsen, D. L., Phillips, E. L., & Wolf, M. M. Achievement place: Experiments in self-government with pre-delinquents. *Journal of Applied Behavior Analysis,* 1973, *6,* 31–47.

Fixsen, D. L., Phillips, E. L., Phillips, E. A., & Wolf, M. M. The teaching-family model of group home treatment. In W. E. Craighead, A. E. Kazdin, & M. J. Mahoney (Eds.), *Behavior modification: Principles, issues, and applications.* Boston: Houghton Mifflin, 1976.

Fjellstedt, N., & Sulzer-Azaroff, B. Reducing the latency of a child's responding to instructions by means of a token system. *Journal of Applied Behavior Analysis,* 1973, *6,* 125–130.

Flanagan, B., Goldiamond, I., & Azrin, N. H. Operant stuttering: The control of stuttering behavior through response-contingent consequences. *Journal of the Experimental Analysis of Behavior,* 1958, *1,* 173–177.

Flowers, J. V. A behavior modification technique to reduce the frequency of unwarranted questions by target students in an elementary school classroom. *Behavior Therapy,* 1974, *5,* 665–667.

Fo, W. S. O., & O'Donnell, C. R. The buddy system: Relationship and contingency conditions in a community intervention program for youth with nonprofessionals as behavior change agents. *Journal of Consulting and Clinical Psychology,* 1974, *42,* 163–169.

Fo, W. S. O., & O'Donnell, C. R. The buddy system: Effect of community intervention on delinquent offenses. *Behavior Therapy*, 1975, *6*, 522–524.

Foreyt, J. P. The punch card token economy program. In R. L. Patterson (Ed.), *Maintaining effective token economies*. Springfield, Ill.: Charles C Thomas, 1976.

Fowler, R. L., & Thomas, E. S. A comparison of the two level and five level grading systems in personalized instruction courses. *Psychological Record*, 1974, *24*, 333–341.

Frazier, J. R., & Williams, B. R. The application of multiple contingencies to rocking behavior in a non-retarded child. *Journal of Behavior Therapy and Experimental Psychiatry*, 1973, *4*, 289–291.

Frederiksen, L. W., & Frederiksen, C. B. Teacher-determined and self-determined token reinforcement in a special education classroom. *Behavior Therapy*, 1975, *6*, 310–314.

Friedman, P. R. Legal regulation of applied behavior analysis in mental institutions and prisons. *Arizona Law Review*, 1975, *17*, 39–104.

Fuller, P. R. Operant conditioning of a vegetative human organism. *American Journal of Psychology*, 1949, *62*, 587–590.

Gambrill, E. D. The use of behavioral methods in short-term detention settings. Paper presented at meeting of the Association for Advancement of Behavior Therapy, Chicago, November, 1974.

Garcia, E. The training and generalization of a conversational speech form in nonverbal retardates. *Journal of Applied Behavior Analysis*, 1974, *7*, 137–149.

Gardner, J. M. Teaching behavior modification to nonprofessionals. *Journal of Applied Behavior Analysis*, 1972, *5*, 517–521.

Gardner, J. M. Training the trainers: A review of research on teaching behavior modification. In R. D. Rubin, J. P. Brady, & J. D. Henderson (Eds.), *Advances in behavior therapy, Volume 4*. New York: Academic Press, 1973.

Gates, J. J. Overspending (stealing) in a token economy. *Behavior Therapy*, 1972, *3*, 152–153.

Gelfand, D. M., Elton, R. H., & Harman, R. E. A videotape-feedback training method to teach behavior modification skills to nonprofessionals. *Research in Education*, 1972, *7*, 15. Document Ed 056 314, U. S. Office of Education. (Abstract)

Gelfand, D. M., Gelfand, S., & Dobson, W. R. Unprogrammed reinforcement of patient's behaviour in a mental hospital. *Behaviour Research and Therapy*, 1967, *5*, 201–207.

Geller, E. S. Prompting anti-litter behaviors. *Proceedings of the 81st Annual Convention of the American Psychological Association*, 1973, *8*, 901–902.

Geller, E. S. Increasing desired waste disposals with instructions. *Man-Environment Systems*, 1975, *5*, 125–128.

Geller, E. S., Chaffee, J. L., & Ingram, R. E. Promoting paper recycling on a university campus. *Journal of Environmental Systems*, 1975, *5*, 39–57.

Geller, E. S., Farris, J. C., & Post, D. S. Prompting a consumer behavior for pollution control. *Journal of Applied Behavior Analysis*, 1973, *6*, 367–376.

Geller, E. S., Wylie, R. G., & Farris, J. C. An attempt at applying prompting and reinforcement toward pollution control. *Proceedings of the 79th Annual Convention of the American Psychological Association*, 1971, *6*, 701–702.

Gericke, O. L. Practical use of operant conditioning procedures in a mental hospital. *Psychiatric Studies and Projects*, 1965, *3*, 2–10.

Gewirtz, J. L., & Baer, D. M. Deprivation and satiation of social reinforcers as drive conditions. *Journal of Abnormal and Social Psychology*, 1958, *57*, 165–172.

Girardeau, F. L., & Spradlin, J. E. Token rewards in a cottage program. *Mental Retardation*, 1964, *2*, 345–351.

Gladstone, B. W., & Sherman, J. A. Developing generalized behavior-modification skills in high-school students working with retarded children. *Journal of Applied Behavior Analysis*, 1975, *8*, 169–180.

Glickman, H., Plutchik, R., & Landau, H. Social and biological reinforcement in an open psychiatric ward. *Journal of Behavior Therapy and Experimental Psychiatry*, 1973, *4*, 121–124.

Glicksman, M., Ottomanelli, G., & Cutler, R. The earn-your-way credit system: Use of a token economy in narcotic rehabilitation. *International Journal of the Addictions*, 1971, *6*, 525–531.

Glynn, E. L. Classroom applications of self-determined reinforcement. *Journal of Applied Behavior Analysis*, 1970, *3*, 123–132.

Glynn, E. L., Thomas, J. D., & Shee, S. M. Behavioral self-control of on-task behavior in an elementary classroom. *Journal of Applied Behavior Analysis*, 1973, *6*, 105–113.

Glynn, E. L., & Thomas, J. D. Effect of cueing on self-control of classroom behavior. *Journal of Applied Behavior Analysis*, 1974, *7*, 299–306.

Goffman, E. *Asylums*. New York: Doubleday, 1961.

Goldfried, M. R., & Merbaum, M. (Eds.), *Behavior change through self-control. New York: Holt, Rinehart & Winston, 1973.*

Goldiamond, I. Toward a constructional approach to social problems: Ethical and constitutional issues raised by applied behavior analysis. *Behaviorism*, 1974, *2*, 1–84.

Golub, C. M. The influence of various demographic variables on the participation of Veterans Administration Day Treatment Center patients in a token economy. Unpublished doctoral dissertation. University of Minnesota, 1969.

Goocher, B. E., & Ebner, M. A behavior modification approach utilizing sequential response targets in multiple settings. Paper presented at Midwestern Psychological Association, Chicago, May, 1968.

Goodman, S. M. Right to treatment: The responsibility of the courts. In D. S. Burris (Ed.), *The right to treatment.* New York: Springer Publishing Company, 1969.

Gorham, D. C., Green, L. W. Caldwell, L. R., & Bartlett, E. R. Effect of operant conditioning techniques on chronic schizophrenics. *Psychological Reports*, 1970, *27*, 223–234.

Gotestam, K. G., & Melin, L. A modified token economy with patients in a methadone maintenance treatment program. *Proceedings of the Fifth National Conference on Methadone Treatment*, 1973, *5*, 1184–1190.

Gotestam, K. G., Melin, L., & Dockens, W. S., III. A behavioral program for intravenous amphetamine addicts. In T. Thompson & W. S. Dockens, III (Eds.), *Applications of behavior modification.* New York: Academic Press, 1975.

Gotestam, K. G., Melin, L., & Ost, L. Behavioral techniques in the treatment of drug abuse: An evaluative review. *Addictive Behaviors*, 1976, in press.

Gouldner, A. Anti-minotaur: The myth of a value-free sociology. *Social Problems*, 1962, *9*, 199–213.

Grandy, G. S., Madsen, C. H., Jr., & De Mersseman, L. M. The effects of individual and interdependent contingencies on inappropriate classroom behavior. *Psychology in the Schools*, 1973, *10*, 488–493.

Grant, M. *Gladiators*. London: Trinity Press, 1967.

Graubard, P. S., Rosenberg, H., & Miller, M. B. Student applications of behavior modification to teachers and environments or ecological approaches to social deviancy. In E. A. Ramp & B. L. Hopkins (Eds.), *A new direction for education: Behavior analysis.* Lawrence, Kansas: Support and Development Center for Follow Through, 1971.

Graubard, P. S., Rosenberg, H., & Miller, M. B. Student applications of behavior modification to teachers and environments or ecological approaches to social deviancy. In R. Ulrich, T. Stachnik, & J. Mabry, *Control of human behavior, Volume 3.* Glenview, Ill.: Scott, Foresman and Company, 1974.

Gray, F., Graubard, P. S., & Rosenberg, H. Little brother is changing you. *Psychology Today*, 1974, *7*, March, 42–46.

Greenberg, D. J., Scott, S. B., Pisa, A., & Friesen, D. D. Beyond the token economy: A comparison of two contingency programs. *Journal of Consulting and Clinical Psychology,* 1975, *43,* 498–503.

Greene, D., & Lepper, M. R. Effects of extrinsic rewards on children's subsequent intrinsic interest. *Child Development,* 1974, *45,* 1141–1145.

Greenspoon, J. The effect of verbal and non-verbal stimuli on the frequency of members of two verbal response classes. Unpublished doctoral dissertation, Indiana University, 1951.

Greenwood, C. R., Hops, H., Delquadri, J., & Guild, J. Group contingencies for group consequences in classroom management: A further analysis. *Journal of Applied Behavior Analysis,* 1974, *7,* 413–425.

Greenwood, C. R., Sloane, H. N., & Baskin, A. Training elementary aged peer-behavior managers to control small group programmed mathematics. *Journal of Applied Behavior Analysis,* 1974, *7,* 103–114.

Griffiths, H., & Craighead, W. E. Generalization in operant speech therapy for misarticulation. *Journal of Speech and Hearing Disorders,* 1972, *37,* 485–494.

Gripp, R. F., & Magaro, P. A. A token economy program evaluation with untreated control ward comparisons. *Behaviour Research and Therapy,* 1971, *9,* 137–149.

Gripp, R. F., & Magaro, P. A. Token economy program in the psychiatric hospital: Review and analysis. *Behaviour Research and Therapy,* 1974, *12,* 205–228.

Grzesiak, R. C., & Locke, B. J. Cognitive and behavioral correlates to overt behavior change with a token economy. *Journal of Consulting and Clinical Psychology,* 1975, *43,* 272.

Guerney, B. G., Jr. (Ed.), *Psychotherapeutic agents: New roles for non-professionals, parents, and teachers.* New York: Holt, Rinehart & Winston, 1969.

Guess, D. A functional analysis of receptive and productive speech: Acquisition of the plural morpheme. *Journal of Applied Behavior Analysis,* 1969, *2,* 55–64.

Guess, D., & Baer, D. M. An analysis of individual differences in generalization between receptive and productive language in retarded children. *Journal of Applied Behavior Analysis,* 1973, *6,* 311–329.

Guess, D., Sailor, W., Rutherford, G., & Baer, D. An experimental analysis of linguistic development: The productive use of the plural morpheme. *Journal of Applied Behavior Analysis,* 1968, *1,* 297–306.

Hall, J., & Baker, R. Token economy systems: Breakdown and control. *Behaviour Research and Therapy,* 1973, *11,* 253–263.

Hall, R. V. Responsive teaching: Focus on measurement and research in the classroom and the home. In E. L. Meyen, G. A. Vergason, & R. J. Whelan (Eds.), *Strategies for teaching exceptional children.* Denver: Love, 1972.

Hall, R. V. Axelrod, S., Foundopoulos, M., Shellman, J., Campbell, R. A., & Cranston, S. S. The effective use of punishment to modify behavior in the classroom. In K. D. O'Leary & S. G. O'Leary (Eds.), *Classroom management: The successful use of behavior modification.* New York: Pergamon Press, 1972.

Hall, R. V., Axelrod, S., Tyler, L., Grief, E., Jones, F. C., & Robertson, R. Modification of behavior problems in the home with a parent as observer and experimenter *Journal of Applied Behavior Analysis,* 1972, *5,* 53–64.

Hall, R. V., & Broden, M. Behavior changes in brain-injured children through social reinforcement. *Journal of Experimental Child Psychology,* 1967, *5,* 463–479.

Hall, R. V., & Copeland, R. E. The Responsive Teaching Model: A first step in shaping school personnel as behavior modification specialists. In F. W. Clark, D. R. Evans, & L. A. Hammerlynck (Eds.), *Implementing behavioral programs for schools and clinics.* Champaign, Ill.: Research Press, 1972.

Hall, R. V., Panyan, M., Rabon, D., & Broden, M. Instructing beginning teachers in reinforcement procedures which improve classroom control. *Journal of Applied Behavior Analysis,* 1968, *1,* 315–322.

Hamblin, R. L., Hathaway, C., & Wodarski, J. Group contingencies, peer tutoring, and accelerating academic achievement. In R. Ulrich, T. Stachnik, & J. Mabry, *Control of human behavior, Volume 3.* Glenview, Ill.: Scott, Foresman and Company, 1974.

Hamilton, J., & Allen, P. Ward programming for severely retarded institutionalized residents. *Mental Retardation,* 1967, *6,* 22–25.

Haring, N. G., & Hauck, M. A. Improved learning conditions in the establishment of reading skills with disabled readers. *Exceptional Children,* 1969, *35,* 341–352.

Harmatz, M. G., & Lapuc, P. Behavior modification of over-eating in a psychiatric population. *Journal of Consulting and Clinical Psychology,* 1968, *32,* 583–587.

Harris, F. R., Johnston, M. K., Kelley, C. S., & Wolf, M. M. Effects of social reinforcement on regressed crawling of a nursery school child. *Journal of Educational Psychology,* 1964, *55,* 35–41.

Harris, F. R., Wolf, M. M., & Baer, D. M. Effects of adult social reinforcement on child behavior. *Young Children,* 1964, *20,* 8–17.

Harris, V. W., Finfrock, S. R., Giles, D. K., Hart, B. M., & Tsosie, P. C. The use of home-based consequences to modify the classroom behavior of institutionalized delinquent youth. *Journal of Applied Behavior Analysis,* in press.

Harris, V. W., & Sherman, J. A. Use and analysis of the "Good Behavior Game" to reduce disruptive classroom behavior. *Journal of Applied Behavior Analysis,* 1973, *6,* 405–417.

Harris, V. W., & Sherman, J. A. Homework assignments, consequences, and classroom performance in social studies and mathematics. *Journal of Applied Behavior Analysis,* 1974, *7,* 505–519.

Harshbarger, D., & Maley, R. F. (Eds.), *Behavior analysis and systems analysis: An integrative approach to mental health programs.* Kalamazoo, Michigan: Behaviordelia, 1974.

Hart, B. M., Allen, K. E., Buell, J. S., Harris, F. R., & Wolf, M. M. Effects of social reinforcement on operant crying. *Journal of Experimental Child Psychology,* 1964, *1,* 145–153.

Hartlage, L. C. Subprofessional therapists' use of reinforcement versus traditional psychotherapeutic techniques with schizophrenics. *Journal of Consulting and Clinical Psychology,* 1970, *34,* 181–183.

Hartmann, D. P., & Atkinson, C. Having your cake and eating it too: A note on some apparent contradictions between therapeutic achievements and design requirements in N = 1 studies. *Behavior Therapy,* 1973, *4,* 589–591.

Hauserman, N. Walen, S. R., & Behling, M. Reinforced racial integration in the first grade: A study in generalization. *Journal of Applied Behavior Analysis,* 1973, *6,* 193–200.

Hauserman, N., Zweback, S., & Plotkin, A. Use of concrete reinforcement to facilitate verbal initiations in adolescent group therapy. *Journal of Consulting and Clinical Psychology,* 1972, *38,* 90–96.

Hayden, T. Osborne, A. E., Hall, S. M., & Hall, R. G. Behavioral effects of price changes in a token economy. *Journal of Abnormal Psychology,* 1974, *83,* 432–439.

Hayes, S. C., & Cone, J. D. Reducing residential electrical energy use: Payments, information, and feedback. *Journal of Applied Behavior Analysis,* in press.

Hayes, S. C., Johnson, V. S., & Cone, J. D. The marked item technique: A practical procedure for litter control. *Journal of Applied Behavior Analysis,* 1975, *8,* 381–386.

Haynes v. Harris, 344 F. 2d 463 (8th Cir. 1965).

Heap, R. F., Boblitt, W. E., Moore, C. H., & Hord, J. E. Behavior-milieu therapy with chronic neuropsychiatric patients. *Journal of Abnormal Psychology,* 1970, *76,* 349–354.

Heller, K. & Marlatt, G. A. Verbal conditioning, behavior therapy, and behavior change: Some problems of extrapolation. In C. M. Franks (Ed.), *Behavior therapy: Appraisal and status.* New York: McGraw-Hill, 1969.

Henderson, J. D. The use of dual reinforcement in an intensive treatment system. In R. D. Rubin & C. M. Franks (Eds.), *Advances in behavior therapy, 1968.* New York: Academic Press, 1969.

Henderson, J. D., & Scoles, P. E. A community-based behavioral operant environment for psychotic men. *Behavior Therapy,* 1970, *1,* 245–251.

Herbert, E. W., & Baer, D. M. Training parents as behavior modifiers: Self-recording of contingent attention. *Journal of Applied Behavior Analysis,* 1972, *5,* 139–149.

Herman, S., & Tramontana, J. Instructions and group versus individual reinforcement in modifying disruptive group behavior. *Journal of Applied Behavior Analysis*, 1971, *4*, 113–119.

Hermann, J. A., de Montes, A. I., Dominguez, B., Montes, F., & Hopkins, B. L. Effects of bonuses for punctuality on the tardiness of industrial workers. *Journal of Applied Behavior Analysis*, 1973, *6*, 563–570.

Hersen, M. Token economies in institutional settings: Historical, political, deprivation, ethical, and generalization issues. *Journal of Nervous and Mental Disease*, 1976, *162*, 206–211.

Hersen, M., & Barlow, D. H. *Single case experimental designs: Strategies for studying behavior change.* New York: Pergamon, 1976.

Hersen, M., & Eisler, R. M. Comments on Heap, Boblitt, Moore, and Hord's "Behavior-milieu therapy with chronic neuropsychiatric patients." *Psychological Reports*, 1971, *29*, 583–586.

Hersen, M., Eisler, R. M., Smith, B. S., & Agras, W. S. A token reinforcement ward for young psychiatric patients. *American Journal of Psychiatry*, 1972, *129*, 142–147.

Hersen, M., Eisler, R. M., Alford, G. S., & Agras, W. S. Effects of token economy on neurotic depression: An experimental analysis. *Behavior Therapy*, 1973, *4*, 392–397.

Hewett, F. M., Taylor, F. D., & Artuso, A. A. The Santa Monica Project: Evaluation of an engineered classroom design with emotionally disturbed children. *Exceptional Children*, 1969, *35*, 523–529.

Higa, W. R. Self-instructional versus direct training in modifying children's impulsive behavior. Unpublished doctoral dissertation, University of Hawaii, 1973.

Higgs, W. J. Effects of gross environmental change upon behavior of schizophrenics: A cautionary note. *Journal of Abnormal Psychology*, 1970, *76*, 421–422.

Hilgard, E. R., & Marquis, D. G. *Conditioning and learning.* New York: Appleton-Century-Crofts, 1940.

Holland, J. G., & Skinner, B. F. *The analysis of behavior.* New York: McGraw-Hill, 1961.

Hollander, M., & Horner, V. Using environmental assessment and operant procedures to build integrated behaviors in schizophrenics. *Journal of Behavior Therapy and Experimental Psychiatry*, 1975, *6*, 289–294.

Hollander, M. A., & Plutchik, R. A reinforcement program for psychiatric attendants. *Journal of Behavior Therapy and Experimental Psychiatry*, 1972, *3*, 297–300.

Hollander, M., Plutchik, R., & Horner, V. Interaction of patient and attendant reinforcement programs: The "piggyback" effect. *Journal of Consulting and Clinical Psychology*, 1973, *41*, 43–47.

Hollingsworth, R., & Foreyt, J. P. Community adjustment of released token economy patients. *Journal of Behavior Therapy and Experimental Psychiatry*, 1975, *6*, 271–274.

Holmes v. Silver Cross Hosp. of Joliet, Illinois, 340 F. Supp. 125 (E.D. Ill. 1972).

Holzman, M. The significance of the value systems of patient and therapist for the outcome of psychotherapy. Unpublished doctoral dissertation, University of Washington, 1961.

Honig, W. K. (Ed.), *Operant Behavior: Areas of research and application.* Appleton-Century-Crofts, 1966.

Hopkins, B. L. Effects of candy and social reinforcement, instructions, and reinforcement schedule learning on the modification and maintenance of smiling. *Journal of Applied Behavior Analysis*, 1968, *1*, 121–129.

Horn, J., & Black, W. A. M. The effect of token reinforcement on verbal participation in a social activity with long stay psychiatric patients. *Australian and New Zealand Journal of Psychiatry*, 1973, *7*, 185–188.

Horner, R. D. Establishing use of crutches by a mentally retarded spina bifida child. *Journal of Applied Behavior Analysis*, 1971, *4*, 183–189.

Horner, R. D., & Keilitz, I. Training mentally retarded adolescents to brush their teeth. *Journal of Applied Behavior Analysis*, 1975, *8*, 301–309.

Hospers, J. *Human conduct: An introduction to the problems of ethics.* New York: Harcourt, Brace & World, 1961.

Hundert, J. The effectiveness of reinforcement, response cost, and mixed programs on class-room behaviors. *Journal of Applied Behavior Analysis*, 1976, *9*, 107.

Hunt, J. G., Fitzhugh, L. C., & Fitzhugh, K. B. Teaching "exit-ward" patients appropriate personal appearance by using reinforcement techniques. *American Journal of Mental Deficiency*, 1968, *73*, 41–45.

Hunt, J. G., & Zimmerman, J. Stimulating productivity in a simulated sheltered workshop setting. *American Journal of Mental Deficiency*, 1969, *74*, 43–49.

Hutt, P. J. Rate of bar pressing as a function of quality and quantity of food reward. *Journal of Comparative and Physiological Psychology*, 1954, *47*, 235–239.

Ingham, R. E., Chaffee, J. L., Rorrer, C. W., & Wellington, C. J. A community-integrated behavior modification approach to facilitate paper recycling. Paper presented at meeting of the Southern Psychological Association, Atlanta, March, 1975.

Ingham, R. J., Andrews, G., & Winkler, R. Stuttering: A comparative evaluation of the short-term effectiveness of four treatment techniques. *Journal of Communication Disorders*, 1972, *5*, 91–117.

Ingham, R. J., & Andrews, G. An analysis of a token economy in stuttering therapy. *Journal of Applied Behavior Analysis*, 1973, *6*, 219–229.

Ingram, R. E., & Geller, E. S. A community-integrated, behavior modification approach to facilitating paper recycling. *Catalog of Selected Documents in Psychology*, 1975, *5*, 327.

Inmates of Boys' Training School v. Affleck, 346 F. Supp. 1354 (D. R. I., 1972).

Isaacs, W., Thomas, J., & Goldiamond, I. Application of operant conditioning to reinstate verbal behavior in psychotics. *Journal of Speech and Hearing Disorders*, 1960, *25*, 8–12.

Ivanov-Smolensky, A. G. Neurotic behavior and the teaching of conditioned reflexes. *American Journal of Psychiatry*, 1927, *84*, 483–488

Iwata, B. A., & Bailey, J. S. Reward versus cost token systems: An analysis of the effects on students and teacher. *Journal of Applied Behavior Analysis*, 1974, *7*, 567–576.

Jackson, D. A., & Wallace, R. F. The modification and generalization of voice loudness in a fifteen-year-old retarded girl. *Journal of Applied Behavior Analysis*, 1974, *7*, 461–471.

Jeffrey, D. B. A comparison of the effects of external-control and self-control on the modification and maintenance of weight. *Journal of Abnormal Psychology*, 1974, *83*, 404–410.

Jenkins, W. O. Witherspoon, A. D., DeVine, M. D., deValera, E. K., Muller, J. B., Barton, M. C., & McKee, J. M. The post-prison analysis of criminal behavior and longitudinal follow-up evaluation of institutional treatment. A report on the Experimental Manpower Laboratory for Corrections, February, 1974.

Jens, K. G., & Shores, R. E. Behavioral graphs as reinforcers for work behavior of mentally retarded adolescents. *Education and Training of the Mentally Retarded*, 1969, *4*, 21–28.

Jesness, C. F. Comparative effectiveness of behavior modification and transactional analysis programs for delinquents. *Journal of Consulting and Clinical Psychology*, 1975, *43*, 758–779.

Jesness, C. F., & DeRisi, W. J. Some variations in techniques of contingency management in a school for delinquents. In J. S. Stumphauzer (Ed.), *Behavior therapy with delinquents*. Springfield, Ill.: Charles C Thomas, 1973.

Jobson vs. Henne, 355 F. 2d 129 (2d Cir. 1966), cited in 61 Calif. L. Rev. at 91.

Johnson, S. M., & Martin, S. Developing self-evaluation as a conditioned reinforcer. In B. Ashem & E. G. Poser (Eds.), *Behavior modification with children*. New York: Pergamon, 1972.

Johnson, S.M., Wahl, G., Martin, S., & Johansson, S. How deviant is the normal child? A behavioral analysis of the preschool child and his family. In R. D. Rubin, J. P. Brady, & J. D. Henderson (Eds.), *Advances in behavior therapy, Volume 4*. New York: Academic Press, 1973.

Johnston, J. M., & Johnston, G. T. Modification of consonant speech-sound articulation in young children. *Journal of Applied Behavior Analysis*, 1972, *5*, 233–246.

Johnston, J. M., & O'Neill, G. The analysis of performance criteria defining course grades as a determinant of college student academic performance. *Journal of Applied Behavior Analysis,* 1973, *6,* 261–268.

Johnston, M. S., Kelley, C. S., Harris, F. R., & Wolf, M. M. An application of reinforcement principles to development of motor skills of a young child. *Child Development,* 1966, *37,* 379–387.

Jones, F. H., & Eimers, R. C. Role playing to train elementary teachers to use a classroom management "skill package." *Journal of Applied Behavior Analysis,* 1975, *8,* 421–433.

Jones, F. H., & Miller, W. H. The effective use of negative attention for reducing group disruption in special elementary school classrooms. *Psychological Record,* 1974, *24,* 435–448.

Jones, R. J., & Azrin, N. H. An experimental application of a social reinforcement approach to the problem of job-finding. *Journal of Applied Behavior Analysis,* 1973, *6,* 345–353.

Jones, R. T., & Kazdin, A. E. Programming response maintenance after withdrawing token reinforcement. *Behavior Therapy,* 1975, *6,* 153–164.

Jones v. Robinson, 440 F 2d 249 (D. C. Cir. 1971).

Kaestle, C. F. (Ed.), *Joseph Lancaster and the monitorial school movement: A documentary history.* New York: Teachers College Press, 1973.

Kagel, J. H., & Winkler, R. C. Behavioral economics: Areas of cooperative research between economics and applied behavior analysis. *Journal of Applied Behavior Analysis,* 1972, *5,* 335–342.

Kaimowitz v. Michigan Department of Mental Health, 42 U. S. L. Week 2063 (Mich. Cir. Ct., Wayne Cty. July 10, 1973).

Kale, R. J., Kaye, J. H., Whelan, P. A., & Hopkins, B. L. The effects reinforcement on the modification, maintenance, and generalization of social responses of mental patients. *Journal of Applied Behavior Analysis,* 1968, *1,* 307–314.

Kallman, W. H., Hersen, M., & O'Toole, D. H. The use of social reinforcement in a case of conversion reaction. *Behavior Therapy,* 1975, *6,* 411–413.

Kanfer, F. H. Verbal conditioning: A review of its current status. In T. R. Dixon & D. L. Horton (Eds.), *Verbal behavior and general behavior theory.* Englewood Cliffs, New Jersey: Prentice-Hall, Inc., 1968.

Kanfer, F. H., & Phillips, J. S. *Learning foundations of behavior therapy.* New York: Wiley & Sons, 1970.

Karacki, L., & Levinson, R. B. A token economy in a correctional institution for youthful offenders. *Howard Journal of Penology and Crime Prevention,* 1970, *13,* 20–30.

Karen, R. L., Eisner, M., & Enders, R. W. Behavior modification in a sheltered workshop for severely retarded students. *American Journal of Mental Deficiency,* 1974, *79,* 338–347.

Kassirer, L. B. Behavior modification for patients and prisoners: Constitutional ramifications of enforced therapy. *The Journal of Psychiatry and Law,* 1974, *2,* 245–302.

Katz, R. C., Johnson, C. A., & Gelfand, S. Modifying the dispensing of reinforcers: Some implications for behavior modification with hospitalized patients. *Behavior Therapy,* 1972, *3,* 579–588.

Kaufman, A., Baron, A., & Kopp, R. E. Some effects of instructions on human operant behavior. *Psychological Monograph Supplements,* 1966, *1,* 243–250.

Kaufman, K. F., & O'Leary, K. D. Reward, cost, and self-evaluation procedures for disruptive adolescents in a psychiatric hospital school. *Journal of Applied Behavior Analysis,* 1972, *5,* 293–309.

Kazdin, A. E. The effect of response cost in suppressing behavior in a pre-psychotic retardate. *Journal of Behavior Therapy and Experimental Psychiatry,* 1971, *2,* 137–140. (a)

Kazdin, A. E. Toward a client administered token reinforcement program. *Education and Training of the Mentally Retarded,* 1971, *6,* 52–55. (b)

Kazdin, A. E. Nonresponsiveness of patients to token economies. *Behaviour Research and Therapy,* 1972, *10,* 417–418. (a)

Kazdin, A. E. Response cost: The removal of conditioned reinforcers for therapeutic change. *Behavior Therapy,* 1972, *3,* 533–546. (b)

Kazdin, A. E. The effect of response cost and aversive stimulation in suppressing punished and nonpunished speech disfluencies. *Behavior Therapy,* 1973, *4,* 73–82. (a)

Kazdin, A. E. The effect of vicarious reinforcement on attentive behavior in the classroom. *Journal of Applied Behavior Analysis,* 1973, *6,* 71–78. (b)

Kazdin, A. E. The effect of vicarious reinforcement on performance in a rehabilitation setting. *Education and Training of the Mentally Retarded, 1973,* 8, 4–11. (c)

Kazdin, A. E. The failure of some patients to respond to token programs. *Journal of Behavior Therapy and Experimental Psychiatry,* 1973, *4,* 7–14. (d)

Kazdin, A. E. Issues in behavior modification with mentally retarded persons. *American Journal of Mental Deficiency,* 1973, *78,* 134–140. (e)

Kazdin, A. E. Methodological and assessment considerations in evaluating reinforcement programs in applied settings. *Journal of Applied Behavior Analysis,* 1973, *6,* 517–531. (f)

Kazdin, A. E. Role of instructions and reinforcement in behavior changes in token reinforcement programs. *Journal of Educational Psychology,* 1973, *64,* 63–71. (g)

Kazdin, A. E. The assessment of teacher training in a reinforcement program. *Journal of Teacher Education,* 1974, *25,* 266–270. (a)

Kazdin, A. E. Reactive self-monitoring: The effects of response desirability, goal setting, and feedback. *Journal of Consulting and Clinical Psychology,* 1974, *42,* 704–716. (b)

Kazdin, A. E. A review of token economy treatment modalities. In D. Harshbarger & R. F. Maley (Eds.), *Behavior analysis and systems analysis: An integrative approach to mental health programs.* Kalamazoo, Michigan: Behaviordelia, 1974. (c)

Kazdin, A. E. Self-monitoring and behavior change. In M. J. Mahoney & C. E. Thoresen (Eds.), *Self-control: Power to the person.* Monterey, California: Brooks Cole, 1974. (d)

Kazdin, A. E. *Behavior modification in applied settings.* Homewood, Illinois: Dorsey Press, 1975. (a)

Kazdin, A. E. Characteristics and trends in applied behavior analysis. *Journal of Applied Behavior Analysis,* 1975, *8,* 332. (b)

Kazdin, A. E. The impact of applied behavior analysis on diverse areas of research. *Journal of Applied Behavior Analysis,* 1975, *8,* 213–229. (c)

Kazdin, A. E. Recent advances in token economy research. In M. Hersen, R. M. Eisler, & P. M. Miller (Eds.), *Progress in behavior modification, Volume 1.* New York: Academic Press, 1975. (d)

Kazdin, A. E. Implementing token programs: The use of staff and patients for maximizing change. In R. L. Patterson (Ed.), *Maintaining effective token economies.* Springfield, Illinois: Charles C Thomas, 1976. (a)

Kazdin, A. E. The modification of "schizophrenic" behavior. In P. A. Magaro (Ed.), *The construction of madness: Emerging conceptions and interventions into the psychotic process.* New York: Pergamon, 1976. (b)

Kazdin, A. E. Statistical analyses for single-case experimental designs. In M. Hersen & D. Barlow, *Single-case experimental designs: Strategies for studying behavior change.* New York: Pergamon, 1976. (c)

Kazdin, A. E. Extensions of reinforcement techniques to socially and environmentally relevant behaviors. In M. Hersen, R. M. Eisler, & P. M. Miller (Eds.), *Progress in behavior modification, Volume 4.* New York: Academic Press, 1977. (a)

Kazdin, A. E., Methodology of applied behavior analysis. In T. A. Brigham & A. C. Catania (Eds.), *The handbook of applied behavior research: Social and instructional processes.* New York: Irvington Press/Halstead Press, in press. (b)

Kazdin, A. E. Vicarious reinforcement and direction of behavior change in the classroom. *Behavior Therapy,* 1977. (c)

Kazdin, A. E., & Bootzin, R. R. The token economy: An evaluative review. *Journal of Applied Behavior Analysis,* 1972, *5,* 343–372.

Kazdin, A. E., & Craighead, W. E. Behavior modification in special education. In L. Mann & D. A. Sabatino (Eds.), *The first review of special education, Volume 2.* Philadelphia: Buttonwood Farms, 1973.

Kazdin, A. E., & Forsberg, S. Effects of group reinforcement and punishment on classroom behavior of retarded children. *Education and Training of the Mentally Retarded,* 1974, *9,* 50–55.

Kazdin, A. E., & Klock, J. The effect of nonverbal teacher approval on student attentive behavior. *Journal of Applied Behavior Analysis,* 1973, *6,* 643–654.

Kazdin, A. E., & Kopel, S. A. On resolving ambiguities of the multiple-baseline design: Problems and recommendations. *Behavior Therapy,* 1975, *6,* 601–608.

Kazdin, A. E., & Moyer, W. Training teachers to use behavior modification. In S. Yen & R. McIntire (Eds.), *Teaching behavior modification.* Kalamazoo, Michigan: Behaviordelia, 1976.

Kazdin, A. E., & Polster, R. Intermittent token reinforcement and response maintenance in extinction. *Behavior Therapy,* 1973, *4,* 386–391.

Kazdin, A. E., Silverman, N. A., & Sittler, J. L. The use of prompts to enhance vicarious effects of nonverbal approval. *Journal of Applied Behavior Analysis,* 1975, *8,* 279–286.

Keep America Beautiful, Inc. *Who litters and why?* 99 Park Avenue, New York, New York, 1968.

Keilitz, I., Tucker, D. J., & Horner, R. D. Increasing mentally retarded adolescents' verbalizations about current events. *Journal of Applied Behavior Analysis,* 1973, *6,* 621–630.

Keith, K. D., & Lange, B. M. Maintenance of behavior change in an institution-wide training program. *Mental Retardation,* 1974, *12,* 34–37.

Kelleher, R. T. A comparison of conditioned and food reinforcement on a fixed ratio schedule in chimpanzees. *Psychology Newsletter,* 1957, *8,* 88–93. (a)

Kelleher, R. T. Conditioned reinforcement in chimpanzees. *Journal of Comparative and Physiological Psychology,* 1957, *50,* 571–575. (b)

Kelleher, R. T., & Gollub, L. R. A review of positive conditioned reinforcement. *Journal of the Experimental Analysis of Behavior,* 1962, *5,* 543–597.

Keller, F. S. *Learning: Reinforcement theory.* New York: Random House, 1954.

Keller, F. S. A personal course in psychology. In R. Ulrich, T. Stachnik, & J. Mabry (Eds.), *Control of human behavior.* Glenview, Illinois: Scott, Foresman and Company, 1966.

Keller, F. S. "Good-bye teacher. . . ." *Journal of Applied Behavior Analysis,* 1968, *1,* 79–89.

Keller, F. S., & Schoenfeld, W. N. *Principles of psychology.* New York: Appleton-Century-Crofts, 1950.

Kelley, K. M., & Henderson, J. D. A community-based operant learning environment II: Systems and procedures. In R. D. Rubin, H. Fensterheim, A. A. Lazarus, & C. M. Franks (Eds.), *Advances in behavior therapy,* New York: Academic Press, 1971.

Kennedy, R. E. Behavior modification in prisons. In W. E. Craighead, A. E. Kazdin, & M. J. Mahoney (Eds.), *Behavior modification: Principles, issues, and applications.* Boston: Houghton Mifflin, 1976.

Kerr, N., Meyerson, L., & Michael, J. A procedure for shaping vocalizations in a mute child. In L. P. Ullmann & L. Krasner (Eds.), *Case studies in behavior modification.* New York: Holt, Rinehart and Winston, 1965.

Kifer, R. E., Lewis, M. A., Green, D. R., & Phillips, E. L. Training predelinquent youths and their parents to negotiate conflict situations. *Journal of Applied Behavior Analysis,* 1974, *7,* 357–364.

Kimble, G. A. *Hilgard and Marquis' conditioning and learning.* New York: Appleton-Century-Crofts, 1961.

Kimmel, H. D. Instrumental conditioning of autonomically mediated behavior. *Psychological Bulletin,* 1967, *67,* 337–345.

Kimmel, H. D. Instrumental conditioning of autonomically mediated responses in human beings. *American Psychologist,* 1974, *29,* 325–335.

Kinkade, K. Commune: A Walden-Two experiment. *Psychology Today,* 1973, *6,* 35–42, 90–93. (a)

Kinkade, K. Commune: A Walden-Two experiment. *Psychology Today,* 1973, *6,* 71–82. (b)

Kinkade, K. *A Walden Two experiment: The first five years of Twin Oaks Community.* New York: William Morrow & Company, 1973. (c)

Kirby, F. D., & Shields, F. Modification of arithmetic response rate and attending behavior in a seventh-grade student. *Journal of Applied Behavior Analysis,* 1972, *5,* 79–84.

Kirigin, K. A., Phillips, E. L., Fixsen, D. L., & Wolf, M. M. Modification of the homework behavior and academic performance of predelinquents with home-based reinforcement. Paper read at the American Psychological Association, Honolulu, Hawaii, 1972.

Kirschner, N. M., & Levin, L. A direct school intervention program for the modification of aggressive behavior. *Psychology in the Schools,* 1975, *12,* 202–208.

Kish, G. B. Studies of sensory reinforcement. In W. K. Honig (Ed.), *Operant behavior: Areas of research and application.* New York: Appleton-Century-Crofts, 1966.

Kittrie, N. N. *The right to be different: Deviance and enforced therapy.* Baltimore, Maryland: The Johns Hopkins Press, 1971.

Klein, R. D., Hapkiewicz, W. G., & Roden, A. H. (Eds.), *Behavior modification in educational settings.* Springfield, Illinois: Charles C Thomas, 1973.

Knapczyk, D. R., & Livingston, G. Self-recording and student teacher supervision: Variables within a token economy structure. *Journal of Applied Behavior Analysis,* 1973, *6,* 481–486.

Knapczyk, D. R., & Yoppi, J. O. Development of cooperative and competitive play responses in developmentally disabled children. *American Journal of Mental Deficiency,* 1975, *80,* 245–255.

Knecht v. Gillman, 488 F. 2d 1136, 1139 (8th Cir. 1973).

Koch, L., & Breyer, N. L. A token economy for the teacher. *Psychology in the Schools,* 1974, *11,* 195–200.

Koegel, R. L., & Rincover, A. Treatment of psychotic children in a classroom environment: I. Learning in a large group. *Journal of Applied Behavior Analysis,* 1974, *7,* 45–59.

Kohlenberg, R., & Phillips, T. Reinforcement and rate of litter depositing. *Journal of Applied Behavior Analysis,* 1973, *6,* 391–396.

Kohlenberg, R., Phillips, T., & Proctor, W. A behavioral analysis of peaking in residential electrical energy consumers. *Journal of Applied Behavior Analysis,* 1976, *9,* 13–18.

Konorski, J. A., & Miller, S. M. On two types of conditioned reflex. *Journal of General Psychology,* 1937, *16,* 264–272.

Kounin, J. S. *Discipline and group management in classrooms.* New York: Holt, Rinehart & Winston, 1970.

Kounin, J. S., & Gump, P. V. The ripple effect in discipline. *Elementary School Journal,* 1958, *59,* 158–162.

Krasner, L. The use of generalized reinforcers in psychotherapy research. *Psychological Reports,* 1955, *1,* 19–25.

Krasner, L. The therapist as a social reinforcement machine. In H. H. Strupp & L. Luborsky (Eds.), *Research in psychotherapy, Volume 2.* Washington, D. C., American Psychological Association, 1962.

Krasner, L. Verbal conditioning and psychotherapy. In L. Krasner & L. P. Ullmann (Eds.), *Research in behavior modification.* New York: Holt, Rinehart and Winston, 1965.

Krasner, L. The behavioral scientist and social responsibility: No place to hide. *Journal of Social Issues,* 1966, *21,* 9–30.

Krasner, L. Assessment of token economy programmes in psychiatric hospitals. In R. Porter (Ed.), *Ciba foundation symposium: The role of learning in psychotherapy.* London: Churchill Ltd., 1968.

Krasner, L., & Ullmann, L. P. *Behavior influence and personality: The social matrix of human action.* New York: Holt, Rinehart & Winston, 1973.

Kruglanski, A. W., Alon, S., & Lewis, T. Retrospective misattribution and task enjoyment. *Journal of Experimental Social Psychology,* 1972, *8,* 493–501.

Kruglanski, A. W., Friedman, I., & Zeevi, G. The effects of extrinsic incentive on some qualitative aspects of task performance. *Journal of Personality,* 1971, *39,* 606–617.

Kubany, E. S., Weiss, L. E., & Sloggett, B. B. The good behavior clock: A reinforcement/time-out procedure for reducing disruptive classroom behavior. *Journal of Behavior Therapy and Experimental Psychiatry,* 1971, *2,* 173–179.

Kuypers, D. S., Becker, W. C., & O'Leary, K. D. How to make a token system fail. *Exceptional Children,* 1968, *11,* 101–108.

Lachenmeyer, C. W. Systematic socialization: Observations on a programmed environment for the habilitation of antisocial retardates. *Psychological Record,* 1969, *19,* 247–257.

Lahey, B., & Drabman, R. Facilitation of the acquisition and retention of sight word vocabulary through token reinforcement. *Journal of Applied Behavior Analysis,* 1974, *7,* 307–312.

Lake v. Cameron, 364 F. 2d. 657 (D. C. 1966).

Lancaster, J. *Improvements in education, as it respects the industrious classes of the community* (Third Edition). London: Darton and Harvey, 1805.

Lanyon, R. I., & Barocas, V. S. Effects of contingent events on stuttering and fluency. *Journal of Consulting and Clinical Psychology,* 1975, *43,* 786–793.

Lattal, K. A. Contingency management of toothbrushing behavior in a summer camp for children. *Journal of Applied Behavior Analysis,* 1969, *2,* 195–198.

Laws, D. R. A patient-managed token economy. Paper presented at the Third Conference on Behavioural Modification. Wexford, Ireland, September, 1971.

Laws, D. R. The failure of a token economy. *Federal Probation,* September, 1974, 33–38.

Lawson, R. B., Greene, R. T., Richardson, J. S., McClure, G., & Padina, R. J. Token economy program in a maximum security correctional hospital. *Journal of Nervous and Mental Diseases,* 1971, *152,* 199–205.

Lazarus, A. A. New methods in psychotherapy: A case study. *South African Medical Journal,* 1958, *32,* 660–664.

Leach, E. Stuttering: Clinical application of response-contingent procedures. In B. B. Gray & G. England (Eds.), *Stuttering and the conditioning therapies.* California: Monterey Institute of Speech and Hearing, 1969.

Lehrer, P., Schiff, L., & Kris, A. The use of a credit card in a token economy. *Journal of Applied Behavior Analysis,* 1970, *3,* 289–291.

Leitenberg, H., Agras, W. S., Thompson, L. D., & Wright, D. E. Feedback in behavior modification: An experimental analysis in two phobic cases. *Journal of Applied Behavior Analysis,* 1968, *1,* 131–137.

Leitenberg, H., Wincze, J., Butz, R., Callahan, E., & Agras, W. Comparison of the effect of instructions and reinforcement in the treatment of a neurotic avoidance response: A single case experiment. *Journal of Behavior Therapy and Experimental Psychiatry,* 1970, *1,* 53–58.

Lent, J. R. Mimosa Cottage: Experiment in hope. *Psychology Today,* 1968, *2*(1), 50–58.

Lepper, M. R., & Greene, D. Overjustification and intrinsic motivation: Some implications for applied behavior analysis. Paper presented at the meeting of the Eastern Psychological Association, Philadelphia, April, 1974.

Lepper, M. R., & Greene, D. On understanding overjustification: A reply to Reiss and Sushinsky. *Journal of Personality and Social Psychology,* 1976, *33,* 25–35.

Lepper, M. R., Greene, D., & Nisbett, R. E. Undermining children's intrinsic interest with extrinsic rewards: A test of the "overjustification" hypothesis. *Journal of Personality and Social Psychology,* 1973, *28,* 129–137.

Lessard v. Schmidt, 349 F. Supp. 1078 (E.D. Wisc. 1972).

Levin, G., & Simmons, J. Response to food and praise by emotionally disturbed boys. *Psychological Reports,* 1962, *2,* 539–546.

Levine, F. M., & Fasnacht, G. Token rewards may lead to token learning. *American Psychologist,* 1974, *29* 816–820.

Libb, J. W., & Clements, C. B. Token reinforcement in an exercise program for hospitalized geriatric patients. *Perceptual and Motor Skills,* 1969, *28,* 957–958.

Liberman, R. P. A view of behavior modification projects in California. *Behaviour Research and Therapy,* 1968, *6,* 331–341.

Liberman, R. P. Reinforcement of social interaction in a group of chronic mental patients. In R. D. Rubin, H. Fensterheim, J. D. Henderson, & L. P. Ullmann (Eds.), *Advances in behavior therapy.* New York: Academic Press, 1972.

Liberman, R. P. Applying behavioral techniques in a community mental health center. In R. D. Rubin, J. P. Brady, & J. D. Henderson (Eds.), *Advances in behavior therapy, Volume 4.* New York: Academic Press, 1973.

Liberman, R. P., & Ferris, C. Antisocial behavior and school performance: Comparison between Welcome Home and Rancho San Antonio. Unpublished report to the California Council on Criminal Justice, 1974.

Liberman, R. P., Ferris, C., Salgado, P., & Salgado, J. Replication of the Achievement Place model in California. *Journal of Applied Behavior Analysis,* 1975, *8,* 287–299.

Liberman, R. P., Teigen, J., Patterson, R., & Baker, V. Reducing delusional speech in chronic, paranoid schizophrenics. *Journal of Applied Behavior Analysis,* 1973, *6,* 57–64.

Liebson, I., Cohen, M., & Faillace, L. Group fines: Technique for behavioral control in a token economy. *Psychological Reports,* 1972, *30,* 895–900.

Lindsley, O. R. Operant conditioning methods applied to research in chronic schizophrenia. *Psychiatric Research Reports,* 1956, *5,* 118–139.

Lindsley, O. R. Characteristics of the behavior of chronic psychotics as revealed by free-operant conditioning methods. *Diseases of the Nervous System,* (Monograph Supplement), 1960, *21,* 66–78.

Lindsley, O. R. Free operant conditioning and psychotherapy. In J. H. Masserman (Ed.), *Current psychiatric therapies, Volume 3.* New York: Grune & Stratton, 1963.

Lindsley, O. R., Skinner, B. F., & Solomon, H. C. Studies in behavior therapy. Metropolitan State Hospital, Waltham, Massachusetts, Status Report I, May, 1953.

Linscheid, T. R., Malosky, P., & Zimmerman, J. Discharge as the major consequence in a hospitalized patient's behavior management program: A case study. *Behavior Therapy,* 1974, *5,* 559–564.

Litow, L. & Pumroy, D. K. A brief review of classroom group-oriented contingencies. *Journal of Applied Behavior Analysis,* 1975, *8,* 341–347.

Lloyd, D. E., & Knutzen, N. J. A self-paced programmed undergraduate course in experimental analysis of behavior. *Journal of Applied Behavior Analysis,* 1969, *2,* 125–133.

Lloyd, K. E., & Abel, L. Performance on a token economy psychiatric ward: A two year summary. *Behaviour Research and Therapy,* 1970, *8,* 1–9.

Lloyd, K. E., & Garlington, W. K. Weekly variations in performance on a token economy psychiatric ward. *Behaviour Research and Therapy,* 1968, *6,* 407–410.

Locke, B. Verbal conditioning with retarded subjects: Establishment or reinstatement of effective reinforcing consequences. *American Journal of Mental Deficiency,* 1969, *73,* 621–626.

Locke, E. A., Cartledge, N., & Koeppel, J. Motivational effects of knowledge of results: A goal setting phenomenon? *Psychological Bulletin,* 1968, *70,* 474–485.

Logan, D. L. A "paper money" token system as a recording aid in institutional settings. *Journal of Applied Behavior Analysis,* 1970, *3,* 183–184.

Logan, D. L. Kinsinger, J., Shelton, G., & Brown, J. M. The use of multiple reinforcers in a rehabilitation setting. *Mental Retardation,* 1971, *9,* 3–6.

Logan, F. A. *Fundamentals of learning and motivation.* Dubuque, Iowa: W. C. Brown, 1969.

London, P. *The modes and morals of psychotherapy.* New York: Holt, Rinehart and Winston, 1964.

London, P. *Behavior control.* New York: Harper & Row, 1969.

Long, J. D., & Williams, R. L. The comparative effectiveness of group and individually contingent free time with inner-city junior high school students. *Journal of Applied Behavior Analysis,* 1973, *6,* 465–474.

Lovaas, O. I. The control of operant responding by rate and content of verbal operants. Paper presented at Western Psychological Association Meeting, Seattle, June, 1961.

Lovaas, O. I., Koegel, R., Simmons, J. Q., & Long, J. S. Some generalization and follow-up measures on autistic children in behavior therapy. *Journal of Applied Behavior Analysis*, 1973, *6*, 131–166.

Lovaas, O. I., Schaeffer, B., & Simmons, J. Q. Building social behavior in autistic children by use of electric shock. *Journal of Experimental Research in Personality*, 1965, *1*, 99–109.

Lovaas, O. I., & Simmons, J. Q. Manipulation of self-destruction in three retarded children. *Journal of Applied Behavior Analysis*, 1969, *2*, 143–157.

Lovitt, T. C., & Curtiss, K. A. Academic response rate as a function of teacher- and self-imposed contingencies. *Journal of Applied Behavior Analysis*, 1969, *2*, 49–53.

Lovitt, T. C., & Esveldt, K. A. The relative effects on math performance of single- versus multiple-ratio schedules: A case study. *Journal of Applied Behavior Analysis*, 1970, *3*, 261–270.

Luborsky, L., & Strupp, H. H. Research problems in psychotherapy: A three-year follow-up. In H. H. Strupp & L. Luborsky (Eds.), *Research in psychotherapy, Volume 2*. Washington, D. C.: American Psychological Association, 1962.

Lucero R. J., Vail, D. J., & Scherber, J. Regulating operant-conditioning programs. *Hospital & Community Psychiatry*, 1968, *19*, 53–54.

Lutzker, J. R., & Sherman, J. Producing generative sentence usage by imitation and reinforcement procedures. *Journal of Applied Behavior Analysis*, 1974, *7*, 447–460.

Luyben, P. D., & Bailey, J. S. Newspaper recycling behavior: The effects of reinforcement versus proximity of containers. Unpublished manuscript, Florida State University, 1975.

MacCubrey, J. Verbal operant conditioning with young institutionalized Down's syndrome children. *American Journal of Mental Deficiency*, 1971, *75*, 676–701.

Mackey v. Procunier, 477 F. 2d. 877 (9th Cir. 1973).

Maconochie, A. *Norfolk Island*. Hobart, Australia: Sullivan's Cove, Publisher, 1973. (Original publication date, 1847.)

Maconochie, A. *The mark system*. London: John Ollivier, 1848.

Madsen, C. H., Becker, W. C., & Thomas, D. R. Rules, praise and ignoring: Elements of elementary classroom control. *Journal of Applied Behavior Analysis*, 1968, *1* 139–150.

Madsen, C. H., Becker, W. C., Thomas, D. R., Koser, L., & Plager, E. An analysis of the reinforcing function of "sit down" commands. In R. K. Parker (Ed.), *Readings in educational psychology*. Boston: Allyn & Bacon, 1970.

Mahoney, M. J., & Mahoney, F. E. A residential program in behavior modification. In R. D. Rubin, J. P. Brady, & J. D. Henderson (Eds.), *Advances in behavior therapy*. New York: Academic Press, 1973.

Mahoney, M. J., Moura, N. G. M., & Wade, T. C. Relative efficacy of self-reward, self-punishment, and self-monitoring techniques for weight loss. *Journal of Consulting and Clinical Psychology*, 1973, *40*, 404–407.

Maley, R. F., Feldman, G. L., & Ruskin, R. S. Evaluation of patient improvement in a token economy treatment program. *Journal of Abnormal Psychology*, 1973, *82*, 141–144.

Maloney, D. M., Harper, T. M., Braukmann, C. J., Fixsen, D. L., Phillips, E. L., & Wolf, M. M. Effects of training pre-delinquent girls on conversation and posture behaviors by teaching-parents and juvenile peers. *Journal of Applied Behavior Analysis*, in press.

Maloney, K. B., & Hopkins, B. L. The modification of sentence structure and its relationship to subjective judgments of creativity in writing. *Journal of Applied Behavior Analysis*, 1973, *6*, 425–433.

Malott, R. W., Hartlep, P. A., Keenan, M., & Michael, J. Groundwork for a student centered experimental college. *Educational Technology*, 1972, *12*, 61.

Mandelker, A. V., Brigham, T. A., & Bushell, D. The effects of token procedures on a teacher's social contacts with her students. *Journal of Applied Behavior Analysis*, 1970, *3*, 169–174.

Mann, J. H. The effect of inter-racial contact on sociometric choices and perceptions. *Journal of Social Psychology,* 1959, *59,* 143–152.

Marholin, D., II, & Gray, D. Effects of group response cost procedures on cash shortages in a small business. *Journal of Applied Behavior Analysis,* 1976, *9,* 25–30.

Marholin, D., II, Plienis, A. J., Harris, S. D., & Marholin, B. L. Mobilization of the community through a behavioral approach: A school program for adjudicated females. *Criminal Justice and Behavior,* 1975, *2,* 130–145.

Marholin, D., II, Siegel, L. J., & Phillips, D. Treatment and transfer: A search for empirical procedures. In M. Hersen, R. M. Eisler, & P. M. Miller (Eds.), *Progress in behavior modification, Vol. 3.* New York: Academic Press, 1976.

Marholin, D., II, Steinman, W. M., McInnis, E. T., & Heads, T. B. The effect of a teacher's presence on the classroom behavior of conduct-problem children. *Journal of Abnormal Child Psychology,* 1975, *3,* 11–25.

Marks, J., Sonoda, B., & Schalock, R. Reinforcement vs. relationship therapy for schizophrenics. *Journal of Abnormal Psychology,* 1968, *73,* 397–402.

Martin, G. L. Teaching operant technology to psychiatric nurses, aides, and attendants. In F. W. Clark, D. R. Evans, & L. A. Hamerlynck (Eds.), *Implementing behavioral programs for schools and clinics.* Champaign, Ill.: Research Press, 1972.

Martin, J. A. Generalizing the use of descriptive adjectives through modeling. *Journal of Applied Behavior Analysis,* 1975, *8,* 208–209.

Matson, F. W. (Ed.) *Without/within: Behaviorism and humanism.* Monterey, California: Brooks/Cole, 1973.

Mattocks, A. L., & Jew, C. Assessment of an aversive treatment program with extreme acting-out patients in a psychiatric facility for criminal offenders. Unpublished manuscript, California Department of Corrections, Undated. (On file with the University of Southern California Law Library, Los Angeles, California.)

McConahey, O. L. A token system for retarded women: Behavior modification, drug therapy, and their combination. In T. Thompson & J. Grabowski (Eds.), *Behavior modification of the mentally retarded.* New York: Oxford University Press, 1972.

McConahey, O. L., & Thompson, T. Concurrent behavior modification and chlorpromazine therapy in a population of institutionalized mentally retarded women. *Proceedings of the 79th Annual Convention of the American Psychological Association,* 1971, *6,* 761–762.

McConahey, O. L., Thompson, T., & Zimmerman, R. A token system for retarded womrn: Behavior therapy, drug administration, and their combination. In T. Thompson & J. Grabowski (Eds.), *Behavior modification of the mentally retarded* (Second Edition). New York: Oxford University Press, 1977.

McDonald, S. The kibitz dimension in teacher consultation. In R. D. Klein, W. G. Hapkiewicz, & A. H. Roden (Eds.), *Behavior modification in educational settings.* Springfield, Ill.: Thomas, 1973.

McInnis, T., Himelstein, H. C., Doty, D. W., & Paul, G. Modification of sampling-exposure procedures for increasing facilities utilization by chronic psychiatric patients. *Journal of Behavior Therapy and Experimental Psychiatry,* 1974, *5,* 119–127.

McKee, J. M. The use of programmed instruction in correctional institutions. *Journal of Correctional Education,* 1970, *22,* 8–12, 28–30.

McKee, J. M. Contingency management in a correctional institution. *Educational Technology,* 1971, *11,* 51–54.

McKee, J. M. The use of contingency management to affect learning performance in adult institutionalized offenders. In R. Ulrich, T. Stachnik, & J. Mabry, *Control of human behavior, Volume 3.* Glenview, Ill.: Scott, Foresman and Company, 1974.

McKee, J. M., & Clements, C. B. A behavioral approach to learning: The Draper model. In H. C. Rickard (Ed.), *Behavioral intervention in human problems.* New York: Pergamon, 1971.

McKenzie, H. S., Clark, M., Wolf, M. M., Kothera, R., & Benson, C. Behavior modification of children with learning disabilities using grades as tokens and allowances as back-up reinforcers. *Exceptional Children,* 1968, *34,* 745–752.

McLaughlin, T. F. A review of applications of group-contingency procedures used in behavior modification in the regular classroom: Some recommendations for school personnel. *Psychological Reports,* 1974, *35,* 1299–1303.

McLaughlin, T. F. The applicability of token reinforcement systems in public school systems. *Psychology in the Schools,* 1975, *12,* 84–89.

McLaughlin, T. F., & Malaby, J. E. Intrinsic reinforcers in a classroom token economy. *Journal of Applied Behavior Analysis,* 1972, *5,* 263–270. (a)

McLaughlin, T. F., & Malaby, J. Reducing and measuing inappropriate verbalizations in a token classroom. *Journal of Applied Behavior Analysis,* 1972, *5,* 329–333. (b)

McLaughlin, T. F., & Malaby, J. E. Increasing and maintaining assignment completion with teacher and pupil controlled individual contingency programs: Three case studies. *Psychology,* 1974, *11,* 1–7. (a)

McLaughlin, T. F., & Malaby, J. E. Note on combined and separate effects of token reinforcement and response cost on completing assignments. *Psychological Reports,* 1974, *35,* 1132. (b)

McLaughlin, T. F., & Malaby, J. E. The utilization of an individual contingency program to control assignment completion in a token classroom: A case study. *Psychology in the Schools,* 1974, *11,* 191–194. (c)

McLaughlin, T. F., & Malaby, J. E. Differential effects of token reinforcement to increase class participation in constructing questions over science films. *Psychological Record,* 1975, *37,* 306. (a)

McLaughlin, T. F., & Malaby, J. E. The effects of various token reinforcement contingencies on assignment completion and accuracy during variable and fixed token exchange schedules. *Canadian Journal of Behavioral Science,* 1975, *7,* 411–419. (b)

McLaughlin, T. F., & Malaby, J. E. Partial component analysis of an inexpensive token system across two classrooms. *Psychological Reports,* 1975, *37,* 362. (c)

McMichael, J. S., & Corey, J. R. Contingency management in an introductory psychology course produces better learning. *Journal of Applied Behavior Analysis,* 1969, *2,* 79–83.

McMurtry, C. A., & Williams, J. E. The evaluation dimension of the affective meaning system of the preschool child. *Developmental Psychology,* 1972, *6,* 238–246.

McNamara, J. R. Teacher and students as a source for behavior modification in the classroom. *Behavior Therapy,* 1971, *2,* 205–213.

McReynolds, W. T., & Coleman, J. Token economy: Patient and staff changes. *Behaviour Research and Therapy,* 1972, *10,* 29–34.

Medland, M. B., & Stachnik, T. J. Good-behavior game: A replication and systematic analysis. *Journal of Applied Behavior Analysis,* 1972, *5,* 45–51.

Meichenbaum, D. H. The effects of instruction and reinforcement on thinking and language behaviour of schizophrenics. *Behaviour Research and Therapy,* 1969, *7,* 101–114.

Meichenbaum, D. H. Cognitive factors in behavior modification: Modifying what clients say to themselves. In R. D. Rubin, J. P. Brady, and J. D. Henderson (Eds.), *Advances in behavior therapy, Volume 4.* New York: Academic Press, 1973.

Meichenbaum, D. H., Bowers, K., & Ross, R. R. Modification of classroom behavior of institutionalized female adolescent offenders. *Behaviour Research and Therapy,* 1968, *6,* 343–353.

Meichenbaum, D. H., & Cameron, R. Training schizophrenics to talk to themselves: A means of developing attentional controls. *Behavior Therapy,* 1973, *4,* 515–534.

Meichenbaum, D. H., & Goodman, J. Training impulsive children to talk to themselves: A means of developing self-control. *Journal of Abnormal Psychology,* 1971, *77,* 115–126.

Melin, L., Andersson, B. E., & Gotestam, K. G. Contingency management in a methadone maintenance treatment program. *Addictive Behaviors,* 1975, *1,* 151–158.

Melin, L., & Gotestam, K. G. A contingency management program on a drug-free unit for intravenous amphetamine addicts. *Journal of Behavior Therapy and Experimental Psychiatry,* 1973, *4,* 331–337.

Mental Health Law Project. *Basic rights of the mentally handicapped.* Washington, D. C.: Mental Health Law Project, 1973.

Metz, J. R. Conditioning generalized imitation in autistic children. *Journal of Experimental Child Psychology,* 1965, *2,* 389–399.

Meyers, A. W., Artz, L. M., & Craighead, W. E. The effects of instructions, incentive, and feedback on a community problem: Dormitory noise. *Journal of Applied Behavior Analysis,* in press.

Milan, M. A., & McKee, J. M. Behavior modification: Principles and applications in corrections. In D. Glaser (Ed.), *Handbook of criminology.* Chicago: Rand McNally, 1974.

Milan, M. A., Wood, Ļ. F., Williams, R. L., Rogers, J. G., Hampton, L. R., & McKee, J. M. *Applied behavior analysis and the Important Adult Felon Project I: The cellblock token economy.* Elmore, Ala.: Rehabilitation Research Foundation, 1974.

Milby, J. B. Modification of extreme social isolation by contingent social reinforcement. *Journal of Applied Behavior Analysis,* 1970, *3,* 149–152.

Milby, J. B. A brief review of token economy treatment. Paper presented at 80th Annual Convention, American Psychological Association, Honolulu, Hawaii, September, 1972.

Milby, J. B., Pendergrass, P. E., & Clarke, C. J. Token economy versus control ward: A comparison of staff and patient attitudes toward ward environment. *Behavior Therapy,* 1975, *6,* 22–29.

Millard v. Cameron, 125 U. S. App. D. C. 383, 373 F. 2d 468 (1966).

Miller, H. R. Identification of effective reinforcers in a token economy for male adolescent retardates. In R. D. Rubin, J. P. Brady, & J. D. Henderson (Eds.), *Advances in behavior therapy, Volume 4.* New York: Academic Press, 1973.

Miller, L. K. The design of better communities through the application of behavioral principles. In W. E. Craighead, A. E. Kazdin, & M. J. Mahoney (Eds.), *Behavior modification: Principles, issues, and applications.* Boston: Houghton Mifflin, 1976.

Miller, L. K., & Feallock, F. A. A behavioral system for group living. In E. Ramp & G. Semb (Eds.), *Behavior analysis and education–1973.* Englewood Cliffs, N. J.: Prentice-Hall, 1975.

Miller, L. K., & Lies, A. A. Everyday behavior analysis: A new direction for applied behavior analysis. *Behavioral Voice,* 1974, *2,* 5–13.

Miller, L. K., & Miller, O. L. Reinforcing self-help group activities of welfare recipients. *Journal of Applied Behavior Analysis,* 1970, *3,* 57–64.

Miller, L. K., & Schneider, R. The use of a token system in project Head Start. *Journal of Applied Behavior Analysis,* 1970, *3,* 213–220.

Miller, L. K., Weaver, F. H., & Semb, G. A procedure for maintaining student progress in a personalized university course. *Journal of Applied Behavior Analysis,* 1974, *7,* 87–91.

Miller, N. E. Learning of visceral and glandular responses, *Science,* 1969, *163,* 434–445.

Miller, P. M. The use of behavioral contracting in the treatment of alcoholism: A case report. *Behavior Therapy,* 1972, *3,* 593–596.

Miller, P. M., & Drennen, W. T. Establishment of social reinforcement as an effective modifier of verbal behavior in chronic psychiatric patients. *Journal of Abnormal Psychology,* 1970 *76,* 392–395.

Miller, P. M., & Eisler, R. M. Alcohol and drug abuse. In W. E. Craighead, A. E. Kazdin, & M. J. Mahoney (Eds.), *Behavior modification: Principles, issues, and applications.* Boston: Houghton Mifflin, 1976.

Miller, P. M., Hersen, M., & Eisler, R. M. Relative effectiveness of instructions, agreements, and reinforcements in behavioral contracts with alcoholics. *Journal of Abnormal Psychology,* 1974, *83,* 548–553.

Miller, P. M., Hersen, M., Eisler, R. M., & Watts, J. G. Contingent reinforcement of lowered blood/alcohol levels in an outpatient chronic alcoholic. *Behaviour Research and Therapy,* 1974, *12,* 261–263.

Miller, S., & Konorski, J. Sur une forme particulière des reflexes conditionnels. *Les Comptes*

Rendus des Séances de la Société de Biologie. Société Polonaise de Biologie, 1928, *XCIX,* 1155. (Translation appears in *Journal of the Experimental Analysis of Behavior,* 1969, *12,* 187–189.)

Minkin, N., Braukmann, C. J., Minkin, B. L., Timbers, G. D., Timbers, B. J., Fixsen, D. L., Phillips, E. L., & Wolf, M. M. The social validation and training of conversational skills. *Journal of Applied Behavior Analysis,* 1976, *9,* 127–139.

Mitchell, W. S., & Stoffelmayr, B. E. Application of the Premack Principle to the behavioral control of extremely inactive schizophrenics. *Journal of Applied Behavior Analysis,* 1973, *6,* 419–423.

Morales v. Turman, 383 F. Supp. 53 (E. D. Tex. 1974).

Morgan, R. R. An exploratory study of three procedures to encourage school attendance. *Psychology in the Schools,* 1975, *12,* 209–215.

Morse, W. H. Intermittent reinforcement. In W. K. Honig (Ed.), *Operant behavior: Areas of research and application.* New York: Appleton-Century-Crofts, 1966.

Morse, W. H., & Kelleher, R. T. Determinants of reinforcement and punishment. In W. K. Honig & J. E. R. Staddon (Eds.), *Handbook of operant behavior.* Englewood Cliffs, New Jersey: Prentice Hall, in press.

Mulligan, W., Kaplan, R. D., & Reppucci, N. D. Changes in cognitive variables among behavior problem elementary school boys treated in a token economy special classroom. In R. D. Rubin, J. P. Brady, & J. D. Henderson (Eds.), *Advances in behavior therapy, Volume 4.* New York: Academic Press, 1973.

Musick, J. K., & Luckey, R. E. A token economy for moderately and severely retarded. *Mental Retardation,* 1970, *8,* 35–36.

Myers, D. V. Extinction, DRO, and response-cost procedures for eliminating self-injurious behavior: A case study. *Behaviour Research and Therapy,* 1975, *13,* 189–191.

Narrol, H. G. Experimental application of reinforcement principles to the analysis and treatment of hospitalized alcoholics. *Quarterly Journal of Studies on Alcohol,* 1967, *28,* 105–115.

Nason v. Superintendent of Bridgewater State Hosp., 233 N. E. 2d 908 (Mass. 1968).

Nay, W. R. A systematic comparison of instructional techniques for parents. *Behavior Therapy,* 1975, *6,* 14–21.

New York City Health and Hosp. Corp. v. Stein, 335 N. Y. S. 2d 461 (Sup. Ct. 1972).

Notz, W. W. Work motivation and the negative effects of extrinsic rewards: A review with implications for theory and practice. *American Psychologist,* 1975, *9,* 884–891.

Novak, M. Is he really a grand inquisitor? In H. Wheeler (Ed.), *Beyond the punitive society.* San Francisco: W. H. Freeman and Co., 1973.

O'Brien, F., & Azrin, H. H. Symptom reduction by functional displacement in a token economy: A case study. *Journal of Behavior Therapy and Experimental Psychiatry,* 1972, *3,* 205–207.

O'Brien, F., Azrin, N. H., & Henson, K. Increased communications of chronic mental patients by reinforcement and response priming. *Journal of Applied Behavior Analysis,* 1969, *2,* 23–29.

O'Brien, F., Bugle, C., & Azrin, N. H. Training and maintaining a retarded child's proper eating. *Journal of Applied Behavior Analysis,* 1972, *5,* 67–72.

O'Brien, J. S., Raynes, A. E., & Patch, V. D. An operant reinforcement system to improve ward behavior in in-patient drug addicts. *Journal of Behavior Therapy and Experimental Psychiatry,* 1971, *2,* 239–242.

O'Dell, S. Training parents in behavior modification: A review. *Psychological Bulletin,* 1974, *81,* 418–433.

O'Leary, K. D. The assessment of psychopathology in children. In H. C. Quay & J. S. Werry (Eds.), *Psychopathological disorders of childhood.* New York: Wiley, 1972.

O'Leary, K. D. Token reinforcement programs in the classroom. In T. A. Brigham & A. C.

Catania (Eds.), *The handbook of applied behavior research: Social and instructional processes.* New York: Irvington Press/Halstead Press, in press.

O'Leary, K. D., & Becker, W. C. Behavior modification of an adjustment class: A token reinforcement program. *Exceptional Children,* 1967, *9,* 637–642.

O'Leary, K. D., Becker, W. C., Evans, M. B., & Saudargas, R. A. A token reinforcement program in a public school: A replication and systematic analysis. *Journal of Applied Behavior Analysis,* 1969, *2,* 3–13.

O'Leary, K. D., & Drabman, R. Token reinforcement programs in the classroom: A review. *Psychological Bulletin,* 1971, *75,* 379–398.

O'Leary, K. D., Drabman, R. S., & Kass, R. E. Maintenance of appropriate behavior in a token program. *Journal of Abnormal Child Psychology,* 1973, *1,* 127–138.

O'Leary, K. D., Kaufman, K. F., Kass, R., & Drabman, R. The effects of loud and soft reprimands on the behavior of disruptive students. *Exceptional Children,* 1970, *37,* 145–155.

O'Leary, K. D., & O'Leary, S. G. (Eds.) *Classroom management: The successful use of behavior modification.* New York: Pergamon, 1972.

O'Leary, K. D., O'Leary, S., & Becker, W. C. Modification of a deviant sibling interaction pattern in the home. *Behaviour Research and Therapy,* 1967, *5,* 113–120.

O'Leary, K. D., Poulos, R. W., & Devine, V. T. Tangible reinforcers: Bonuses or bribes? *Journal of Consulting and Clinical Psychology,* 1972, *38,* 1–8.

Olson, R. P., & Greenberg, D. J. Effects of contingency-contracting and decision-making groups with chronic mental patients. *Journal of Consulting and Clinical Psychology,* 1972, *38,* 376–383.

Opton, E. M., Jr. Psychiatric violence against prisoners: When therapy is punishment. *Mississippi Law Journal,* 1974, *45,* 605–644.

Osborne, J. G. Free time as a reinforcer in the management of classroom behavior. *Journal of Applied Behavior Analysis,* 1969, *2,* 113–118.

Packard, R. G. The control of "classroom attention": A group contingency for complex behavior. *Journal of Applied Behavior Analysis,* 1970, *3,* 13–28.

Palmer, A. B., & Wohl, J. Voluntary-admission forms: Does the patient know what he's signing? *Hospital & Community Psychiatry,* 1972, *23,* 250–252.

Palmer, M. H., Lloyd, M. E., & Lloyd, K. E. An experimental analysis of electricity conservation procedures. *Journal of Applied Behavior Analysis,* in press.

Panek, M. Token economies on a shoestring: Successes and failures. Unpublished research report, Northern State Hospital, Sedro Wooley, Washington, May, 1969.

Panek, D. M. Word association learning by chronic schizophrenics on a token economy ward under conditions of reward and punishment. *Journal of Clinical Psychology,* 1970, *26,* 163–167.

Panyan, M., Boozer, H., & Morris, N. Feedback to attendants as a reinforcer for applying techniques. *Journal of Applied Behavior Analysis,* 1970, *3,* 1–4.

Parker, H. C. Contingency management and concomitant changes in elementary-school students' self-concepts. *Psychology in the Schools,* 1974, *11,* 70–79.

Parrino, J. J., George, L., & Daniels, A. C. Token control of pill-taking behavior in a psychiatric ward. *Journal of Behavior Therapy and Experimental Psychiatry,* 1971, *2,* 181–185.

Parsonson, B. S., Baer, A. M., & Baer, D. M., The application of generalized correct social contingencies by institutional staff: An evaluation of the effectiveness and durability of a training program. *Journal of Applied Behavior Analysis,* 1974, *7,* 427–437.

Patterson, G. R. An application of conditioning techniques to the control of a hyperactive child. In L. P. Ullmann & L. Krasner (Eds.), *Case studies in behavior modification.* New York: Holt, Rinehart and Winston, 1965.

Patterson, G. R. Interventions for boys with conduct problems: Multiple settings, treatments, and criteria. *Journal of Consulting and Clinical Psychology,* 1974, *42,* 471–481.

Patterson, G. R., Cobb, J. A., & Ray, R. S. Direct intervention in the classroom: A set of procedures for the aggressive child. In F. W. Clark, D. R. Evans, & L. A. Hamerlynck (Eds.), *Implementing behavioral programs for schools and clinics.* Champaign, Illinois: Research Press, 1972.

Patterson, G. R., Cobb, J. A., & Ray, R. S. A social engineering technology for retraining families of aggressive boys. In H. E. Adams & I. P. Unikel (Eds.), *Issues and trends in behavior therapy.* Springfield, Illinois: Charles C Thomas, 1973.

Patterson, G. R., & Reid, J. B. Intervention for families of aggressive boys: A replication study. *Behaviour Research and Therapy,* 1973, *11,* 1–12.

Patterson, G. R., Shaw, D. A., & Ebner, M. J. Teachers, peers, and parents as agents of change in the classroom. In F. A. M. Benson (Ed.), *Modifying deviant social behaviors in various classroom settings.* Eugene, Oregon: University of Oregon, 1969.

Patterson, R. L. *Maintaining effective token economies.* Springfield, Ill.: Charles C Thomas, 1976.

Patterson, R., & Teigen, J. Conditioning and post-hospital generalization of nondelusional responses in a chronic psychotic patient. *Journal of Applied Behavior Analysis,* 1973, *6,* 65–70.

Paul, G. L. Chronic mental patient: Current status–future directions. *Psychological Bulletin,* 1969, *71,* 81–94.

Paul, G. L., & McInnis, T. L. Attitudinal changes associated with two approaches to training mental health technicians in milieu and social-learning programs. *Journal of Consulting and Clinical Psychology,* 1974, *42,* 21–33.

Paul, G. L., McInnis, T. L., & Mariotto, M. J. Objective performance outcomes associated with two approaches to training mental health technicians in milieu and social-learning programs. *Journal of Abnormal Psychology,* 1973, *82,* 523–532.

Paul, G. L., Tobias, L. T., & Holly, B. L. Maintenance psychotropic drugs with chronic mental patients in the presense of active treatment programs: A "triple-blind" withdrawal study. *Archives of General Psychiatry,* 1972, *27,* 106–115.

Pavlov, I. P. The reply of a physiologist to psychologists. *Psychological Review,* 1932, *39,* 91–127.

Pedalino, E., & Gamboa, V. U. Behavior modification and absenteeism: Intervention in one industrial setting. *Journal of Applied Psychology,* 1974, *59,* 694–698.

Peek v. Ciccone, 288 F. Supp. 329 (E. D. Mo. 1968).

Peine, H. A., & Munro, B. C. Behavioral management of parent training programs. *Psychological Record,* 1973, *23,* 459–466.

Peniston, E. Reducing problem behaviors in the severely and profoundly retarded. *Journal of Behavior Therapy and Experimental Psychiatry,* 1975, *6,* 295–299.

People ex rel. Blunt v. Narcotic Addiction Control Comm'n, 295 N. Y. S. 2d 276 (Sup. Ct.), aff'd mem., 296 N. Y. S. 2d 533 (App. Div. 1968).

People ex rel. Stutz v. Conboy, 300 N. Y. S. 2d 453 (Sup. Ct. 1969).

Phillips, E. L. Achievement Place: Token reinforcement procedures in a home-style rehabilitation setting for "pre-delinquent" boys. *Journal of Applied Behavior Analysis,* 1968, *1,* 213–223.

Phillips, E. L., Phillips, E. A., Fixsen, D. L., & Wolf, M. M. Achievement Place: Modification of the behaviors of pre-delinquent boys within a token economy. *Journal of Applied Behavior Analysis,* 1971,·*4,* 45–59.

Phillips, E. L., Phillips, E. A., Fixsen, D. L., & Wolf, M. M. Behavior shaping works for delinquents. *Psychology Today,* 1973, *7,* January, 75–79.

Phillips, E. L., Phillips, E. A., Wolf, M. M., & Fixsen, D. L. Achievement Place: Development of the elected manager system. *Journal of Applied Behavior Analysis,* 1973, *6,* 541–561.

Pierce, C. H., & Risley, T. R. Improving job performance of Neighborhood Youths Corps aides in an urban recreation program. *Journal of Applied Behavior Analysis,* 1974, *7,* 207–215.

Platt, J. R. The Skinnerian revolution. In H. Wheeler (Ed.), *Beyond the punitive society.* San Francisco: W. H. Freeman and Co., 1973.

Pomerleau, O. F., Bobrove, P. H., & Harris, L. C. Some observations on a controlled social environment for psychiatric patients. *Journal of Behavior Therapy and Experimental Psychiatry,* 1972, *3,* 15–21.

Pomerleau, O. F., Bobrove, P. H., & Smith, R. H. Rewarding psychiatric aides for the behavioral improvement of assigned patients. *Journal of Applied Behavior Analysis,* 1973, *6,* 383–390.

Pommer, D. A., & Streedbeck, D. Motivating staff performance in an operant learning program for children. *Journal of Applied Behavior Analysis,* 1974, *7,* 217–221.

Powers, R. B., Osborne, J. G., & Anderson, E. G. Positive reinforcement of litter removal in the natural environment. *Journal of Applied Behavior Analysis,* 1973, *6,* 579–586.

Premack, D. Toward empirical behavior laws: I. Positive reinforcement. *Psychological Review,* 1959, *66,* 219–233.

Premack, D. Reversibility of the reinforcement relation. *Science,* 1962, *136,* 255–257.

Premack, D. Reinforcement theory. In D. Levine (Ed.), *Nebraska symposium on motivation.* Lincoln: University of Nebraska Press, 1965.

Quay, H. C., & Hunt, W. A. Psychopathy, neuroticism and verbal conditioning: A replication and extension. *Journal of Consulting Psychology,* 1965, *29,* 283.

Quay, H. C., Sprague, R. L., Werry, J. S., & McQueen, M. M. Conditioning visual orientation of conduct problem children in the classroom. *Journal of Experimental Child Psychology,* 1967, *5,* 512–517.

Quilitch, H. R. A comparison of three staff-management procedures. *Journal of Applied Behavior Analysis,* 1975, *8,* 59–66.

Rachlin, H. (Ed.), *Introduction to modern behaviorism.* San Francisco: W. H. Freeman & Co., 1970.

Rachman, S., & Teasdale, J. *Aversion therapy and behaviour disorders.* Coral Gables: University of Miami, 1969.

Ravitch, D. *The great school wars, New York City, 1805–1973: A history of the public schools as battlefield of social change.* New York: Basic Books, 1974.

Redd, W. H. Effects of mixed reinforcement contingencies on adults' control of children's behavior. *Journal of Applied Behavior Analysis,* 1969, *2,* 249–254.

Redd, W. H., & Birnbrauer, J. S. Adults as discriminative stimuli for different reinforcement contingencies with retarded children. *Journal of Experimental Child Psychology,* 1969, *7,* 440–447.

Reid, D. H., Luyben, P. L., Rawers, R. J., & Bailey, J. S. The effects of prompting and proximity of containers on newspaper recycling behavior. *Environment and Behavior,* in press.

Reisinger, J. J. The treatment of "anxiety-depression" via positive reinforcement and response cost. *Journal of Applied Behavior Analysis,* 1972, *5,* 125–130.

Reiss, S., & Redd, W. H. Suppression of screaming behavior in an emotionally disturbed, retarded child. *Proceedings of the 78th Annual Convention of the American Psychological Association,* 1970, *5,* 741–742.

Reiss, S., & Sushinsky, L. W. Overjustification, competing responses, and the acquisition of intrinsic interest. *Journal of Personality and Social Psychology,* 1975, *31,* 1116–1125. (a)

Reiss, S., & Sushinsky, L. W. Undermining extrinsic interest. *American Psychologist,* 1975, *30,* 782–783. (b)

Reiss, S., & Sushinsky, L. W., The competing response hypothesis of decreased play effects: A reply to Lepper and Greene. *Journal of Personality and Social Psychology,* 1976, *33,* 233–244.

Renelli v. Department of Mental Hygiene, 340, N. Y. S. 2d 498 (Sup. Ct. 1973).

Repp, A. C., Klett, S. Z., Sosebee, L. H., & Speir, N. C. Differential effects of four token conditions on rate and choice of responding in a matching to sample task. *American Journal of Mental Deficiency,* 1975, *80,* 51–56.

Resnick, P. A., Forehand, R., & Peed, S. Prestatement of contingencies: The effects on acquisition and maintenance of behavior. *Behavior Therapy,* 1974, *5,* 642–647.

Reynolds, G. S. *A primer of operant conditioning.* Glenview, Illinois: Scott, Foresman & Co., 1968.

Ribes-Inesta, E., Duran, L., Evans, B., Felix, G., Rivera, G., & Sanchez, S. An experimental evaluation of tokens as conditioned reinforcers in retarded children. *Behaviour Research and Therapy,* 1973, *11,* 125–128.

Rice, R. D. Educo-therapy: A new approach to delinquent behavior. *Journal of Learning Disabilities, 1970, 3,* 16–23.

Rickard, H. C., Dignam, P. J., & Horner, R. F. Verbal manipulation in a psychotherapeutic relationship. *Journal of Clinical Psychology,* 1960, *16,* 364–367.

Rickard, H. C., Melvin, K. B., Creel, J., & Creel, L. The effects of bonus tokens upon productivity in a remedial classroom for behaviorally disturbed children. *Behavior Therapy,* 1973, *4,* 378–385.

Rickard, H. C., & Mundy, M. B. Direct manipulation of stuttering behavior: An experimental-clinical approach. In L. P. Ullmann & L. Krasner (Eds.), *Case studies in behavior modification.* New York: Holt, Rinehart and Winston, 1965.

Rincover, A., & Koegel, R. L. Setting generality and stimulus control in autistic children. *Journal of Applied Behavior Analysis,* 1975, *8,* 235–246.

Ringer, V. M. The use of a "token helper" in the management of classroom behavior problems and in teacher training. *Journal of Applied Behavior Analysis,* 1973, *6,* 671–677.

Risley, T. R. Behavior modification: An experimental-therapeutic endeavor. In L. A. Hamerlynck, P. O. Davidson, & L. E. Acker (Eds.), *Behavior modification and ideal mental health services.* Calgary, Alberta, Canada: University of Calgary Press, 1970.

Risley, T. R., & Wolf, M. Establishing functional speech in echolalic children. In H. N. Sloane, Jr., & B. D. MacAulay (Eds.), *Operant procedures in remedial speech and language training.* Boston: Houghton Mifflin, 1968.

Robin, A. L., Armel, S., & O'Leary, K. D. The effects of self-instruction on writing deficiencies. *Behavior Therapy,* 1975, *6,* 178–187.

Roe, A. Man's forgotten weapon. *American Psychologist,* 1959, *14,* 261–266.

Rogers, C. R., & Skinner, B. F. Some issues concerning the control of human behavior: A symposium. *Science,* 1956, *124,* 1057–1066.

Rollins, H. A., McCandless, B. R., Thompson, M., & Brassell, W. R. Project success environment: An extended application of contingency management in inner-city schools. *Journal of Educational Psychology,* 1974, *66,* 167–178.

Rose, S. D., Sundel, M., Delange, J., Corwin, L., & Palumbo, A. The Hartwig project: A behavioral approach to the treatment of juvenile offenders. In R. Ulrich, T. Stachnik, & J. Mabry, *Control of human behavior, Volume 2.* Glenview, Illinois: Scott, Foresman and Company, 1970.

Rosenbaum, A., O'Leary, K. D., & Jacob, R. G. Behavioral intervention with hyperactive children: Group consequences as a supplement to individual contingencies. *Behavior Therapy,* 1975, *6,* 315–323.

Rosenfeld, G. W. Some effects of reinforcement on achievement and behavior in a regular classroom. *Journal of Educational Psychology.* 1972, *63,* 189–193.

Rosenthal, D. Changes in some moral values following psychotherapy. *Journal of Consulting Psychology,* 1955, *19,* 431–436.

Ross, J. A. The use of contingency contracting in controlling adult nail-biting. *Journal of Behavior Therapy and Experimental Psychiatry,* 1974, *5,* 105–106. (a)

Ross, J. A. Use of teachers and peers to control classroom thumbsucking. *Psychological Reports,* 1974, *34,* 327–330. (b)

Ross, S. A. Effects of intentional training in social behavior on retarded children. *American Journal of Mental Deficiency,* 1969, *73,* 912–919.

Rotenstreich, N. Skinner and "Freedom and Dignity." In H. Wheeler (Ed.), *Beyond the punitive society.* San Francisco: W. H. Freeman and Co., 1973.

Rouse v. Cameron, 373 F. 2d 451 (D. C. Cir. 1966).

Rule, S. A. Comparison of three different types of feed back on teachers' performance. In G. Semb (Ed.), *Behavior analysis and education.* Lawrence: University of Kansas, 1972.

Ruskin, R. S., & Maley, R. F. Item preference in a token economy ward store. *Journal of Applied Behavior Analysis,* 1972, *5,* 373–376.

Ryan, B. A. *Keller's Personalized System of Instruction: An appraisal.* Washington, D. C.: American Psychological Association, 1972.

Rybolt, G. A. Token reinforcement therapy with chronic psychiatric patients: A three-year evaluation. *Journal of Behavior Therapy and Experimental Psychiatry,* 1975, *6,* 188–191.

Sachs, D. A. WISC changes as an evaluative procedure within a token economy. *American Journal of Mental Deficiency,* 1971, *76,* 230–234.

Sachs, D. A. Behavioral techniques in a residential nursing home facility. *Journal of Behavior Therapy and Experimental Psychiatry,* 1975, *6,* 123–127.

Sacks, A. S., Moxley, R. A., Jr., & Walls, R. T. Increasing social interaction of preschool children with "mixies." *Psychology in the Schools,* 1975, *12,* 74–79.

Salmon, D. *Joseph Lancaster.* London: British and Foreign School Society, 1904.

Salzberg, B. H., Wheeler, A. A., Devar, L. T., & Hopkins, B. L. The effect of intermittent feedback and intermittent contingent access to play on printing of kindergarten children. *Journal of Applied Behavior Analysis,* 1971, *4,* 163–171.

Santogrossi, D. A., O'Leary, K. D., Romanczyk, R. G., & Kaufman, K. F. Self-evaluation by adolescents in a psychiatric hospital school token program. *Journal of Applied Behavior Analysis,* 1973, *6,* 277–287.

Saunders, A. G., Milstein, B. M., & Roseman, R. *Motion for partial summary judgment.* Filed in U. S. District Court for the Western District of Missouri, Southern Division, January 7, 1974. National Prison Project, Washington, D. C.

Schaefer, H. H., & Martin, P. L. Behavioral therapy for "apathy" of hospitalized schizophrenics. *Psychological Reports,* 1966, *19,* 1147–1158.

Scheff, T. J. *Being mentally ill.* Chicago: Aldine, 1966.

Schlosberg, H. The relationship between success and the laws of conditioning. *Psychological Review,* 1937, *44,* 379–394.

Schmidt, G. W., & Ulrich, R. E. Effects of group contingent events upon classroom noise. *Journal of Applied Behavior Analysis,* 1969, *2,* 171–179.

Schnelle, J. F. A brief report on invalidity of parent evaluations of behavior change. *Journal of Applied Behavior Analysis,* 1974, *7,* 341–343.

Schnelle, J. F., & Frank, L. J. A quasi-experimental retrospective evaluation of a prison policy change. *Journal of Applied Behavior Analysis,* 1974, *7,* 483–494.

Schroeder, S. R. Parametric effects of reinforcement frequency, amount of reinforcement, and required response force on sheltered workshop behavior. *Journal of Applied Behavior Analysis,* 1972, *5,* 431–441.

Schumaker, J., & Sherman, J. A. Training generative verb usage by imitation and reinforcement procedures. *Journal of Applied Behavior Analysis,* 1970, *3,* 273–287.

Schwartz, J., & Bellack, A. S. A comparison of a token economy with standard inpatient treatment. *Journal of Consulting and Clinical Psychology,* 1975, *43,* 107–108.

Schwartz, M. L., & Hawkins, R. P. Application of delayed reinforcement procedures to the behavior of an elementary school child. *Journal of Applied Behavior Analysis,* 1970, *3,* 85–96.

Schwitzgebel, R. *Street-corner research: An experimental approach to the juvenile delinquent.* Cambridge, Mass.: Harvard University Press, 1964.

Schwitzgebel, R. K. Limitations on the coercive treatment of offenders. *Criminal Law Bulletin,* 1972, *8,* 267–320.

Schwitzgebel, R. K. A contractual model for the protection of the rights of institutionalized mental patients. *American Psychologist, 1975, 8,* 815–820.

Schwitzgebel, R. L. Short-term operant conditioning of adolescent offenders on socially relevant variables. *Journal of Abnormal Psychology, 1967, 72,* 134–142.

Schwitzgebel, R. L., & Kolb, D. A. Inducing behavior change in adolescent delinquents. *Behaviour Research and Therapy, 1964, 1,* 297–304.

Scott, R. W., Peters, R. D., Gillespie, W. J., Blanchard, E. B., Edmunson, E. D., & Young, L. D. The use of shaping and reinforcement in the operant acceleration and deceleration of heart rate. *Behaviour Research and Therapy, 1973, 11,* 179–185.

Scott, S. B., Pisa, A., & Friesen, D. D. Beyond the token economy: A comparison of two contingency programs. *Journal of Consulting and Clinical Psychology, 1975, 43,* 498–503.

Seaver, W. B., & Patterson, A. H. Decreasing fuel oil consumption through feedback and social commendation. *Journal of Applied Behavior Analysis, 1976, 9,* 147–152.

Semb, G. The effects of mastery criteria and assignment length on college student test performance. *Journal of Applied Behavior Analysis, 1974, 7,* 61–69.

Sewell, E., McCoy, J. F., & Sewell, W. R. Modification of an antagonistic social behavior using positive reinforcement for other behavior. *Psychological Record, 1973, 23,* 499–504.

Seymour, F. W., & Stokes, T. F. Self-recording in training girls to increase work and evoke staff praise in an institution for offenders. *Journal of Applied Behavior Analysis, 1976, 9,* 41–54.

Shean, J. D., & Zeidberg, A. Token reinforcement therapy: A comparison of matched groups. *Journal of Behavior Therapy and Experimental Psychiatry, 1971, 2,* 95–105.

Sheppard, W. C., & MacDermot, H. G. Design and evaluation of a programmed course in introductory psychology. *Journal of Applied Behavior Analysis, 1970, 3,* 5–11.

Sherman, J. A. Reinstatement of verbal behavior in a psychotic by reinforcement methods. *Journal of Speech and Hearing Disorders, 1963, 28,* 398–401.

Sherman, J. A. Use of reinforcement and imitation to reinstate verbal behavior in mute psychotics. *Journal of Abnormal Psychology, 1965, 70,* 155–164.

Sherman, J. A., & Baer, D. M. Appraisal of operant therapy techniques with children and adults. In C. M. Franks (Ed.), *Behavior therapy: Appraisal and status.* New York: McGraw-Hill, 1969.

Sidman, M. *Tactics of scientific research.* New York: Basic Books, 1960.

Siegel, G. M., Lenske, J., & Broen, P. Suppression of normal speech disfluencies through response cost. *Journal of Applied Behavior Analysis, 1969, 2,* 265–276.

Sirota, A. D., Schwartz, G. E., & Shapiro, D. Voluntary control of human heart rate: Effect on reaction to aversive stimulation. *Journal of Abnormal Psychology, 1974, 83,* 261–267.

Skinner, B. F. Two types of conditioned reflex and a pseudo type. *Journal of General Psychology, 1935, 12,* 66–77.

Skinner, B. F. *The behavior of organisms.* New York: Appleton-Century-Crofts, 1938.

Skinner, B. F. *Walden Two.* New York: The Macmillan Company, 1948.

Skinner, B. F. *Science and human behavior.* New York: The Macmillan Company, 1953. (a)

Skinner, B. F. Some contributions of an experimental analysis of behavior to psychology as a whole. *American Psychologist, 1953, 8,* 69–78. (b)

Skinner, B. F. A new method for the experimental analysis of the behavior of psychotic patients. *Journal of Nervous and Mental Disease, 1954, 120,* 403–406.

Skinner, B. F. A case history in scientific methods. *American Psychologist, 1956, 11,* 221–233.

Skinner, B. F. *Verbal behavior.* New York: Appleton-Century-Crofts, 1957.

Skinner, B. F. The design of cultures. *Daedalus, 1961, 90,* 534–546.

Skinner, B. F. What is the experimental analysis of behavior? *Journal of the Experimental Analysis of Behavior, 1966, 9,* 213–218.

Skinner, B. F. *Beyond freedom and dignity.* New York: Knopf, 1971.

Skinner, B. F. Answers for my critics. In H. Wheeler (Ed.), *Beyond the punitive society.* San Francisco: W. H. Freeman, 1973.

Skinner, B. F., Solomon, H. C., Lindsley, O. R., & Richards, M. E. Studies in behavior therapy. Metropolitan State Hospital, Waltham, Massachusetts, Status Report II, May, 1954.

Slack, C. Experimenter-subject psychotherapy. *Mental Hygiene*, 1960, *44*, 238–256.

Sloane, H. N., Jr., Johnston, M. K., & Harris, F. R. Remedial procedures for teaching verbal behavior to speech deficient or defective young children. In H. N. Sloane, Jr., & B. D. MacAulay (Eds.), *Operant procedures in remedial speech and language training*. Boston: Houghton Mifflin, 1968.

Smith, D. E. P., Brethower, D., & Cabot, R. Increasing task behavior in a language arts program by providing reinforcement. *Journal of Experimental Child Psychology*, 1969, *8*, 45–62.

Sobell, L. C., Schaefer, H. H., Sobell, M. B., & Kremer, M. E. Food priming: A therapeutic tool to increase the percentage of meals bought by chronic mental patients. *Behaviour Research and Therapy*, 1970, *8*, 339–345.

Solomon, R. W., & Wahler, R. G. Peer reinforcement control of classroom problem behavior. *Journal of Applied Behavior Analysis*, 1973, *6*, 49–56.

In re Spadafora, 54 Misc. 2d 123, 281 N. Y. S. 2d 923 (Sup. Ct. 1967).

Spradlin, J. E. Effects of reinforcement schedules on extinction in severely mentally retarded children. *American Journal of Mental Deficiency*, 1962, *66*, 634–640.

Spradlin, J. E., & Girardeau, F. L. The behavior of moderately and severely retarded persons. In N. R. Ellis (Ed.), *International review of research in mental retardation, Volume 1*. New York: Academic Press, 1966.

Staats, A. W. A general apparatus for the investigation of complex learning in children. *Behaviour Research and Therapy*, 1968, *6*, 45–50. (a)

Staats, A. W. *Learning, language, and cognition*. New York: Holt, Rinehart and Winston, 1968. (b)

Staats, A. W. Reinforcer systems in the solution of human problems. In G. A. Fargo, C. Behrns, & P. Nolen (Eds.), *Behavior modification in the classroom*. Belmont, California: Wadsworth, 1970.

Staats, A. W., & Butterfield, W. H. Treatment of nonreading in a culturally deprived juvenile delinquent: An application of learning principles. *Child Development*, 1965, *4*, 925–942.

Staats, A. W., Finley, J. R., Minke, K. A., & Wolf, M. M. Reinforcement variables in the control of unit reading responses. *Journal of the Experimental Analysis of Behavior*, 1964, *7*, 139–149.

Staats, A. W., Minke, K. A., & Butts, P. A token-reinforcement remedial reading program administered by black therapy techniques to problem black children. *Behavior Therapy*, 1970, *1*, 331–353.

Staats, A. W., Minke, K. A., Finley, J. R., Wolf, M., & Brooks, L. O. A reinforcer system and experimental procedure for the laboratory study of reading acquisition. *Child Development*, 1964, *35*, 209–231.

Staats, A. W., Minke, K. A., Goodwin, W., & Landeen, J. Cognitive behavior modification: "Motivated learning" reading treatment with subprofessional therapy technicians. *Behaviour Research and Therapy*, 1967, *5*, 283–299.

Staats, A. W., Staats, C. K., Schultz, R. E., & Wolf, M. The conditioning of textual responses using "extrinsic" reinforcers. *Journal of the Experimental Analysis of Behavior*, 1962, *5*, 33–40.

Stahl, J. R., Thomson, L. E., Leitenberg, H., & Hasazi, J. E. Establishment of praise as a conditioned reinforcer in socially unresponsive psychiatric patients. *Journal of Abnormal Psychology*, 1974, *83*, 488–496.

Staw, B. M. *Intrinsic and extrinsic motivation*. Morristown, N. J.: General Learning Press, 1975.

Stayer, S. J., & Jones, F. Ward 108: Behavior modification and the delinquent soldier. Paper presented at Behavioral Engineering Conference, Walter Reed General Hospital, 1969.

Stedman, J. M., Peterson, T. L., & Cardarelle, J. Application of a token system in a preadolescent boys' group. *Journal of Behavior Therapy and Experimental Psychiatry*, 1971, *2*, 23–29.

Steffy, R. A., Hart, J., Craw, M., Torney, D., & Marlett, N. Operant behaviour modification techniques applied to severely regressed and aggressive patients. *Canadian Psychiatric Association Journal,* 1969, *14,* 59–67.

Stein, T. J. Some ethical considerations of short-term workshops in the principles and methods of behavior modification. *Journal of Applied Behavior Analysis,* 1975, *8,* 113–115.

Steinman, C. The case of the frightened convict. *The Nation,* 1973, *217,* 590–593.

Stokes, T. F., & Baer, D. M. An implicit technology of generalization. *Journal of Applied Behavior Analysis,* in press.

Stokes, T. F., & Baer, D. M. Preschool peers as mutual generalization-facilitating agents. *Behavior Therapy,* 1976, *7,* 549–556.

Stokes, T. F., Baer, D. M., & Jackson, R. L. Programming the generalization of a greeting response in four retarded children. *Journal of Applied Behavior Analysis, 1974,* 7, 599–610.

Stolz, S. B. Ethical issues in behavior modification. In G. Bermant & H. Kelman (Eds.), *Ethics of social intervention.* Washington, D. C.: Hemisphere, 1976.

Stolz, S. B. Ethics of social and educational interventions: Historical context and a behavioral analysis. In T. A. Brigham & A. C. Catania (Eds.), *The handbook of applied behavior research: Social and instructional processes.* New York: Irvington Press/Halstead Press, in press.

Stolz, S. B., Wienckowski, L. A., & Brown, B. S. Behavior modification: A perspective on critical issues. *American Psychologist,* 1975, *11,* 1027–1048.

Stuart, R. B. Operant-interpersonal treatment for marital discord. *Journal of Consulting and Clinical Psychology,* 1969, *33,* 675–682. (a)

Stuart, R. B. Token reinforcement in marital treatment. In R. D. Rubin & C. M. Franks (Eds.), *Advances in behavior therapy, 1968.* New York: Academic Press, 1969. (b)

Stumphauzer, J. S. *Behavior therapy with delinquents.* Springfield, Ill.: Charles C Thomas, 1973.

Subcommittee on Constitutional Rights, Committee of the Judiciary, United States Senate, Ninety-Third Congress, *Individual rights and the federal role in behavior modification.* Washington, D. C.: U. S. Government Printing Office, 1974.

Suchotliff, L., Greaves, S., Stecker, H., & Berke, R. Critical variables in the token economy. *Proceedings of the 78th Annual Convention of the American Psychological Association,* 1970, *5,* 517–518.

Sulzbacher, S. I., & Houser, J. E. A tactic to eliminate disruptive behaviors in the classroom: Group contingent consequences. *American Journal of Mental Deficiency,* 1968, *73,* 88–90.

Sulzer, B., Hunt, S., Ashby, E., Koniarski, C., & Krams, M. Increasing rate and percentage correct in reading and spelling in a fifth grade public school class of slow readers by means of a token system. In E. Ramp & B. L. Hopkins (Eds.), *A new direction for education: Behavior analysis 1971.* Lawrence: The University of Kansas, 1971.

Sulzer, E. S. Behavior modification in adult psychiatric patients. *Journal of Counseling Psychology,* 1962, *9,* 271–276.

Surratt, P. R., & Hopkins, B. L. The effects of access to a playroom on the speed and quality of arithmetic work of first and second grade children: A replication and systematic analysis. Paper read at the American Educational Research Association meeting, Minneapolis, 1970.

Surratt, P. R., Ulrich, R. E., & Hawkins, R. P. An elementary student as a behavioral engineer. *Journal of Applied Behavior Analysis,* 1969, *2,* 85–92.

Szasz, T. S. The myth of mental illness. *American Psychologist,* 1960, *15,* 113–118.

Taffel, C. Anxiety and the conditioning of verbal behavior. *Journal of Abnormal and Social Psychology,* 1955, *51,* 496–501.

Tanner, B. A., Parrino, J. J., & Daniels, A. C. A token economy with "automated" data collection. *Behavior Therapy,* 1975, *6,* 111–118.

Tharp, R. G., & Wetzel, R. J. *Behavior modification in the natural environment*. New York: Academic Press, 1969.

Thomas, D. R. Self-monitoring as a technique for modifying teacher behaviors. Unpublished doctoral dissertation, University of Illinois, 1972.

Thomas, E. J., & Walter, C. W. Guidelines for behavioral practice in the open community agency: Procedure and evaluation. *Behaviour Research and Therapy*, 1973, *11*, 193–205.

Thompson, T., & Grabowski, J. (Eds.), *Behavior modification of the mentally retarded*. New York: Oxford University Press, 1972.

Thomson, N., Fraser, D., & McDougall, A. The reinstatement of speech in near-mute chronic schizophrenics by instructions, imitative prompts, and reinforcement. *Journal of Behavior Therapy and Experimental Psychiatry*, 1974, *5*, 77–80.

Thoresen, C. E., & Mahoney, M. J. *Behavioral self-control*. New York: Holt, Rinehart and Winston, 1974.

Thorndike, E. L. *Animal intelligence*. New York: Macmillan, 1911.

Timbers, G. D., Timbers, B. J., Fixsen, D. L., Phillips, E. L., & Wolf, M. M. Achievement Place for pre-delinquent girls: Modification of inappropriate emotional behaviors with token reinforcement and instructional procedures. Paper read at the American Psychological Association, Montreal, Canada, 1973.

Tracey, D. A., Briddell, D. W., & Wilson, G. T. Generalization of verbal conditioning to verbal and nonverbal behavior: Group therapy with chronic psychiatric patients. *Journal of Applied Behavior Analysis*, 1974, *7*, 391–402.

Tribby v. Cameron, Id. at 328, 379 F. 2d at 105.

Truax, C. B. Reinforcement and non-reinforcement in Rogerian psychotherapy. *Journal of Abnormal Psychology*, 1966, *71*, 1–9.

Trudel, G., Boisvert, J., Maruca, F., & Leroux, P. Unprogrammed reinforcement of patients' behaviors in wards with and without token economy. *Journal of Behavior Therapy and Experimental Psychiatry*, 1974, *5*, 147–149.

Tsosie, P., & Giles, D. Intermountain Youth Center program description. Unpublished manuscript, Intermountain Youth Center, 1973.

Turkewitz, H., O'Leary, K. D., & Ironsmith, M. Generalization and maintenance of appropriate behavior through self-control. *Journal of Consulting and Clinical Psychology*, 1975, *43*, 577–583.

Turton, B. K., & Gathercole, C. E. Token economies in the U. K. and Eire. *Bulletin of the British Psychological Society*, 1972, *25*, 83–87.

Twardosz, S., & Baer, D. M. Training two severely retarded adolescents to ask questions. *Journal of Applied Behavior Analysis*, 1973, *6*, 655–661.

Twardosz, S., & Sajwaj, T. Multiple effects of a procedure to increase sitting in a hyperactive, retarded boy. *Journal of Applied Behavior Analysis*, 1972, *5*, 73–78.

Tyler, V. O. Application of operant token reinforcement to academic performance of an institutionalized delinquent. *Psychological Reports*, 1967, *21*, 249–260.

Tyler, V. O., & Brown, G. D. Token reinforcement of academic performance with institutionalized delinquent boys. *Journal of Educational Psychology*, 1968, *59*, 164–168.

Ullmann, L. P., & Krasner, L. *A psychological approach to abnormal behavior* (Second Edition). Englewood Cliffs, N. J.: Prentice-Hall, 1975.

Ulman, J. D., & Klem, J. L. Communication. *Journal of Applied Behavior Analysis*, 1975, *8*, 210.

Ulmer, R. A. Relationships between objective personality test scores, schizophrenics' history and behavior in a mental hospital token economy ward. *Psychological Reports*, 1971, *29*, 307–312.

Ulrich, R. Behavior control and public concern. *Psychological Record*, 1967, *17*, 229–234.

Ulrich, R., Stachnik, T., & Mabry, J. (Eds.), *Control of human behavior, Volume 1*. Glenview, Ill.: Scott, Foresman and Company, 1966.

Ulrich, R., Stachnik, T., & Mabry, J. (Eds.), *Control of human behavior: From cure to prevention, Volume 2.* Glenview, Ill.: Scott, Foresman and Company, 1970.

Ulrich, R., Stachnik, T., & Mabry, J. (Eds.), *Control of human behavior: Behavior modification in education, Volume 3.* Glenview, Ill.: Scott, Foresman and Company, 1974.

Upper, D. A "ticket" system for reducing ward rule violations on a token economy program. *Journal of Behavior Therapy and Experimental Psychiatry,* 1973, *4,* 137–140.

Upper, D., & Newton, J. G. A weight-reduction program for schizophrenic patients on a token economy unit: Two case studies. *Journal of Behavior Therapy and Experimental Psychiatry,* 1971, *2,* 113–115.

Vah Houten, R., Morrison, E., Jarvis, R., & McDonald, M. The effects of explicit timing and feedback on compositional response rate in elementary school children. *Journal of Applied Behavior Analysis,* 1974, *7,* 547–555.

Van Houten, R., & Sullivan, K. Effects of an audio cueing system on the rate of teacher praise. *Journal of Applied Behavior Analysis,* 1975, *8,* 197–201.

Verplanck, W. S. The control of the content of conversation: Reinforcement of statements of opinion. *Journal of Abnormal and Social Psychology,* 1955, *51,* 668–676.

Wagner, R. F., & Guyer, B. P. Maintenance of discipline through increasing children's span of attending by means of a token economy. *Psychology in the Schools,* 1971, *8,* 285–289.

Wahler, R. G. Behavior therapy for oppositional children: Love is not enough. Paper read at Eastern Psychological Association, Washington, D. C., April, 1968.

Wahler, R. G. Oppositional children: A quest for parental reinforcement control. *Journal of Applied Behavior Analysis,* 1969, *2,* 159–170.

Wahler, R. G. Some ecological problems in child behavior modification. In S. W. Bijou & E. Ribes-Inesta (Eds.), *Behavior modification: Issues and extensions.* New York: Academic Press, 1972.

Wahler, R. G. Some structural aspects of deviant child behavior. *Journal of Applied Behavior Analysis,* 1975, *8,* 27–42.

Walker, H. M., & Buckley, N. K. Programming generalization and maintenance of treatment effects across time and across settings. *Journal of Applied Behavior Analysis,* 1972, *5,* 209–224.

Walker, H. M., & Hops, H. Use of normative peer data as a standard for evaluating classroom treatment effects. *Journal of Applied Behavior Analysis,* 1976, *9,* 159–168.

Walker, H. M., Hops, H., & Fiegenbaum, E. Deviant classroom behavior as a function of combinations of social and token reinforcement and cost contingency. *Behavior Therapy,* 1976, *7,* 76–88.

Walker, H. M., Hops, H., & Johnson, S. M. Generalization and maintenance of classroom treatment effects. *Behavior Therapy,* 1975, *6,* 188–200.

Walker, H. M., Mattson, R. H., & Buckley, N. K. The functional analysis of behavior within an experimental class setting. In W. C. Becker (Ed.), *An empirical basis for change in education.* Chicago: Science Research Associates, 1971.

Wallace, C. J., & Davis, J. R. Effects of information and reinforcement on the conversational behavior of chronic psychiatric patient dyads. *Journal of Consulting and Clinical Psychology,* 1974, *42,* 656–662.

Wallace, C. J., Davis, J. R., Liberman, R. P., & Baker, V. Modeling and staff behavior. *Journal of Consulting and Clinical Psychology,* 1973, *41,* 422–425.

Ward, M. H., & Baker, B. L. Reinforcement therapy in the classroom. *Journal of Applied Behavior Analysis,* 1968, *1,* 323–328.

Wasik, B. H. Janus House for delinquents: An alternative to training schools. New Treatment Approaches to Juvenile Delinquency, University of Tennessee, Memphis, 1972.

Watson, L. S. Shaping and maintaining behavior modification skills in staff using contingent reinforcement techniques. In R. L. Patterson (Ed.), *Maintaining effective token economies.* Springfield, Ill.: Charles C Thomas, 1976.

Wehman, P. Maintaining oral hygiene skills in geriatric retarded women. *Mental Retardation,* 1974, *12,* 20.

Weisberg, P., Lieberman, C., & Winter, K. Reduction of facial gestures through loss of token reinforcement. *Psychological Reports,* 1970, *26,* 227–230.

Welch, M. W., & Gist, J. W. *The open token economy system: A handbook for a behavioral approach to rehabilitation.* Springfield, Ill.: Charles C Thomas, 1974.

West, J. *The history of Tasmania.* Launceston, Tasmania: H. Dowling, 1852.

Wexler, D. B. Token and taboo: Behavior modification, token economies, and the law. *California Law Review,* 1973, *61,* 81–109.

Wexler, D. B. Behavior modification and other behavior change procedures: The emerging law and the proposed Florida Guidelines. *Criminal Law Bulletin,* 1975, *11,* 600–616. (a)

Wexler, D. B. Reflections on the legal regulation of behavior modification in institutional settings. *Arizona Law Review,* 1975, *17,* 132–143. (b)

Wheeler, A. J., & Sulzer, B. Operant training and generalization of a verbal response form in a speech-deficient child. *Journal of Applied Behavior Analysis,* 1970, *3,* 139–147.

Wheeler, H. (Ed.), *Beyond the punitive society.* San Francisco: W. H. Freeman and Co., 1973.

Whitman, T. L., Mercurio, J. R., & Caponigri, V. Development of social responses in two severely retarded children. *Journal of Applied Behavior Analysis,* 1970, *3,* 133–138.

Whitree v. State, 56 Misc. 2d 693, 290 N. Y. S. 2d 486 (Ct. Cl. 1968).

Wildman, R. W., II, & Wildman, R. W. The generalization of behavior modification procedures: A review with special emphasis on classroom applications. *Psychology in the Schools,* 1975, *12,* 432–448.

Williams, C. D. The elimination of tantrum behaviors by extinction procedures. *Journal of Abnormal and Social Psychology,* 1959, *59,* 269.

Williams, J. E., & Morland, J. K. *Race, color, and the young.* Chapel Hill: University of North Carolina Press, 1976.

Williams, J. L. *Operant learning: Procedures for changing behavior.* Monterey, California: Brooks/Cole, 1973.

Williams, M. G., & Harris, V. W. Program description. Unpublished manuscript, Southwest Indian Youth Center, 1973.

Wilson, C. W., & Hopkins, B. L. The effects of contingent music on the intensity of noise in junior high home economics classes. *Journal of Applied Behavior Analysis,* 1973, *6,* 269–276.

Wilson, M. D., & McReynolds, L. V. A procedure for increasing oral reading rate in hard of hearing children. *Journal of Applied Behavior Analysis,* 1973, *6,* 231–239.

Wiltz, N. A., & Gordon, S. B. Parental modification of a child's behavior in an experimental residence. *Journal of Behavior Therapy and Experimental Psychiatry,* 1974, *5,* 107–109.

Wincze, J. P., Leitenberg, H., & Agras, W. S. The effects of token reinforcement and feedback on the delusional verbal behavior of chronic paranoid schizophrenics. *Journal of Applied Behavior Analysis,* 1972, *5,* 247–262.

Winett, R. A., & Nietzel, M. T. Behavioral ecology: Contingency management of consumer energy use. *American Journal of Community Psychology,* 1975, *3,* 123–133.

Winett, R. A., Richards, C. S., & Krasner, L. Child-monitored token reading program. *Psychology in the Schools,* 1971, *8,* 259–262.

Winett, R. A., & Roach, E. M. The effects of reinforcing academic performance on social behavior. *Psychological Record,* 1973, *23,* 391–396.

Winett, R. A., & Winkler, R. C. Current behavior modification in the classroom: Be still, be quiet, be docile. *Journal of Applied Behavior Analysis,* 1972, *5,* 499–504.

Winkler, R.C. Management of chronic psychiatric patients by a token reinforcement system. *Journal of Applied Behavior Analysis,* 1970, *3,* 47–55.

Winkler, R. C. Reinforcement schedules for individual patients in a token economy. *Behavior Therapy,* 1971, *2,* 534–537. (a)

Winkler, R. C. The relevance of economic theory and technology of token reinforcement systems. *Behaviour Research and Therapy,* 1971, *9,* 81–88. (b)

Winkler, R. C. A theory of equilibrium in token economies. *Journal of Abnormal Psychology,* 1972, *79,* 169–173.

Winkler, R. C. An experimental analysis of economic balance, savings and wages in a token economy. Behavior Therapy, 1973, 4, 22–40. (a)

Winkler, R. C. A reply to Fethke's comment on "The relevance of economic theory and technology to token reinforcement systems." Behaviour Research and Therapy, 1973, 11, 223–224. (b)

Winkler, R. C., Kagel, J. H., Battalio, R. C., Fisher, E. F., Basmann, R. L., & Krasner, L. Methodological and conceptual issues in testing economic theory of consumer demand in a token economy. Paper presented at the Addiction Research Foundation, Toronto, Canada, October, 1973.

Winkler, R. C., & Krasner, L. The relevance of economics to token economies. Paper presented at the meeting of the Eastern Psychological Association, New York, April, 1971.

Winters v. Miller, 446 F. 2d 65 (2d Cir. 1970), reversing 306 F. Supp. 1158 (E.D. N.Y. 1969), cert. denied 404 U.S. 985, 92 S. Ct. 450. 30L. Ed. 2d 369.

Witmer, J. F., & Geller, E. S. Facilitating paper recycling: Effects of prompts, raffles, and contests. Journal of Applied Behavior Analysis, in press.

Wodarski, J. S. The reduction of electrical energy consumption: The application of behavioral analysis. Ninth Annual Meeting, Association for Advancement of Behavior Therapy, San Francisco, December, 1975.

Wodarski, J. S., Hamblin, R. L., Buckholdt, D. R., & Ferritor, D. C. The effects of different reinforcement contingencies on cooperative behaviors exhibited by fifth graders. In R. D. Rubin, J. P. Brady, & J. D. Henderson (Eds.), Advances in behavior therapy, Volume 4. New York: Academic Press, 1973.

Wolf, M. M., Birnbrauer, J. S., Williams, T., & Lawler, J. A note on apparent extinction of the vomiting behavior of a retarded child. In L. P. Ullmann & L. Krasner (Eds.), Case studies in behavior modification. New York: Holt, Rinehart and Winston, 1965.

Wolf, M. M., Giles, D. K., & Hall, R. V. Experiments with token reinforcement in a remedial classroom. Behaviour Research and Therapy, 1968, 6, 51–64.

Wolf, M. M., Hanley, E. L., King, L. A., Lachowicz, J., & Giles, D. K. The timer-game: A variable interval contingency for the management of out-of-seat behavior. Exceptional Children, 1970, 37, 113–117.

Wolf, M. M., Risley, T., & Mees, H. Application of operant conditioning procedures to the behaviour problems of an autistic child. Behaviour Research and Therapy, 1964, 1, 305–312.

Wolf, M. M., Sidman, M., & Miller, L. K. Methodological and assessment considerations in applied settings: Reviewers' comments. Journal of Applied Behavior Analysis, 1973, 6, 532–539.

Wolfe, J. B. Effectiveness of token-rewards for chimpanzees. Comparative Psychological Monographs, 1936, 12, No. 60.

Wooley, S. C., & Blackwell, B. A behavioral probe into social contingencies on a psychosomatic ward. Journal of Applied Behavior Analysis, 1975, 8, 337–339.

Wyatt v. Stickney, 344 F. Supp. 373, 344 F. Supp. 387 (M.D. Ala. 1972) affirmed sub nom. Wyatt v. Aderholt, 503 F. 2d 1305 (5th Cir. 1974).

Yen, S., & McIntire, R. W. (Eds.), Teaching behavior modification. Kalamazoo, Michigan: Behaviordelia, 1976.

Zifferblatt, S. M. The effectiveness of modes and schedules of reinforcement on work and social behavior in occupational therapy. Behavior Therapy, 1972, 3, 567–578.

Zimmerman, E. H., & Zimmerman, J. The alteration of behavior in a special classroom situation. Journal of the Experimental Analysis of Behavior, 1962, 5, 59–60.

Zimmerman, E. H. Zimmerman, J., & Russell, C. D. Differential effects of token reinforcement on instruction-following behavior in retarded students instructed as a group. Journal of Applied Behavior Analysis, 1969, 2, 101–112.

Zimmerman, J., Overpeck, C., Eisenberg, H., & Garlick, B. Operant conditioning in a sheltered workshop. *Rehabilitation Literature,* 1969, *30,* 326–334.

Zimmerman, J., Stuckey, T. E., Garlick, B. J., & Miller, M. Effects of token reinforcement on productivity in multiple handicapped clients in a sheltered workshop. *Rehabilitation Literature,* 1969, *30,* 34–41.

AUTHOR INDEX

SUBJECT INDEX